The Materials Research Society Series

The Materials Research Society Series covers the multidisciplinary field of materials research and technology, publishing across chemistry, physics, biology, and engineering. The Series focuses on premium textbooks, professional books, monographs, references, and other works that serve the broad materials science and engineering community worldwide. Connecting the principles of structure, properties, processing, and performance and employing tools of characterization, computation, and fabrication the Series addresses established, novel, and emerging topics.

More information about this series at http://www.springer.com/series/16655

Ray LaPierre

Introduction to Quantum Computing

Ray LaPierre
Department of Engineering Physics
McMaster University
Hamilton, ON, Canada

ISSN 2730-7360   ISSN 2730-7379   (electronic)
The Materials Research Society Series
ISBN 978-3-030-69320-6   ISBN 978-3-030-69318-3   (eBook)
https://doi.org/10.1007/978-3-030-69318-3

This Springer imprint is published by the registered company Springer Nature Switzerland AG
The registered company address is: Gewerbestrasse 11, 6330 Cham, Switzerland

Preface

My interest in quantum computing began 20 years ago with my research career in semiconductor nanowires, which are being investigated as a potential platform for quantum computing. In 2020, I offered a senior undergraduate course in quantum computing at McMaster University (Hamilton, Ontario, Canada). The course assumed only an introductory background in quantum mechanics and was intended for undergraduate students in their third or fourth year of a four-year bachelor's degree program. While researching the content for the course, I found several excellent books but none of them gave a completely satisfactory understanding of the topic at an introductory level. The book most often cited in this field is *Quantum Computation and Quantum Information* (Cambridge University Press 2017) by Nielson and Chuang. I once heard someone say that "you're not doing quantum computing unless it's in Nielson and Chuang." However, the book by Nielson and Chuang is probably not the best for the undergraduate beginner. Conversely, other books are available for a more general audience without assuming much background in physics, but I found these books did not delve into the topic deeply enough. Lastly, many textbooks on quantum computing are written for computer scientists with an emphasis on algorithms rather than the hardware that may be of more interest to engineers. Quantum computing is now ramping up in the private sector [1], requiring more graduates with engineering- and physics-based knowledge in the field. Therefore, the present book includes a physics- and engineering-based approach to quantum computing.

The book begins in Chap. 1 with the double-slit experiment and the principle of superposition, which contain some of the essential features of quantum mechanics. From there, the book moves onto energy quantization in Chap. 2 and spins in Chap. 3, from which the concept of qubits in Chap. 4 can be understood. Chapters 5 and 6 introduce entanglement, the Bell inequality, and applications of entanglement. Quantum gates are introduced in Chap. 7, followed by typical circuits (teleportation, superdense coding) in Chap. 8. These ideas are refined in Chaps. 9 and 10 with tensor products and computational complexity. Three famous algorithms (Deutsch, Grover, and Shor) are covered in Chaps. 11 to 13. Concepts underlying the physical implementation of quantum gates are treated in Chaps. 14 to 18. Finally, the different physical platforms for quantum computing are described in detail in Chaps. 19 to 26, including quantum error correction in Chap. 25.

Quantum computing is a fascinating and exciting topic, but also a complicated one. Understanding quantum computing requires the synthesis of knowledge across many disciplines of physics and computer science, including electrodynamics, electronics, condensed matter physics, and quantum mechanics, posing a challenge to any beginner. I hope this book offers a good beginning to your journey.

I am grateful to McMaster University, the Department of Engineering Physics, my many colleagues, my students, and my family for permitting me the time and inspiration to write this book.

Reference

1.   Physics Today 73 (2020) 22.

How to Use This Book

This book is intended for a single semester (~12 weeks) elective course on quantum computing, comprised of approximately 36 one-hour lectures (3 hours per week). A suggested lecture schedule is as follows:

Lecture 1–2: Chapter 1—Superposition
Lecture 3–4: Chapter 2—Quantization
Lecture 5–6: Chapter 3—Spin
Lecture 7–8: Chapter 4—Qubits
Lecture 9–10: Chapter 5—Entanglement
Lecture 11: Chapter 6—Quantum Key Distribution
Lecture 12–13: Chapter 7—Quantum Gates
Lecture 14: Chapter 8—Teleportation
Lecture 15: Chapter 9—Tensor Products
Lecture 16: Chapter 10—Quantum Parallelism and Computational Complexity
Lecture 17: Chapter 11—Deutsch Algorithm
Lecture 18: Chapter 12—Grover Algorithm
Lecture 19: Chapter 13—Shor Algorithm
Lecture 20–21: Chapter 14—Precession
Lecture 22: Chapter 15—Electron Spin Resonance
Lecture 23: Chapter 16—Two-State Dynamics
Lecture 24: Chapter 17—Implementing Two-Qubit Gates
Lecture 25: Chapter 18—DiVincenzo Criteria
Lecture 26: Chapter 19—Nuclear Magnetic Resonance
Lecture 27–28: Chapter 20—Solid-State Spin Qubits
Lecture 29: Chapter 21—Trapped Ion Quantum Computing
Lecture 30–31: Chapter 22—Superconducting Qubits
Lecture 32: Chapter 23—Adiabatic Quantum Computing
Lecture 33: Chapter 24—Optical Quantum Computing
Lecture 34–35: Chapter 25—Quantum Error Correction

Lecture 36: Chapter 26—Topological Quantum Computing

The book assumes that students have successfully completed an introductory course in quantum mechanics, which is typically in the second year of a four-year undergraduate program in science, engineering, or related disciplines. Thus, this book is intended for the third or fourth year of an undergraduate program or the entry level of a graduate program.

The book is divided into three main parts. Chapters 1–13 focus on the basic principles underlying quantum computing and an understanding of quantum algorithms. Chapters 14–18 introduce the principles underlying the physical implementation of single-qubit and two-qubit gates. Finally, Chaps. 19–26 present specific physical platforms for quantum computers, as well as quantum error correction.

Each chapter is intended to be taught consecutively. Chapter 6 on quantum key distribution and Chap. 8 on teleportation, are given as sample applications of entanglement and may be considered optional (although students typically enjoy this material). Chapter 25 on quantum error correction, although of importance to quantum computing, may also be considered optional or quickly skimmed. Instructors who wish to emphasize quantum algorithms may choose to focus on Chaps. 1–13, while those more interested in hardware can focus on Chaps. 14–26.

Each chapter includes exercises which can be completed by the student as homework assignments or used for tutorial instruction. A solutions manual is available for qualified instructors. Each chapter also includes references for more advanced study, and further reading is listed at the end of the book.

Hamilton, Canada   Ray LaPierre

Contents

1 Superposition 1
   1.1   Classical Physics 1
   1.2   Wave Nature of Light 1
   1.3   Particle Nature of Light (Photons) 5
   1.4   Probability Amplitude 6
   1.5   Mach–Zehnder Interferometer 8
   1.6   Superposition 9
   1.7   "Which Path?" and Measurement Collapse 9
   1.8   What is Matter? 10
   1.9   Wave-Particle Duality 12
   1.10   The Meaning of Quantum Mechanics 14
   1.11   The Basic Idea Behind Quantum Computing 15
   References 16

2 Quantization 19
   2.1   Probability Amplitude 19
   2.2   Time-Dependent Schrodinger Equation 20
   2.3   The Free Particle 21
   2.4   The Hamiltonian 21
   2.5   Time-Independent Schrodinger Equation 22
   2.6   Observables 23
   2.7   Infinite Quantum Well or "Particle in a Box" 23
   2.8   Quantization of Energy 26
   2.9   Orthonormal States 27
   2.10   Basis States 27
   2.11   Time Dependence 28
   2.12   The Hydrogen Atom 31
   2.13   Expectation Value 33
   2.14   Fundamental Postulates of Quantum Mechanics 33
   References 34

3   Spin   35
   3.1   Orbital Angular Momentum   35
   3.2   Magnetic Dipole Moment   37
   3.3   Stern-Gerlach Experiment   38
   3.4   Electron Spin   42
   3.5   Intrinsic Magnetic Moment   43
   3.6   Combinations of SG Apparatus   44
   3.7   Mathematics of Spin   46
   3.8   Vector Representation   48
   3.9   Spin Operators   48
   3.10   Pauli Spin Matrices   50
   3.11   Arbitrary Spin Direction   51
   3.12   Bloch Sphere   54
   References   55

4   Qubits   57
   4.1   Bits   57
   4.2   Qubits   59
   4.3   Dirac Ket   59
   4.4   Computational Basis   60
   4.5   Dirac Bra   61
   4.6   Inner Product   62
   4.7   Outer Product   64
   4.8   Projection Operators   65
   4.9   Bloch Sphere   66
   4.10   Two Qubits   66
   4.11   Partial Measurement Rule   69
   4.12   Three Qubits   70
   4.13   Multiple Qubits   71

5   Entanglement   73
   5.1   Composite States   73
   5.2   Separable States   74
   5.3   Entanglement   74
   5.4   Bell States   75
   5.5   Quantum Eraser   77
   5.6   Quantum Metrology   78
   5.7   EPR Paradox and Hidden Variables   78
   5.8   The Bell Test   79
   5.9   Measurements Along One Direction   79
   5.10   Measurements Along Two Directions   81
   5.11   Measurements Along Three Directions   82
   5.12   CHSH Inequality   86
   5.13   Testing Bell's Inequality   88
   5.14   Loopholes   89
   References   90

6   Quantum Key Distribution 91
   6.1   Cryptography 91
   6.2   One-Time Pad 93
   6.3   Polarization Basis 93
   6.4   BB84 95
   6.5   No Cloning Theorem 96
   6.6   Technology 97
   6.7   Other QKD Schemes 97
   6.8   Quantum Random Number Generator 97
   6.9   Quantum Money 98
   6.10   Outlook 98
   References 98

7   Quantum Gates 101
   7.1   Classical Gates 101
   7.2   Quantum Gates 101
   7.3   Circuit or Gate Model of Quantum Computing 102
   7.4   Linear Transformations 103
   7.5   Unitary Transformations 104
   7.6   Reversibility 105
   7.7   Hermitian Operators 106
   7.8   Single-Qubit (Unary) Gates 108
   7.9   NOT (X) Gate 108
   7.10   Y Gate 109
   7.11   Z Gate 110
   7.12   Pauli Gates 111
   7.13   Hadamard Gate 111
   7.14   Identity Operator 112
   7.15   Phase Gate 113
   7.16   Two-Qubit (Binary) Gates 114
   7.17   Controlled NOT (CNOT) Gate 115
   7.18   Other Control Gates 117
   7.19   SWAP Gate 118
   7.20   Fredkin Gate 119
   7.21   Toffoli Gate 120
   7.22   Measurement 121
   7.23   Bell State Circuit 122
   7.24   Greenberger-Horne-Zeilinger (GHZ) State 123
   7.25   Universal Gates 124
   7.26   No Cloning Theorem 124
   References 125

8   Teleportation 127
   8.1   Teleporting Quantum States 127
   8.2   Teleportation Circuit 127
   8.3   Superdense Coding 129
   References 131

9   Tensor Products 133
   9.1   Tensor Product of Qubits 133
   9.2   Tensor Product of Operators 135

10   Quantum Parallelism and Computational Complexity 139
   10.1   Achieving Superpositions 139
   10.2   Quantum Parallelism 142
   10.3   The Measurement Problem 143
   10.4   The Control Problem 144
   10.5   The Challenge 144
   10.6   Computational Complexity 144
   10.7   Big O Notation 145
   10.8   Church-Turing Thesis 147
   10.9   Classes of Quantum Algorithms 147
   10.10 Quantum Advantage 148
   References 148

11   Deutsch Algorithm 149
   11.1   Classical Unary Operator 149
   11.2   Deutsch's Problem 152
   11.3   Deutsch's Quantum Circuit 154
   11.4   Phase Kick-Back 156
   11.5   General Analysis 158
   11.6   Deutsch-Jozsa Algorithm 159
   Reference 161

12   Grover Algorithm 163
   12.1   Searching a Database 163
   12.2   Grover Algorithm 164
   12.3   Initialization 165
   12.4   The Oracle 165
   12.5   Amplitude Amplification 167
   12.6   Diffusion Transform 168
   12.7   Circuit for the Diffusion Transform 169
   12.8   Solving NP Problems 172
   12.9   Geometrical Interpretation 172
   References 176

13 Shor Algorithm 177
  13.1 RSA Encryption 177
  13.2 Shor Algorithm 178
  13.3 Finding Prime Factors 179
  13.4 Birthday Paradox 180
  13.5 Discrete Fourier Transform 180
  13.6 Quantum Fourier Transform 182
  13.7 QFT Matrix 183
  13.8 Period Finding 185
  13.9 QFT Circuit 187
  13.10 Computational Complexity of QFT 188
  13.11 Shor Circuit 189
  13.12 Outlook 191
  References 192

14 Precession 193
  14.1 Rotation Matrices 193
  14.2 Time Evolution of a Quantum System 197
  14.3 Classical Description of Precession 198
  14.4 Hamiltonian of an Electron in a Magnetic Field 201
  14.5 Zeeman Effect 202
  14.6 Quantum Description of Precession 205
  References 206

15 Electron Spin Resonance 209
  15.1 ESR 209
  15.2 Rigorous Derivation of ESR 213
  References 216

16 Two-State Dynamics 217
  16.1 Hamiltonian of a Two-Level System 217
  16.2 Spin Qubits 219
  16.3 Charge Qubits 223
  16.4 Avoided Crossing 224
  16.5 Rabi Formula 227
  16.6 Driven Rabi Oscillations 229
  References 231

17 Implementing Two-Qubit Gates 233
  17.1 CNOT Gate 233
  17.2 Spin Qubit Implementation 233
  17.3 Another Way to Implement Two-Qubit Gates 237

18   DiVincenzo Criteria 241
   18.1   A Scalable Physical System with Well-Defined Qubits 241
   18.2   Initialization of Qubit Register 242
   18.3   Read-Out of Qubits 242
   18.4   Ability to Implement Universal Quantum Gates 242
   18.5   Long Qubit Lifetimes 243
   18.6   Extended DiVincenzo Criterion 243
   References 244

19   Nuclear Magnetic Resonance 245
   19.1   Nuclear Spin 245
   19.2   Two-Level System 247
   19.3   NMR Spectrometer 250
   19.4   Single-Qubit Gates 250
   19.5   Read-Out 252
   19.6   Free Induction Decay 253
   19.7   Relaxation Time 253
   19.8   Decoherence 254
   19.9   NMR Spectrum 255
   19.10 Magnetic Resonance Imaging (MRI) 255
   19.11 Algorithms 256
   19.12 Outlook 256
   References 257

20   Solid-State Spin Qubits 259
   20.1   Electron Confinement 259
   20.2   Quantum Wells, Wires, and Dots 259
   20.3   Coulomb Blockade 262
   20.4   Single Electron Transistor 264
   20.5   Spin Qubits 266
   20.6   Spin Read-Out 267
   20.7   Electrostatic Quantum Dots 267
   20.8   Other Spin Qubits 270
   20.9   Decoherence 270
   20.10 Outlook 272
   References 272

21   Trapped Ion Quantum Computing 275
   21.1   Paul Trap 275
   21.2   Ion Production 277
   21.3   Qubits 277
   21.4   Ion Cooling 277
   21.5   Initialization 279
   21.6   Single-Qubit Operations 279
   21.7   Two-Qubit Operations 279
   21.8   Read-Out 281

21.9   Neutral Atom Quantum Computing 282
21.10   Outlook 282
References 282

22 **Superconducting Qubits** 285
22.1   Superconductivity 285
22.2   BCS Theory 286
22.3   Flux Quantization 288
22.4   Quantum Harmonic Oscillator 289
22.5   Josephson Junction 296
22.6   Potential Energy of a Josephson Junction 300
22.7   Flux Quantization in a Josephson Junction 302
22.8   DC-SQUID (Tunable Josephson Junction) 303
22.9   General Superconducting Circuit 304
22.10   Phase Qubit 305
22.11   Flux Qubit 308
22.12   Charge Qubit 311
22.13   Complementary Variables 315
22.14   Decoherence 315
22.15   Transmon Qubit 316
22.16   Two-Qubit Operations 317
22.17   Circuit QED 318
22.18   Outlook 320
References 321

23 **Adiabatic Quantum Computing** 323
23.1   The Adiabatic Theorem 323
23.2   Ising Model 323
23.3   Outlook 325
References 325

24 **Optical Quantum Computing** 327
24.1   Phase Shifter 327
24.2   Path Encoding 328
24.3   Mirror 329
24.4   Beam Splitter 331
24.5   Single-Qubit Gate 332
24.6   NOT Gate 333
24.7   Two-Qubit Gate 333
24.8   Outlook 335
References 335

25 **Quantum Error Correction** 337
25.1   Leakage 337
25.2   Longitudinal Relaxation 337
25.3   Pure Dephasing 338
25.4   Transverse Relaxation 339

25.5   Noise Mitigation Strategies 341
25.6   Quantum Error Correction 341
25.7   Redundant Encoding 342
25.8   Correcting Bit Flips 343
25.9   Correcting Sign (Phase) Flips 344
25.10   Correcting Bit Flips and Phase Flips 344
25.11   Quantum Threshold Theorem 345
References 345

26 Topological Quantum Computing 347
26.1   Exchange Statistics: Bosons, Fermions and Anyons 347
26.2   Topological Quantum Computing 349
26.3   Fractional Quantum Hall Effect 350
26.4   Majorana Fermions 352
26.5   Outlook 355
References 357

Further Reading 359

Index 361

Chapter 1
Superposition

One of the defining characteristics of quantum mechanics is the possibility of a superposition of states. This is what gives quantum computers their power by enabling a superposition of digital 0 and 1. This chapter explores the principle of quantum superposition.

1.1 Classical Physics

Classical physics encompasses Newtonian mechanics, Maxwell's equations, and thermodynamics. In the early 1900s, it was thought that classical physics could explain everything. Consequently, a scientist (usually attributed to Lord William Thomson Kelvin) expressed the common sentiment that "there is nothing new to be discovered in physics now. All that remains is more and more precise measurement." However, it was these precise measurements and further experiments on the nature of light and matter that would soon reveal new physics.

1.2 Wave Nature of Light

Isaac Newton (Fig. 1.1) believed that light was made of particles. Newton's "Optiks", which appeared in 1704, dealt with his "corpuscular" (particle) theory of light. This was revised to waves with the double slit interference experiment of Thomas Young (Fig. 1.1) in 1801.

Young's double slit interference experiment, illustrated in Fig. 1.2, involves the superposition of two light waves resulting in an interference pattern on a screen or detector. The two light waves are produced by a coherent light source, such as a laser, placed behind two slits with a separation, a, on the order of the wavelength of light, λ. If slit 2 is blocked, a light intensity profile I_1 is observed on the screen due

© The Author(s), under exclusive license to Springer Nature Switzerland AG 2021
R. LaPierre, *Introduction to Quantum Computing*, The Materials Research Society Series,
https://doi.org/10.1007/978-3-030-69318-3_1

Fig. 1.1 From left: Isaac Newton (1643–1727) [1] and Thomas Young (1773–1829) [2]. Each held different views on the nature of light. *Credit* Wikimedia Commons [1, 2]

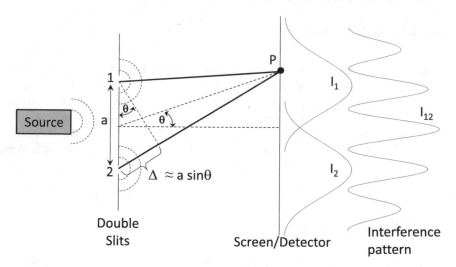

Fig. 1.2 Superposition of two light waves on a screen. If slit 2 is blocked (leaving only slit 1 open), the light intensity profile I_1 is observed. If slit 1 is blocked (leaving only slit 2 open), the light intensity profile I_2 is observed. With both slits open, I_{12} is observed

to slit 1 only. If slit 1 is blocked, a light intensity profile I_2 is observed on the screen due to slit 2 only. I_1 and I_2 are essentially shadows of each slit but broadened due to diffraction. If both slits are open, then an interference pattern I_{12} is observed.

Referring to Fig. 1.2, the interference pattern observed at some point P on the screen at angle θ can be explained by the interference of the two light waves from slit 1 and 2. If the distance between the screen and the slits is much larger than the slit separation, then the two optical paths in Fig. 1.2 are almost parallel. In this case, the path length difference between the two light waves can be approximated as:

$$\Delta \approx a \sin \theta \tag{1.1}$$

This path length difference can be converted to a phase difference, δ, by using the wavevector, k ($k = 2\pi/\lambda$) as a conversion factor:

$$\delta = k\Delta = \frac{2\pi}{\lambda}\Delta = \frac{2\pi}{\lambda}a \sin \theta \tag{1.2}$$

Constructive interference of the two light waves will occur at any point on the screen when the phase difference is an integer multiple of 2π:

$$\delta = 2\pi m, \, m = 0, \pm 1, \pm 2, \ldots \tag{1.3}$$

On the other hand, destructive interference occurs when:

$$\delta = 2\pi (m + 1/2), \, m = 0, \pm 1, \pm 2, \ldots \tag{1.4}$$

Now let us calculate the intensity profile of the interference pattern on the screen. The electric field at point P on the screen due to light from slit 1 can be written as:

$$\mathbf{E}_1 = \mathbf{E}_{01} \, e^{i(kx - \omega t)} \tag{1.5}$$

where k is the wavevector:

$$k = 2\pi/\lambda \tag{1.6}$$

and ω is the angular frequency:

$$\omega = 2\pi \nu \tag{1.7}$$

$\mathbf{E}_{01}$ is the electric field amplitude and x is the distance from slit 1 to point P on the screen. Similarly, with only slit 2 open, we have:

$$\mathbf{E}_2 = \mathbf{E}_{02} \, e^{i(kx - \omega t + \delta)} \tag{1.8}$$

Note that $\mathbf{E}_2$ contains the phase difference of δ from Eq. (1.2) due to the optical path length difference between the two light paths.

Photodetectors (and screens) measure light intensity, not electric field. From classical physics, we know that light intensity is proportional to the square of electric field:

$$I_1 \alpha \, |\mathbf{E}_1(x)|^2 \tag{1.9}$$

$$I_2 \alpha \, |\mathbf{E}_2(x)|^2 \tag{1.10}$$

With both slits open, the total intensity is:

$$I_{12} \alpha \, |\mathbf{E}_1 + \mathbf{E}_2|^2 = I_1 + I_2 + 2\sqrt{I_1 I_2} \cos \delta \tag{1.11}$$

The last term in Eq. (1.11) is the "interference term" which produces the intensity oscillations illustrated in Fig. 1.2. Note that $I_{12} \neq I_1 + I_2$ due to the interference term.

Exercise 1.1 Derive Eq. (1.11).

Young's double slit experiment provided definitive proof of the wave nature of light. The wave theory of light culminated in Maxwell's theory (Fig. 1.3) where the wave nature of light is derived from electrostatics. The issue seemed to be settled—light is a wave. However, this was not the end of the story. What happens if you turn down the intensity of light in Young's double slit experiment?

Fig. 1.3 James Clerk Maxwell (1831–1879). *Credit* Wikimedia Commons [3]

1.3 Particle Nature of Light (Photons)

Let us replace the screen in Young's double slit experiment with a CCD (charge coupled device) camera. If we turn down the intensity of light, individual pixels on the CCD will "light up", as seen in Fig. 1.4a, as though the light were composed of particles like bullets. Furthermore, only light of sufficient frequency will trigger the pixel to light up. This can be explained by interpreting the light particle as a packet of energy which is proportional to the frequency of light:

$$E = h\nu = hc/\lambda \tag{1.12}$$

where h is Planck's constant ($h = 6.626 \times 10^{-34}$ Js), c is the speed of light (3×10^8 m/s), ν is the frequency of light, and λ is the wavelength of light. The CCD pixel is triggered only when the light energy (related to frequency, according to Eq. (1.12)) exceeds the bandgap energy of the CCD semiconductor (usually silicon), as shown in Fig. 1.4b.

Similarly, in 1905, based on the photoelectric effect, Albert Einstein (Fig. 1.5) postulated that light itself is quantized; i.e., made of particles or packets of energy (light quanta) like bullets [4]. Each light particle travels at the speed of light (c) with energy given by Eq. (1.12) and momentum given by:

$$p = E/c = h/\lambda \tag{1.13}$$

where the first term ($E = pc$) can be derived from special relativity and the second term ($p = h/\lambda$) is derived by substitution of Eq. (1.12). E and p can be related to ω and k from Eqs. (1.6) and (1.7). Thus, we can also write:

$$E = \hbar\omega \tag{1.14}$$

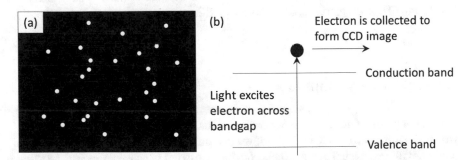

Fig. 1.4 a CCD image where individual pixels "light up", one at a time, under low intensity light. **b** An electron is excited across the semiconductor bandgap, from valence band to conduction band, if the incident light exceeds the semiconductor bandgap energy

Fig. 1.5 Albert Einstein
(1879–1955); Nobel Prize in
Physics in 1921. *Credit*
Wikimedia Commons [5]

$$p = \hbar k \tag{1.15}$$

where $\hbar$ is the reduced Planck constant ($\hbar = h/2\pi$).

In 1926, the chemist, Gilbert Lewis, coined the term "photon" for the particles of light. The concept of the photon solved many problems in physics including the spectrum of blackbody radiation, the photoelectric effect, and the Compton effect.

Let us return to Young's double slit experiment. The intensity of light can be reduced to the point where only a single photon at a time is incident on the double slits. Thus, only one particle at a time is passing through either slit 1 or slit 2. In this case, we might expect a light intensity profile like that for bullets in Fig. 1.6a. Instead, an interference pattern I_{12} is formed when both slits are open, as in Fig. 1.6b. The CCD images of Fig. 1.7 show how the interference pattern builds up photon-by-photon, slowly over time. Although the photons are detected as single particles on the CCD, an interference pattern is still observed from many photons. Single photon experiments of this type introduce many questions. How is an interference pattern possible if photons are particles like bullets? How do individual particles, passing one at a time through either slit 1 or slit 2, interfere with each other? What exactly is interfering? How do particles, like photons, exhibit wave properties? What exactly is waving?

1.4 Probability Amplitude

We cannot predict with certainty where individual particles (like bullets or photons) will strike the screen or CCD. Instead, we can only talk about the probability that a particle will be detected at a particular position. Also, it is an experimental fact that quantum mechanical particles like photons exhibit wave properties, as seen in the double slit interference experiment. In this sense, quantum mechanical particles are not like bullets. If we increase the number of possible paths that a photon can travel

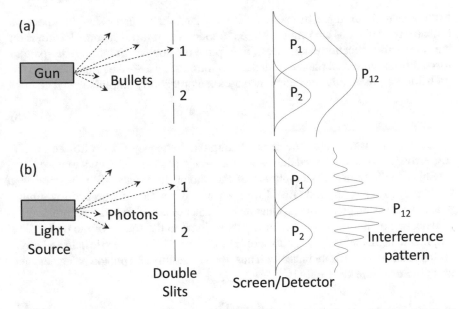

Fig. 1.6 **a** The probability of a bullet hitting the screen is P_1 from slit 1 and P_2 from slit 2. If both slits are open, the probability of a bullet hitting the screen is $P_{12} = P_1 + P_2$. **b** If photons are particles like bullets, we might expect the probability of a photon hitting the screen to be like that for bullets. Instead, an interference pattern is formed. Obviously, $P_{12} \neq P_1 + P_2$

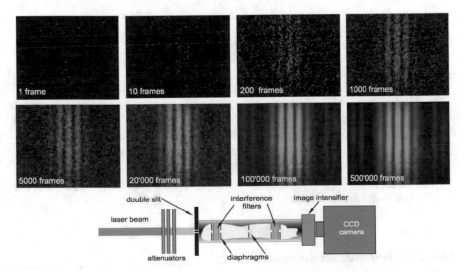

Fig. 1.7 Gradual formation of double slit interference pattern in a CCD image, built up photon-by-photon. Reproduced from T.L. Dimitrova and A. Weis, Am. J. Phys. 76 (2008) 137, https://doi.org/10.1119/1.2815364, with the permission of the American Association of Physics Teachers [6]

to the screen (for example, by opening slit 1 and then slit 2), then we might expect the probability of photons striking the screen to increase. In fact, for some locations on the screen, the probability decreases! To explain the double slit interference pattern formed from individual photons, we need to introduce wave-like properties into the probabilities. To do this, we define the so-called probability amplitude:

$$\Psi = \Psi_0 e^{i(kx-\omega t)} \tag{1.16}$$

where k is the wavevector and ω is the angular frequency (Eqs. (1.6) and (1.7), respectively). Ψ is also called the "wavefunction" or "state" of the photon.

Suppose the probability amplitude of the photon from each slit is Ψ_1 and Ψ_2, analogous to Eqs. (1.5) and (1.6). However, we do not measure probability amplitudes; we measure probabilities. Quantum mechanics postulates that the probability of detecting a photon on the screen is proportional to the square of the probability amplitude. In this sense, probability amplitude is analogous to electric field and probability is analogous to light intensity. Thus, the probability of a photon being detected on the screen from slit 1 only is:

$$P_1 = |\Psi_1|^2 \tag{1.17}$$

and the probability from slit 2 only is:

$$P_2 = |\Psi_2|^2 \tag{1.18}$$

The probability when both slits are open is a superposition:

$$P_{12} = |\Psi_1 + \Psi_2|^2 \tag{1.19}$$

Note that $P_{12} \neq P_1 + P_2$ in the same way that $I_{12} \neq I_1 + I_2$. If P_{12} were to equal $P_1 + P_2$, then no interference pattern would be possible. Instead, P_{12} is given by Eq. (1.19), which will produce an interference term identical to that in Eq. (1.11). What exactly is waving? It is the probability amplitude that is waving. What exactly is interfering? It is the probability amplitudes associated with multiple paths of the single photon that are interfering.

1.5 Mach–Zehnder Interferometer

The Mach–Zehnder (MZ) interferometer (Fig. 1.8) is another way of performing the double slit interference experiment. A 50:50 beam splitter (BS1) sends an incident photon either along path 1 with 50% probability or path 2 with 50% probability. The mirrors M1 and M2 redirect the photon to another 50:50 beam splitter, BS2, which then sends the photon to one of the two detectors, D1 or D2. Only one detector (either D1 or D2) registers a signal at a time, but never both simultaneously. We can

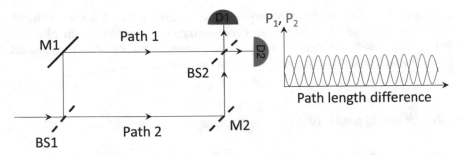

Fig. 1.8 Mach-Zehnder interferometer. The probability of detecting a single photon at detector D1 (P_1, red) or D2 (P_2, blue) oscillates as the path length difference between path 1 and path 2 changes (for example, by moving the mirrors and beam splitters)

change the path length difference between path 1 and path 2 by moving, for example, the position of the mirrors or beam splitters. At each position, we measure many photons and determine the probability, P_1 and P_2, of detecting a photon at D1 and D2, respectively. This is called a correlation experiment. Remarkably, an interference pattern is observed in P_1 and P_2 as a function of the path length difference between path 1 and path 2. We find that $P_1 + P_2 = 1$ because each photon must be observed at either D1 *or* D2, so the two probabilities must add to 1. An interference pattern is observed because the probability amplitude of each individual photon is put into a superposition of the two paths, identical to the double slit interference experiment.

1.6 Superposition

Each individual photon, represented by Eq. (1.16), appears to travel both paths simultaneously in the double slit interference experiment (or MZ interferometer), resulting in interference. This is known as "superposition", as expressed in Eq. (1.19). In quantum mechanics, particles take all possible paths simultaneously (superposition of paths) and interfere. It is this ability of particles to exist in superpositions, and to undergo interference, that gives quantum computers their power.

1.7 "Which Path?" and Measurement Collapse

What happens if you try to determine which path the photon took in the double slit experiment or MZ interferometer? When a measurement is made, we say the wavefunction "collapses" into one of its states. This is called the Copenhagen interpretation of quantum mechanics. For example, we may have a superposition of a particle in path 1 *and* path 2 from the double slits or the MZ interferometer. If we try to determine which path is taken by the photons, the wavefunction collapses into either path 1 *or* path 2, and the superposition is destroyed. This is called "collapse of the wavefunction". The interference pattern disappears. Quantum mechanics does not tell us how this collapse happens upon a measurement. It is one of the fundamental postulates and enduring mystery of quantum mechanics.

1.8 What is Matter?

In 1897, J.J. Thomson (Fig. 1.9) measured the charge to mass ratio (q/m) for the electron, using electric and magnetic fields to steer the electron. Electrons were believed to be particles, like bullets.

In 1924, Louis de Broglie (Fig. 1.10) hypothesized in his Ph.D. thesis that matter also has wave properties. He almost failed his Ph.D. exam! He postulated that matter particles, like photons, have a wavelength. The wavelength, called the de Broglie wavelength, is given by:

$$\lambda = \frac{h}{p} = \frac{h}{\sqrt{2mE}} \tag{1.20}$$

where the momentum, p, is determined by the kinetic energy ($E = p^2/2m$).

In 1927, Davisson and Germer (Fig. 1.11) showed that electrons scattered from a Ni crystal produced constructive interference as the angle θ of the detector changed [9]. An accident in the lab led to these interesting results! The vacuum system broke

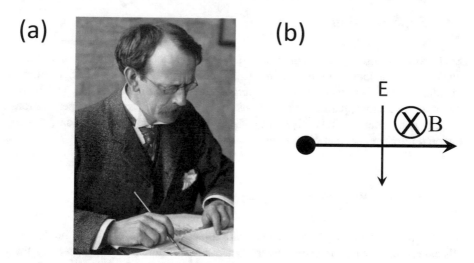

Fig. 1.9 **a** Joseph John "J.J." Thomson (1856–1940); Nobel Prize in Physics in 1906. *Credit* Wikimedia Commons [7]. **b** J.J. Thomson's experiment involved forces from an electric field (E) and magnetic field (B) on an electron that can cancel each other, resulting in a straight electron path. In this way, the charge-to-mass ratio of the electron can be determined

Fig. 1.10 Louis de Broglie
(1892–1987); Nobel Prize in
Physics in 1929. *Credit*
Wikimedia Commons [8]

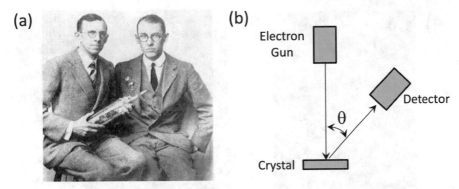

Fig. 1.11 a Left: Clinton Davisson (1881–1958). Nobel Prize in Physics in 1937. Right: Lester
Germer (1896–1971). *Credit* Wikimedia Commons [10]. **b** The Davisson-Germer experiment

while the Ni target was hot, thereby oxidizing the target. While trying to change the
nickel oxide back to nickel by annealing it, they incidentally changed the target from
polycrystalline into a few large crystals needed to observe the electron interference
pattern.

G.P. Thomson, together with Clinton Davisson (Fig. 1.12), also proved the wave
properties of electrons using diffraction experiments [11]. G.P. Thomson was son of
J.J. Thomson who showed electrons were particles! The experiments of Thomson
and Davisson laid the foundations for modern transmission electron microscopy.

In 1961, C. Jönsson (Fig. 1.13a) succeeded in showing double slit interference
for electrons (Fig. 1.13b) by constructing very narrow slits [14, 15]. What happens
if you shoot electrons, one at a time, through the slits? Like the photon interference
pattern of Fig. 1.7, an interference pattern is built up electron-by-electron (Fig. 1.14)
[16]. What is each electron interfering with? If electrons are waves, what is waving?
Once again, the answer is given by Eqs. (1.16)–(1.19) with the wavelength given by
Eq. (1.20). It is the probability amplitude of each electron path that is interfering.

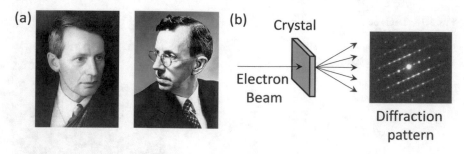

Fig. 1.12 a Left: George Paget Thomson (1892–1975). *Credit* Wikimedia Commons [12]. Right: Clinton Davisson (1881–1958). *Credit* Wikimedia Commons [13]. Both won the Nobel Prize in Physics in 1937. **b** Electron diffraction from a crystal, demonstrating constructive interference (diffraction spots)

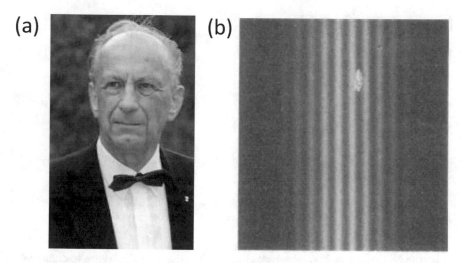

Fig. 1.13 a Claus Jönsson. *Credit* Wikimedia Commons [17]. **b** Electron interference pattern from two slits. Reproduced from C. Jönsson, Am. J. Phys. 42 (1974) 4, https://doi.org/10.1119/1.198 7592, with the permission of the American Association of Physics Teachers [15]

1.9 Wave-Particle Duality

Experiments demonstrate that light and matter exhibit both wave and particle properties—a so-called "wave-particle duality". Interference and diffraction experiments show that light and matter exhibit wave properties. Detection (like electrons or photons hitting a detector) shows that light and matter exhibit particle properties. In the two-slit problem, the photons propagate as waves but are detected as particles. This is the "Copenhagen interpretation"—a particle does not have a location until an experiment localizes it and collapses the wavefunction. This wave-particle

Fig. 1.14 Double slit
interference pattern built up
electron-by-electron. The
total number of electrons is
a 11, **b** 200, **c** 6000,
d 40,000, and **e** 140,000.
Reproduced from A.
Tonomura, J. Endo, T.
Matsuda and T. Kawasaki,
Am. J. Phys. 57 (1989) 117,
https://doi.org/10.1119/1.
16104, with the permission
of the American Association
of Physics Teachers [16]

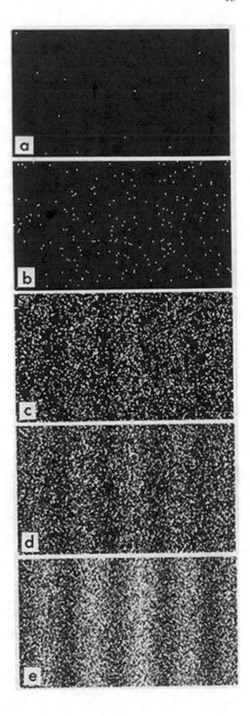

duality is a result of the complementarity of observables, as described by the Heisenberg uncertainty principle. One cannot simultaneously measure both the position and
momentum of a particle with absolute precision. We will have more to say about this
when we discuss the process of measurement in Chap. 7.

Wave properties have been shown for neutrons, atoms, α particles, and molecules
as large as carbon-70. In 2019, a superposition of organic molecules containing
~2000 atoms was demonstrated [18]. How large can a particle be and still exhibit
wave properties? This was the topic of Schrodinger's famous thought experiment in
which a cat is closed inside a box [19]. A radioactive decay triggers a hammer that
breaks a poisonous vial. Until you open the box and observe it (i.e., collapse the
wavefunction), the cat is in a superposition of being alive and dead simultaneously.
This sounds ridiculous, but it illustrates an important question: At what size or under
what conditions does quantum mechanics transition to classical physics? Maintaining
a superposition of quantum objects is one of the challenges of quantum computing.
Maintaining something as big as a cat in a superposition of "alive" and "dead" is
impossible since the cat is constantly interacting with its environment, essentially
resulting in a measurement. This causes "decoherence" of the superposition into one
state or the other ("alive" or "dead"). We will have more to say about decoherence
in Chap. 25.

1.10 The Meaning of Quantum Mechanics

The meaning and interpretation of quantum mechanics is still debated today. The
famous physicist, Richard Feynman (Fig. 1.15), once said: "If you are not confused

Fig. 1.15 Richard Feynman
(1918–1988); Nobel Prize in
Physics in 1965. *Credit*
Wikimedia Commons [20]

by quantum physics, then you haven't really understood it." He also said: "I think I can safely say that nobody understands quantum mechanics." Although we may not *understand* quantum mechanics, we know how to *do* quantum mechanics. To *do* quantum mechanics, we use the postulates of quantum mechanics such as Eqs. (1.16)–(1.19). Another physicist, David Mermin, once said: "Shut up and calculate."

1.11   The Basic Idea Behind Quantum Computing

Quantum computing relies on quantum superpositions of two (binary) states called "qubits". It is this superposition of states that gives an exponential speedup for some computations (Fig. 1.16). For example, suppose we have an input to a classical computer of 4 bits of information (a bit represents information by a 0 or 1). There are 16 possible states of these 4 bits (0000, 0001, 0010, 0011, 0100, 0101, 0110, 0111, 1000, 1001, 1010, 1011, 1100, 1101, 1110, 1111). In a classical computer, this sequence can only be input sequentially. In a quantum computer, a superposition of these states is possible so that all 16 states can be input simultaneously. This is known as quantum parallelism. Thus, quantum algorithms can outperform classical algorithms for certain problems such as Grover's algorithm for database search [21] and Shor's algorithm for factoring large numbers [22, 23]. The "Quantum Algorithm Zoo" website contains a compendium of quantum algorithms [24]. Quantum computers have many other applications, such as:

- the simulation of quantum systems (usually credited to Richard Feynman [25]);
- solving optimization problems (e.g., the traveling salesperson problem);

Fig. 1.16 The difference between classical and quantum computers. With 4 bits (0 or 1), we have 16 possible states. In classical computing, these states are input sequentially. In a quantum computer, we can form a superposition of the states and input them simultaneously. This is known as quantum parallelism

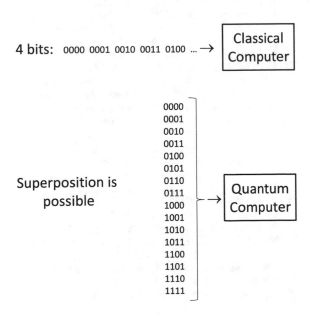

- quantum chemistry: computing the behaviour of molecules and materials;
- designing new pharmaceuticals;
- modeling climate change; and
- machine learning.

In this book, we will study quantum algorithms as well as quantum hardware used to implement the algorithms.

References

1. File: Sir Isaac Newton (1643–1727).jpg. (2020, November 16). *Wikimedia Commons, the free media repository*. Retrieved 10:11, December 7, 2020 from https://commons.wikimedia.org/w/index.php?title=File:Sir_Isaac_Newton_(1643-1727).jpg&oldid=512898315
2. File: Thomas Young by Briggs.jpg. (2020, November 16). *Wikimedia Commons, the free media repository*. Retrieved 10:17, December 7, 2020 from https://commons.wikimedia.org/w/index.php?title=File:Thomas_Young_by_Briggs.jpg&oldid=512878100
3. File: James Clerk Maxwell.png. (2020, July 8). *Wikimedia Commons, the free media repository*. Retrieved 10:20, December 7, 2020 from https://commons.wikimedia.org/w/index.php?title=File:James_Clerk_Maxwell.png&oldid=431814026
4. A. Einstein, Ann. Phys. **17**, 132 (1905)
5. File: Einstein 1921 by F Schmutzer - restoration.jpg. (2020, September 26). *Wikimedia Commons, the free media repository*. Retrieved 10:26, December 7, 2020 from https://commons.wikimedia.org/w/index.php?title=File:Einstein_1921_by_F_Schmutzer_-_restoration.jpg&oldid=471895083
6. T.L. Dimitrova, A. Weis, Am. J. Phys. **76**, 137 (2008)
7. File: J.J Thomson.jpg. (2018, May 11). *Wikimedia Commons, the free media repository*. Retrieved 11:01, December 7, 2020 from https://commons.wikimedia.org/w/index.php?title=File:J.J_Thomson.jpg&oldid=300699582
8. File: Broglie Big.jpg. (2020, September 28). *Wikimedia Commons, the free media repository*. Retrieved 11:04, December 7, 2020 from https://commons.wikimedia.org/w/index.php?title=File:Broglie_Big.jpg&oldid=474406761.
9. C. Davisson, L.H. Germer, Phys. Rev. **30**, 705 (1927)
10. File: Davisson and Germer.jpg. (2015, January 15). *Wikimedia Commons, the free media repository*. Retrieved 11:07, December 7, 2020 from https://commons.wikimedia.org/w/index.php?title=File:Davisson_and_Germer.jpg&oldid=146640797
11. G.P. Thomson, Proc. Roy. Soc. A **117**, 600 (1928)
12. File: George Paget Thomson.jpg. (2020, September 18). *Wikimedia Commons, the free media repository*. Retrieved 11:09, December 7, 2020 from https://commons.wikimedia.org/w/index.php?title=File:George_Paget_Thomson.jpg&oldid=463378222
13. File: Clinton Davisson.jpg. (2020, March 9). *Wikimedia Commons, the free media repository*. Retrieved 11:10, December 7, 2020 from https://commons.wikimedia.org/w/index.php?title=File:Clinton_Davisson.jpg&oldid=402856815
14. C. Jönsson, Z. Phys. **161**, 454 (1961)
15. C. Jönsson, Am. J. Phys. **42**, 4 (1974)
16. A. Tonomura, J. Endo, T. Matsuda, T. Kawasaki, Am. J. Phys. **57**, 117 (1989)
17. File: Prof. Dr. Claus Jönsson 2008.jpg. (2020, October 30). *Wikimedia Commons, the free media repository*. Retrieved 11:15, December 7, 2020 from https://commons.wikimedia.org/w/index.php?title=File:Prof._Dr._Claus_J%C3%B6nsson_2008.jpg&oldid=507805690
18. Y.Y. Fein et al., Nat. Phys. **15**, 1242 (2019)
19. J.D. Trimmer, Proc. Am. Philos. Soc. **124**, 323 (1980)

20. File: RichardFeynman-PaineMansionWoods1984 copyrightTamikoThiel bw.jpg. (2020, October 25). *Wikimedia Commons, the free media repository*. Retrieved 11:27, December 7, 2020 from https://commons.wikimedia.org/w/index.php?title=File:RichardFeynman-PaineM ansionWoods1984_copyrightTamikoThiel_bw.jpg&oldid=501086478.
21. L. Grover, in *Proceedings of 28th Annual ACM Symposium on the Theory of Computation* (ACM Press, New York, 1996), p. 212
22. P. Shor, in *Proceedings of 35th Annual Symposium on the Foundations of Computer Science*, ed. by S Goldwasser (IEEE Computer Society, Los Alamos, NM, 1994), p. 124796
23. P.W. Shor, SIAM J. Comput. **2**, 1484 (1997)
24. https://quantumalgorithmzoo.org/
25. R.P. Feynman, Int. J. Theor. Phys. **21**, 467 (1982)

Chapter 2
Quantization

Quantum computers represent information (digital bits, 0 and 1) using two different states of a quantum system. One possibility is to represent 0 and 1 using two different quantized energy levels; e.g., 0 as the ground state and 1 as an excited state. This chapter examines the quantization of measurables, especially energy, in quantum mechanics.

2.1 Probability Amplitude

As we saw in Chap. 1, it is an experimental fact that particles (photons, electrons, etc.) exhibit wave properties. The wave nature is described by the probability amplitude:

$$\Psi(x, t) = \Psi_0 e^{i(kx - \omega t)} \tag{2.1}$$

where k is the wavevector and ω is the angular frequency. Ψ is also called the "wavefunction" or the "state" of the particle.

In Chap. 1, we saw that the probability (what we measure) is related to the square of the probability amplitude. More concisely, $|\Psi(x,t)|^2$ is called the probability density. In one dimension, $|\Psi(x,t)|^2 \, dx$ is the probability of finding the particle in the region x to $x + dx$ at time t. In three dimensions, $|\Psi(x,y,z,t)|^2 \, dxdydz$ is the probability of finding the particle in the volume element $dV = dxdydz$ at time t. This is known as the Born interpretation or Born rule, developed by the physicist Max Born (Fig. 2.1).

Classical physics is deterministic. If all particle positions (r_i) and velocities (v_i) are known at some time $(t = 0)$, then subsequent positions and velocities can be determined at some other time, t: $r_i(t = 0), v_i(t = 0) \rightarrow r_i(t), v_i(t)$. For classical systems, we can trace a particle's path in space. Conversely, quantum physics is probabilistic. In quantum mechanics, it is not possible to trace the path of a particle because we can only predict probabilities. If the same measurement is performed on several identically prepared systems, one cannot in general expect the same outcome. This is not

© The Author(s), under exclusive license to Springer Nature Switzerland AG 2021
R. LaPierre, *Introduction to Quantum Computing*, The Materials Research Society Series,
https://doi.org/10.1007/978-3-030-69318-3_2

because we lack information about the system (like in statistical mechanics); rather, it is because the outcome of a measurement is inherently probabilistic.

2.2 Time-Dependent Schrodinger Equation

If particles are waves, we need a wave equation. The wave equation for matter particles was developed by Erwin Schrodinger (Fig. 2.2) [2]. The time-dependent Schrodinger equation states that a particle with potential energy $V(x, t)$ has a probability amplitude described by:

$$-\frac{\hbar^2}{2m}\frac{\partial^2 \Psi(x, t)}{\partial x^2} + V(x, t)\Psi(x, t) = i\hbar\frac{\partial \Psi(x, t)}{\partial t} \tag{2.2}$$

No derivation of Schrodinger's equation is possible; it is a fundamental postulate of quantum mechanics. All non-relativistic quantum mechanics is contained in Schrodinger's equation. The relativistic version of Schrodinger's equation is called the Klein–Gordon equation, and the equation including spin is called the Dirac equation. We will deal only with the Schrodinger equation.

2.3 The Free Particle

For example, consider a free particle, meaning the potential energy is $V(x, t) = 0$. By setting $V(x, t) = 0$ in Eq. (2.2), the time-dependent Schrodinger equation simplifies to:

$$-\frac{\hbar^2}{2m}\frac{\partial^2 \Psi(x, t)}{\partial x^2} = i\hbar\frac{\partial \Psi(x, t)}{\partial t} \tag{2.3}$$

The solution to this differential equation is Eq. (2.1), which is easily verified by substituting Eqs. (2.1) into (2.3), resulting in:

$$\frac{\hbar^2 k^2}{2m}\Psi = \hbar\omega\Psi \tag{2.4}$$

Since the momentum is $p = \hbar k$ (Eq. 1.15), the left-hand side of Eq. (2.4) contains the kinetic energy of the particle, $p^2/2m$. The right-hand side of Eq. (2.4) is the total energy of the particle, $E = \hbar\omega$, identical to Eq. (1.14). Since the potential energy is zero for the free particle, Eq. (2.4) simply states that the total energy is equal to the kinetic energy of the particle.

2.4 The Hamiltonian

A more compact way of writing the time-dependent Schrodinger equation is:

$$\widehat{H}\Psi = i\hbar\frac{\partial \Psi}{\partial t} \tag{2.5}$$

where $\widehat{H}$ is called the Hamiltonian:

$$\widehat{H} = -\frac{\hbar^2}{2m}\frac{\partial^2}{\partial x^2} + V(x, t) \tag{2.6}$$

The "hat" symbol (^) indicates that $\widehat{H}$ is an operator. This means that $\widehat{H}$ operates on the wavefunction $\Psi(x, t)$. Equation (2.5) describes the dynamical behavior of a quantum system. It is similar, in a sense, to Newton's equations of motion in classical mechanics. The term "Hamiltonian" is derived from the name of the mathematician Rowan Hamilton who made significant contributions to the theory of mechanics. The term Hamiltonian is usually applied in mechanics if the total energy is expressed in terms of momentum and position variables, as opposed to position and velocity. As in classical mechanics, determining the Hamiltonian is usually the first step in understanding the dynamical behavior of a quantum system.

2.5   Time-Independent Schrodinger Equation

Let us solve the time-dependent Schrodinger equation, Eq. (2.2), for the case where V is independent of time but still dependent on x: $V = V(x)$. As a solution, let us try a separation of variables, which is a common technique for solving differential equations:

$$\Psi(x, t) = \psi(x)\psi(t) \tag{2.7}$$

Substituting Eqs. (2.7) into (2.2), gives:

$$-\frac{\hbar^2}{2m}\psi(t)\frac{d^2\psi(x)}{dx^2} + V(x)\psi(t)\psi(x) = i\hbar\psi(x)\frac{d\psi(t)}{dt} \tag{2.8}$$

Dividing $\psi(x)\psi(t)$ into both sides of Eq. (2.8) gives:

$$-\frac{1}{\psi(x)}\frac{\hbar^2}{2m}\frac{d^2\psi(x)}{dx^2} + V(x) = i\hbar\frac{1}{\psi(t)}\frac{d\psi(t)}{dt} \tag{2.9}$$

The left-hand side of Eq. (2.9) depends on x only, while the right-hand side depends on t only. The only possible solution is provided when both sides of Eq. (2.9) is equal to a constant, which we will call E:

$$-\frac{1}{\psi(x)}\frac{\hbar^2}{2m}\frac{d^2\psi(x)}{dx^2} + V(x) = i\hbar\frac{1}{\psi(t)}\frac{d\psi(t)}{dt} = E \tag{2.10}$$

After some simple rearrangement, Eq. (2.10) yields two equations:

$$-\frac{\hbar^2}{2m}\frac{d^2\psi(x)}{dx^2} + V(x)\psi(x) = E\psi(x) \tag{2.11}$$

and:

$$i\hbar\frac{d\psi(t)}{dt} = E\psi(t) \tag{2.12}$$

As will be shown below, the constant E is the total energy of the particle. We will deal with Eq. (2.11) first, and Eq. (2.12) in Sect. 2.11.

Equation (2.11) is known as the time-independent Schrodinger equation for obvious reasons (it contains no time dependence). If we set $V(x) = 0$, as we did for the free particle, and substitute Eq. (2.1) into (2.11), then we recover Eq. (2.4) with $\hbar\omega$ replaced by E. Hence, we see that E is the total energy of the particle. Therefore, Eq. (2.11) can be interpreted as a statement on the conservation of energy where the first term on the left-hand side is the kinetic energy (as we saw with the free particle), the second term is the potential energy, and these sum to the total energy on the right-hand side of Eq. (2.11). As before, we can simplify Eq. (2.11) using the Hamiltonian:

$$\widehat{H}\psi = E\psi \tag{2.13}$$

which is simply a more compact form of the time-independent Schrodinger equation.

2.6   Observables

In quantum mechanics, all observables are described by an eigenvalue equation. This is another of the fundamental postulates of quantum mechanics. An eigenvalue equation takes the form:

$$\widehat{A}\Psi = a\Psi \tag{2.14}$$

where $\widehat{A}$ is an operator or eigenfunction, and a is the eigenvalue that is the result of a measurement. Examples of operators include the Hamiltonian, $\widehat{H} = -\frac{\hbar^2}{2m}\frac{\partial^2}{\partial x^2} + V(x,t)$ as in Eq. (2.13); linear momentum, $\widehat{p} = \frac{\hbar}{i}\nabla$; angular momentum, $\widehat{L} = \frac{\hbar}{i}(\mathbf{r} \times \nabla)$; and many others.

2.7   Infinite Quantum Well or "Particle in a Box"

The "particle in a box" or "infinite quantum well" (Fig. 2.3) is a popular problem in introductory quantum mechanics, not only because it is one of the few problems that can be easily solved analytically, but also because of its practical significance. For example, quantum wells are used in many optoelectronic devices [4]. The problem consists of a particle (for example, an electron) trapped in the domain between $x = 0$ and $x = L$ due to an infinite potential energy, V, outside this domain. This problem is similar to the free particle but with boundary conditions. We can solve the time-independent Schrodinger equation, Eq. (2.11), inside the well:

Fig. 2.3 Potential energy of
the infinite quantum well

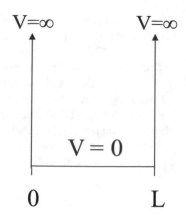

$$-\frac{\hbar^2}{2m}\frac{d^2\psi(x)}{dx^2} = E\psi(x), \text{ inside the well where } V = 0 \qquad (2.15)$$

Rearranging, we get:

$$\frac{d^2\psi(x)}{dx^2} = -\frac{2mE}{\hbar^2}\psi(x) \qquad (2.16)$$

or

$$\frac{d^2\psi(x)}{dx^2} = -k^2\psi(x) \qquad (2.17)$$

where k is the wavevector (since $E = \hbar^2 k^2/2m$).

The solution to Eq. (2.17) is:

$$\psi(x) = A\sin(kx) \qquad (2.18)$$

which is easily verified by substituting Eq. (2.18) into (2.17). Equation (2.18) contains two unknowns: A and k. To solve for k, we use the boundary conditions. The state $\psi(x)$ cannot exist outside the potential well; otherwise, the potential energy of the particle would be infinite. Therefore, we must satisfy the following boundary condition:

$$\psi(L) = 0 \qquad (2.19)$$

(The boundary condition $\psi(0) = 0$ simply yields $k = 0$ or $A = 0$, which is a trivial solution where ψ vanishes). Equation (2.19) yields:

$$k_n = n\pi/L \qquad (2.20)$$

where $n = 1, 2, 3, \ldots$. Hence, not all values of k are allowed. Instead, k is quantized to certain discrete values, k_n, and Eq. (2.18) becomes:

$$\psi_n(x) = A \sin\left(\frac{n\pi}{L}x\right) \tag{2.21}$$

where we have added a subscript n to ψ to indicate the discrete solutions.

To solve for A, we use the "normalization" condition. The probability of finding the particle somewhere in the box should equal 1:

$$\int_0^L |\psi_n(x)|^2 dx = 1 \tag{2.22}$$

Substituting Eq. (2.21), we get:

$$\int_0^L A^2 \sin^2\left(\frac{n\pi}{L}x\right) dx = 1 \tag{2.23}$$

To solve Eq. (2.23), a trigonometric substitution is made:

$$A^2 \int_0^L \frac{1}{2}\left[1 - \cos\left(\frac{2n\pi}{L}x\right)\right] dx = 1 \tag{2.24}$$

Equation (2.24) is easily solved, giving:

$$A = \sqrt{\frac{2}{L}} \tag{2.25}$$

Thus, the final solution is:

$$\psi_n(x) = \sqrt{\frac{2}{L}} \sin\left(\frac{n\pi}{L}x\right) \tag{2.26}$$

These $\psi_n(x)$ are known as "stationary states", meaning the probability is independent of time. These are also known as "bound states". The first three states (ψ_1, ψ_2, ψ_3) for the infinite quantum well are illustrated in Fig. 2.4.

According to the Born rule, $|\psi_n(x)|^2 dx$ is the probability of finding the particle between x and $x + dx$. The probability density is:

$$|\psi_n(x)|^2 = \frac{2}{L} \sin^2\left(\frac{n\pi}{L}x\right) \tag{2.27}$$

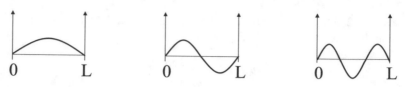

Fig. 2.4 The probability amplitude, $\psi_n(x)$, for a few of the states of the infinite quantum well

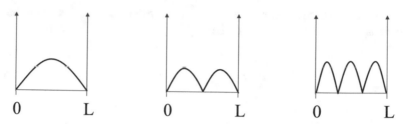

Fig. 2.5 The probability density, $|\psi_n(x)|^2$, for a few of the states of the infinite quantum well

The probability density for the first three states of the infinite quantum well is depicted in Fig. 2.5.

2.8 Quantization of Energy

As we saw above, the boundary conditions of the infinite quantum well led to quantization of the particle momentum (Eq. 2.20) and, thus, quantization of the energy:

$$p = \hbar k_n = n\hbar\pi/L \tag{2.28}$$

$$E_n = p^2/2m = \hbar^2 k_n^2/2m = n^2\hbar^2\pi^2/2mL^2 \tag{2.29}$$

This leads to a "ladder" of energy states as depicted in Fig. 2.6. The lowest possible energy, E_1, is called the ground state energy, and the higher energy states (E_2, E_3, ...) are called the excited states.

Why don't we notice the discretization of energy in everyday objects (classical particles)? If we substitute, for example, $L = 1$ m and $m = 1$ kg into Eq. (2.29) we get $E = (5.5 \times 10^{-68}$ J$)\, n^2$. This is an extremely small separation between quantized energy levels. This means that energy appears to be continuous, not discrete, for large objects. On the other hand, if we substitute the mass of an electron ($m = 9.11 \times 10^{-31}$ kg) into a quantum well of width on the order of 1 nm, then we get energy separations on the order of 1 eV, which is easily measurable (it can be measured by the emission spectrum of light when electrons transition between energy levels).

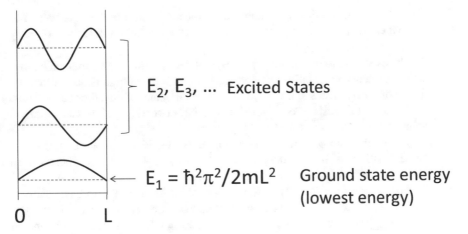

 $E_2, E_3, \ldots$ Excited States

$E_1 = \hbar^2\pi^2/2mL^2$ Ground state energy
(lowest energy)

Fig. 2.6 Discretization of energy in the infinite quantum well

2.9 Orthonormal States

An important property of ψ_n is that these states are "orthonormal", meaning:

$$\int_{-\infty}^{+\infty} \psi_i^* \psi_j \, dx = \delta_{ij} \tag{2.30}$$

where ψ_i^* is the complex conjugate and δ_{ij} is the Kronecker delta defined as:

$$\delta_{ij} = \begin{cases} 0, i \neq j \\ 1, i = j \end{cases} \tag{2.31}$$

When $i = j$, Eq. (2.30) is simply the normalization condition of Eq. (2.22). When $i \neq j$, the two states are said to be orthogonal.

Exercise 2.1 Show that the first two states (two lowest energy levels) of the infinite quantum well are orthonormal.

2.10 Basis States

The ψ_n form "basis" states. Basis states satisfy the following properties:

1. The basis states are the eigenfunctions associated with the possible values of some measurable property of the system; for example, states $\psi_n(x)$ with energy E_n, where $\widehat{H}\psi_n = E_n\psi_n$.

2. The basis states are orthogonal: $\int_{-\infty}^{+\infty}\psi_i^*\psi_j dx = \delta_{ij}$. This means that each basis state represents a mutually exclusive possibility. If the system is observed to be in one of the basis states, it will not be observed in any of the other basis states.

3. The basis states must be "complete" in that they cover all possible measurement values.

4. Any possible state ψ of the system can be expressed as a linear combination (superposition) of basis states: $\psi = c_1\psi_1 + c_2\psi_2 + \cdots c_n\psi_n$ where c_n are complex coefficients (probability amplitudes).

5. The basis states are linearly independent, which means that no basis state in the set can be expressed as a linear combination of other basis states in the set: $\psi_i \neq \sum_{n\neq i} c_n\psi_n$. If $c_1\psi_1 + c_2\psi_2 + \cdots c_n\psi_n = 0$ for complex coefficients c_i and basis states ψ_i, then $c_i = 0$ for all i. The basis vectors are said to "span" the state space.

6. The eigenvalues of the basis states are the result of a measurement. Upon measurement, the state, $\psi = \sum c_n\psi_n$, will collapse with probability $|c_n|^2$ to one of the basis states ψ_n. The probabilities must sum to unity: $\sum |c_n|^2 = 1$

2.11 Time Dependence

What does all this have to do with quantum computing? In computing, information is represented digitally by 0's and 1's. In quantum computing, 0 and 1 are represented by two different quantum states. For example, 0 may be represented by a particle (e.g., an electron) in the ground state of a quantum well, and 1 may be represented by an electron in the first excited state (Fig. 2.7).

To do computing, something needs to change in time. We now need to consider Eq. (2.12) which arose from the separation of variables in the solution to the time-dependent Schrodinger equation. After a simple rearrangement of Eq. (2.12), and replacing E with E_n for a basis state, we get:

$$i\hbar\frac{d\psi(t)}{dt} = E_n\psi(t) \qquad (2.32)$$

Fig. 2.7 Representation of 0 and 1 by an electron in the ground and first excited state, respectively, of an infinite quantum well

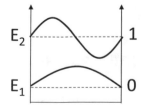

The solution to Eq. (2.32) is:

$$\psi(t) = \psi(0)e^{-iE_n t/\hbar} \tag{2.33}$$

where $\psi(0)$ is the initial state at time $t = 0$. Equation (2.33) is easily verified as a solution by substitution into (2.32). Combining the spatial dependence of Eq. (2.26) and the time dependence of Eq. (2.33) gives the complete solution for a basis state:

$$\Psi_n(x, t) = \left[\sqrt{\frac{2}{L}} \sin\left(\frac{n\pi}{L}x\right)\right] e^{-iE_n t/\hbar} \tag{2.34}$$

$$= \left[\sqrt{\frac{2}{L}} \sin\left(\frac{n\pi}{L}x\right)\right] e^{-i(n^2 \hbar \pi^2/2mL^2)t} \tag{2.35}$$

Note that the probability density is $|\Psi(x, t)|^2 = \Psi * \Psi = |\psi_n(x)|^2$ where Ψ^* is the complex conjugate of Ψ; i.e., Ψ is independent of time as required for a stationary state (by definition, stationary states do not change in time).

Equation (2.35) are the basis states for a particle in an infinite quantum well. According to the properties of basis states, any state Ψ can be expressed as a linear combination (superposition) of basis states:

$$\Psi(x, t) = \sum_{n=1}^{\infty} c_n \Psi_n = \sum_{n=1}^{\infty} c_n \left[\sqrt{\frac{2}{L}} \sin\left(\frac{n\pi}{L}x\right)\right] e^{-i(n^2 \hbar \pi^2/2mL^2)t} \tag{2.36}$$

The superposition of Eq. (2.36) is a solution of the time-dependent Schrodinger equation because the Schrodinger equation is a linear differential equation. The c_n are the complex probability amplitudes, and $|c_n|^2$ are the probabilities of measuring state Ψ_n with energy E_n. The net probability must sum to unity: $\sum_{n=1}^{\infty} |c_n|^2 = 1$. After measurement, we obtain some energy E_n and the state collapses to Ψ_n with the probability $|c_n|^2$.

As an example, consider the superposition of an electron in the two lowest energy levels of an infinite quantum well (superposition of 0 and 1):

$$\Psi(x, t) = c_1 \Psi_1 + c_2 \Psi_2 = \frac{1}{\sqrt{2}} e^{-iE_1 t/\hbar} \sqrt{\frac{2}{L}} \sin\left(\frac{\pi}{L}x\right)$$

$$+ \frac{1}{\sqrt{2}} e^{-iE_2 t/\hbar} \sqrt{\frac{2}{L}} \sin\left(\frac{2\pi}{L}x\right) \tag{2.37}$$

The probability amplitudes are $c_1 = \frac{1}{\sqrt{2}}$ for the electron in the ground state with energy E_1, and $c_2 = \frac{1}{\sqrt{2}}$ for an electron in the first excited state with energy E_2. The probability of measuring the electron with energy E_1 is $|c_1|^2 = 1/2$, and the probability of measuring energy E_2 is $|c_2|^2 = 1/2$. Note that the probabilities add to unity as required. Thus, Eq. (2.37) is properly normalized.

The probability density for the superposition of Eq. (2.37) is:

$$|\Psi(x,t)|^2 = \Psi\Psi^* = \frac{1}{L}\sin^2\left(\frac{\pi}{L}x\right) + \frac{1}{L}\sin^2\left(\frac{2\pi}{L}x\right)$$
$$+ \frac{2}{L}\sin\left(\frac{\pi}{L}x\right)\sin\left(\frac{2\pi}{L}x\right)\cos(\omega t) \tag{2.38}$$

where

$$\omega = (E_2 - E_1)/\hbar \tag{2.39}$$

Equation (2.38) exhibits a time dependence! The probability distribution oscillates back and forth inside the quantum well (Fig. 2.8) with the angular frequency, $\omega = (E_2 - E_1)/\hbar$. We can consider this as a type of clock. Only the individual eigenstates or basis states (Ψ_n) of Eq. (2.35) are time-independent stationary states. Superpositions of Ψ_n are not time-independent and are not stationary states. In general, we can express a superposition of two stationary states as:

$$\Psi(x,t) = \psi_1(x)e^{-\frac{iE_1 t}{\hbar}} + \psi_2(x)e^{-\frac{iE_2 t}{\hbar}} \tag{2.40}$$

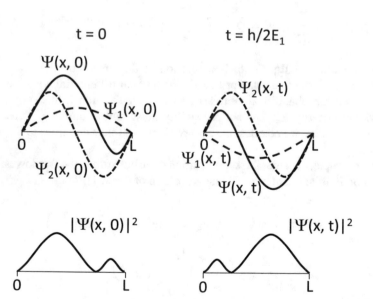

Fig. 2.8 A superposition of a particle in the ground and first excited state of an infinite quantum well at $t = 0$ and $t = h/2E_1$. This produces a time-dependent state where the probability density oscillates between the left and right side of the well

The probability density is:

$$|\Psi(x,t)|^2 = \Psi \, \Psi^* = |\psi_1(x)|^2 + |\psi_2(x)|^2 + 2|\psi_1(x)||\psi_2(x)|\cos(\omega t) \qquad (2.41)$$

Does Eq. (2.41) look familiar? It is analogous to Eq. (1.11) for the interference of two photons.

2.12 The Hydrogen Atom

The solution of the time-independent Schrodinger equation for the H atom is considerably more complicated, but it is another of the few examples that can be solved exactly. The potential energy is the Coulomb potential for the electron bound by a proton (Fig. 2.9). The solution can be found in any introductory quantum mechanics textbook, resulting in a quantization of energy E_n, orbital angular momentum (L), and orbital angular momentum projected along a particular direction, say z (L_z):

$$E_n = 13.6 \, \text{eV}/n^2, n = 1, 2, 3, \ldots \qquad (2.42)$$

$$L^2 = l(l+1)\hbar^2, \quad l = 0, 1, 2, \ldots, n-1 \qquad (2.43)$$

$$L_z = m\hbar, m = 0, \pm 1, \pm 2, \ldots, \pm l \text{ or } m = -l, -l+1, \ldots +l \qquad (2.44)$$

There are three quantum numbers: n, l, and m. n is called the principal quantum number, l is called the orbital angular momentum quantum number, and m is called the magnetic or azimuthal quantum number. There are n values of l, and $2l + 1$ values of m. In spherical coordinates, the stationary states of the H atom take the form:

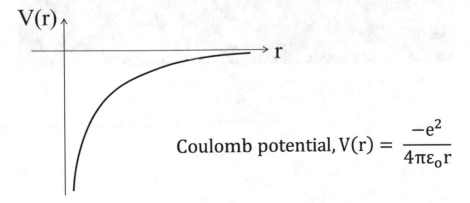

Fig. 2.9 The Coulomb potential for the H atom

$$\psi_{n,l,m}(r, \theta, \phi) = R_{n,l}(r)P_{l,m}(\theta)e^{im\phi} \qquad (2.45)$$

where $R_{n,l}(r)$ contains the radial dependence, $P_{l,m}(\theta)$ contains the polar angle dependence, and $e^{im\phi}$ contains the azimuthal angle dependence. $\Psi_{n,l,m}$ is a simultaneous eigenstate of E_n, L and L_z. Plots showing surfaces of constant $\Psi_{n,l,m}(r, \theta, \phi)$ are known as "charge clouds" or orbitals, as shown in Fig. 2.10. The values of $l = 0$, 1, 2 and 3 are also called s, p, d and f states in spectroscopic notation. Atomic orbitals are responsible for chemistry.

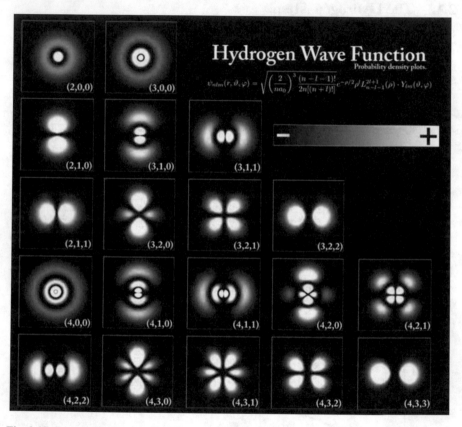

Fig. 2.10 Depiction of $\psi_{n,l,m}$ orbitals for different states of the H atom. The brighter areas represent a higher probability of finding the electron. *Credit* Wikimedia Commons [5]

2.13 Expectation Value

The average value of an observable, called the expectation value in quantum mechanics, is given by:

$$\langle A \rangle = \frac{\int_{-\infty}^{+\infty} \Psi^* \hat{A} \Psi \, dx}{\int_{-\infty}^{+\infty} \Psi^* \Psi \, dx} \qquad (2.46)$$

where Ψ is the wavefunction and $\hat{A}$ is an operator corresponding to the observable. For example, if Ψ is an eigenstate of the operator $\hat{A}$, then $\hat{A}\,\Psi = a\Psi$ where a is the result of a measurement. Thus,

$$\langle A \rangle = \frac{\int_{-\infty}^{+\infty} \Psi^* a \Psi \, dx}{\int_{-\infty}^{+\infty} \Psi^* \Psi \, dx} \qquad (2.47)$$

The observed value, a, is a constant and can be moved outside the integral, giving:

$$\langle A \rangle = \frac{a \int_{-\infty}^{+\infty} \Psi^* \Psi \, dx}{\int_{-\infty}^{+\infty} \Psi^* \Psi \, dx} = a \qquad (2.48)$$

Since Ψ is an eigenstate corresponding to the measurable, a, its average value is a, as expected. If Ψ is a superposition of eigenstates, $\Psi = c_1 \Psi_1 + c_2 \Psi_2 + \cdots c_n \Psi_n$ where each Ψ_i are eigenstates with corresponding eigenvalues a_i, then Eq. (2.46) gives the average value:

$$\langle A \rangle = |c_1|^2 a_1 + |c_2|^2 a_2 + \cdots + |c_n|^2 a_n \qquad (2.49)$$

where $|c_i|^2$ are the probability for each measurable a_i and the probabilities sum to unity ($|c_1|^2 + |c_2|^2 + \cdots + |c_n|^2 = 1$).

Exercise 2.2 Prove Eq. (2.49).

2.14 Fundamental Postulates of Quantum Mechanics

Let us summarize the fundamental postulates of quantum mechanics covered in Chaps. 1 and 2. The postulates are:

1. A quantum mechanical system is completely specified by a state $\Psi(r, t)$.

2. Every observable, *a*, has a corresponding operator, $\hat{A}$, in quantum mechanics. The operator $\hat{A}$ is Hermitian (more on Hermitian operators in Chap. 7).

3. Observables associated with operator $\hat{A}$ satisfy the equation $\hat{A}\Psi = a\Psi$ where *a* are the eigenvalues that are the result of the measurement and Ψ are the eigenstates. The state can be represented as a superposition of eigenstates with complex coefficients representing the probability amplitudes of each state. The square of the modulus of the probability amplitude represents the probability of measuring each state.

4. The average value of the observable corresponding to operator $\hat{A}$ is given by
$$\langle A \rangle = \frac{\int_{-\infty}^{+\infty} \Psi^* \hat{A} \Psi \, dx}{\int_{-\infty}^{+\infty} \Psi^* \Psi \, dx}$$

5. The wavefunction evolves in time according to the time-dependent Schrödinger equation.

References

1. File: Max Born.jpg. (2017, October 26). *Wikimedia Commons, the free media repository.* Retrieved 11:48, December 7, 2020 from https://commons.wikimedia.org/w/index.php?title= File:Max_Born.jpg&oldid=264524855
2. E. Schrödinger, Phys. Rev. **28**, 1049 (1926)
3. File: Erwin Schrödinger (1933).jpg. (2020, October 28). *Wikimedia Commons, the free media repository.* Retrieved 11:52, December 7, 2020 from https://commons.wikimedia.org/w/index. php?title=File:Erwin_Schr%C3%B6dinger_(1933).jpg&oldid=505588599
4. R. Dingle, W. Wiegmann, C.H. Henry, Phys. Rev. Lett. **33**, 827 (1974)
5. File: Hydrogen Density Plots.png. (2020, December 1). *Wikimedia Commons, the free media repository.* Retrieved 12:06, December 7, 2020 from https://commons.wikimedia.org/w/index. php?title=File:Hydrogen_Density_Plots.png&oldid=516077335

Chapter 3
Spin

In Chap. 2, we represented "0" by an electron in a lower energy level (ground state) and "1" by an electron in an excited state. Another method of implementing quantum computing is by using the spin of electrons or protons to represent 0 and 1. For example, we can represent spin "down" as 0 and spin "up" as 1 (or vice versa), and we can have superpositions of these two spin states. Let us examine spin in more detail (Fig. 3.1).

3.1 Orbital Angular Momentum

First, let us examine orbital angular momentum, such as that of an electron in a hydrogen atom (Fig. 3.2), which we explored briefly at the end of Chap. 2. Angular momentum is defined as:

$$\mathbf{L} = \mathbf{r} \times \mathbf{p} \tag{3.1}$$

where $\mathbf{r}$ is the position vector of the electron relative to the origin (nucleus of the H atom), and $\mathbf{p}$ is the linear momentum vector of the electron. In classical mechanics, $\mathbf{L}$ is a constant of the motion, meaning $\mathbf{L}$ is conserved and it can take a continuum of values. Conversely, in quantum mechanics, $\mathbf{L}$ is quantized for bound states in two- or three-dimensional problems (angular momentum is not possible in only one dimension).

Quantization of $\mathbf{L}$ was first proposed by Bohr (Fig. 3.3) in his model of the H atom. In a semi-classical model, Bohr proposed that an integer number of electron wavelengths must fit within one complete circular orbit of the electron around the nucleus, resulting in a standing wave pattern (Fig. 3.4).

$$2\pi r = m\lambda = mh/p, \ m = 1, 2, 3, \ldots \tag{3.2}$$

© The Author(s), under exclusive license to Springer Nature Switzerland AG 2021
R. LaPierre, *Introduction to Quantum Computing*, The Materials Research Society Series,
https://doi.org/10.1007/978-3-030-69318-3_3

Fig. 3.1 Representation of 0 and 1 by particle spin

spin down
"0"

spin up
"1"

Fig. 3.2 Classical model of angular momentum (**L**) and magnetic moment (**μ**) of an electron in a hydrogen atom. The electron velocity (**v**), current (*I*) and area encircled by the electron (*A*) are indicated

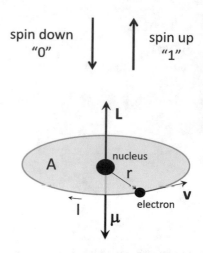

Fig. 3.3 Niels Bohr (1885–1962); Nobel Prize in Physics in 1922. *Credit* Wikimedia Commons [1]

Fig. 3.4 Bohr's model where an integer number of wavelengths must fit within the circumference, $2\pi r$, of the circular electron orbits. Here, the circumference is stretched out as a line for clarity

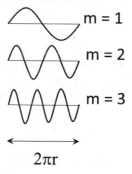

$m = 1$

$m = 2$

$m = 3$

$2\pi r$

In a sense, this condition is necessary to prevent the electron from destructively interfering with itself in its orbit around the nucleus. Alternatively, we say that the wavefunction must be "single-valued". Thus, Fig. 3.4 depicts the allowed orbits of the electron. For a circular orbit, **r** and **p** are always perpendicular (Fig. 3.2). Hence, according to the definition of the cross product, Eq. (3.1) gives the magnitude of *L*:

$$L = rp \tag{3.3}$$

Rearranging Eq. (3.2) and substituting into Eq. (3.3) gives:

$$L = rp = mh/2\pi = m\hbar, \quad m = 1, 2, 3, \ldots \tag{3.4}$$

The direction of **L** is given by the right-hand rule (**r** × **p**) with the direction as shown in Fig. 3.2.

Although Bohr's model correctly predicts the quantization of angular momentum along z (L_z), the planetary model of the atom where the electron orbits the nucleus is impossible because, according to classical electromagnetism, the electron would radiate away its energy and thus spiral into the nucleus due to its centripetal motion. Today, we know that the correct solution is given by the Schrodinger equation where the total orbital angular momentum (L) for the H atom is given by Eq. (2.43) of Chap. 2, and the corresponding component of orbital angular momentum along z (L_z) is given by Eq. (2.44).

3.2 Magnetic Dipole Moment

Classically, the electron orbit also produces a magnetic dipole moment along z defined as:

$$\mu_z = IA \tag{3.5}$$

where I is the electron current and A is the area defined by the electron orbit. The electron current is:

$$I = q/t \tag{3.6}$$

where

$$t = 2\pi r/v \tag{3.7}$$

t is the time required for the electron to complete one circular orbit, and v is the velocity given by:

$$v = p/m_e \tag{3.8}$$

where m_e is the electron mass, and p is the linear momentum:

$$p = L_z/r \tag{3.9}$$

Combining Eqs. (3.5)–(3.9) gives:

$$\mu_z = \frac{q}{2m_e} L_z \tag{3.10}$$

Thus, the magnetic moment is proportional to the angular momentum. Note that the charge on the electron (q) is negative; therefore, μ_z and L_z point in opposite directions for the electron as shown in Fig. 3.2. This is also obvious from the definitions of L_z and μ_z: the current, I (used to define μ_z) is in a direction opposite that of the electron velocity and linear momentum (used to define L_z). Thus, atoms can act as tiny magnets. This has made it possible to use magnetic fields to perform experiments on the electron, such as the Stern-Gerlach experiment described below.

3.3   Stern-Gerlach Experiment

In 1922, Otto Stern (Fig. 3.5) and Walther Gerlach tried to demonstrate the quantization of orbital angular momentum of electrons in the atom [2]. In the Stern-Gerlach (SG) experiment (Fig. 3.6), a beam of Ag atoms is produced by an oven and a collimating slit. The beam of atoms is passed through a non-uniform magnetic field produced by the shaped poles of a permanent magnet. This field will interact with the magnetic dipole moment of the atom, if any, and deflect it. The atoms then impinge onto a screen (a glass plate) where the Ag atoms deposit to form a pattern. Note that we cannot use charged particles, such as a beam of electrons, for the SG experiment since the Lorentz force due to the electron velocity would overshadow the SG effects. Therefore, neutral atoms are used. Incidentally, a bad cigar may have led to the success of the SG experiment! [3].

An inhomogeneous magnetic field is required to deflect the magnetic moment of the atom. This is easy to understand by consideration of the potential energy, U, of a magnetic dipole moment in a magnetic field, B, given by classical electromagnetism:

Fig. 3.5 Otto Stern (1888–1969); Nobel Prize in Physics in 1943. *Credit* Wikimedia Commons [4]

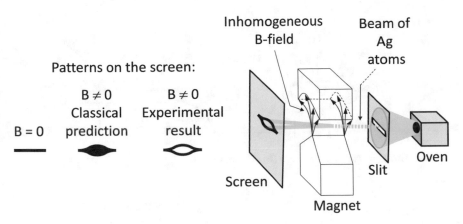

Fig. 3.6 Illustration of the Stern-Gerlach experiment

$$U = -\mu \cdot B \tag{3.11}$$

If the magnetic field is predominantly along the z direction, then:

$$U = -\mu_z B_z \tag{3.12}$$

The dipole moment will be deflected in the direction that decreases its potential energy. According to Eq. (3.12), a magnetic moment parallel to the magnetic field can decrease its energy by moving to regions of higher magnetic field, while a magnetic moment pointing opposite the direction of the magnetic field can decrease its energy by moving to regions of lower magnetic field. Alternatively, we can understand the atomic deflection by considering the force on the magnetic moment:

$$F_z = -\frac{\partial U}{\partial z} = \mu_z \frac{\partial B_z}{\partial z} \tag{3.13}$$

If the B field becomes weaker as we move along $+z$, as depicted in Fig. 3.6, then $\frac{\partial B_z}{\partial z}$ is negative. If μ_z is positive, meaning it points along $+z$, then the force is negative according to Eq. (3.13). A third way of understanding the deflection of the magnetic moment is by consideration of the Lorenz force ($F = qv \times B$) on a current loop constituting the magnetic moment (Fig. 3.7). The horizontal forces on the current loop cancel, while the vertical forces combine to deflect the magnetic moment downwards. Of course, all three methods give the same result.

Classically, we would expect a continuous distribution of Ag atoms on the "screen" (glass plate) in the SG experiment because a continuous distribution of magnetic moment vectors (all possible orientations) will exit the oven. For example, a magnetic moment pointing down in Fig. 3.8 (angular momentum pointing up) will be deflected up (towards regions of lower magnetic field) to reduce its energy as discussed above. A magnetic moment perpendicular to the magnetic field will not be deflected. A

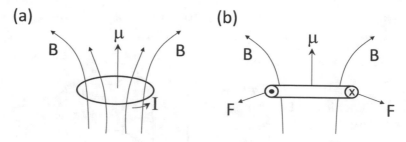

Fig. 3.7 Tilted view (**a**) and side view (**b**) of a current loop

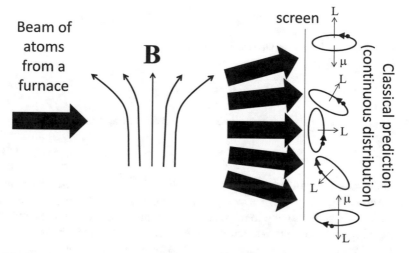

Fig. 3.8 Classical prediction of Ag atom distribution on the screen. A random distribution of magnetic moment enters the inhomogeneous magnetic field. We expect a continuous distribution of Ag atoms on the screen

magnetic moment pointing up in Fig. 3.8 (angular momentum pointing down) will be deflected down (towards regions of higher magnetic field) to reduce its energy. A tilted magnetic moment will be deflected according to the component of magnetic moment along z.

Rather than a continuous distribution of Ag atoms hitting the screen, two lines were observed, indicating a quantization of angular momentum (Fig. 3.9). Today, we know that Ag atoms contain 47 electrons surrounding the nucleus, of which 46 form a closed inner core of total angular momentum equal to zero. The one remaining electron also has zero orbital angular momentum since it resides in an s orbital with $l = 0$. Thus, the deflection of Ag atoms in the SG experiment was not due to the orbital angular momentum of the electrons. So, what causes the deflection of the Ag atoms?

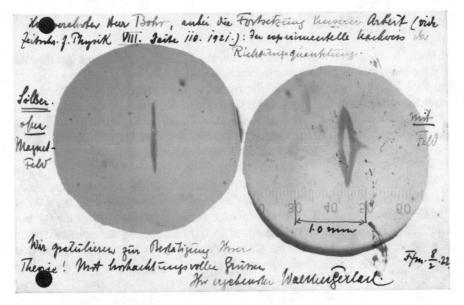

Fig. 3.9 A postcard from Gerlach to Bohr congratulating him on the success of his theory
Reproduced with permission from the Niels Bohr Archive, Copenhagen

Fig. 3.10 Left: George Uhlenbeck (1900–1988). Right: Samuel Goudsmit (1902–1978). *Credit*
Wikimedia Commons [6, 7]

3.4 Electron Spin

In 1925, George Uhlenbeck and Samuel Goudsmit (Fig. 3.10) proposed that the electron itself has an intrinsic angular momentum, called spin, and therefore a magnetic moment. The 46 core electrons of the Ag atom form pairs with opposite spin, so the net angular momentum and magnetic moment of these core electrons is zero. The only source of any magnetic moment is that due to the intrinsic spin of the 47th electron. Therefore, the SG experiment was actually measuring the intrinsic spin of a single electron, not orbital angular momentum of electrons. A history of spin is available in Ref. [5].

We might model the electron as a spinning ball of charge. The magnetic moment could then be derived by integrating the magnetic moment contributions from all the current loops constituting the rotating ball. However, it is found that the electron would have to spin faster than the speed of light to explain the measured value of magnetic moment. Besides, the electron is believed to be a point particle with diameter less than 10^{-17} m based on the results of particle accelerator experiments. How can a point particle have angular momentum? For this reason, spin is considered to be an entirely quantum phenomenon.

By analogy to Eq. (2.43), the spin angular momentum is:

$$S^2 = s(s+1)\hbar^2, s = 0, \frac{1}{2}, 1, \frac{3}{2}, \dots \tag{3.14}$$

where s is the spin quantum number with half-integer values allowed. By analogy to Eq. (2.44), the spin angular momentum along z (projection along z) is:

$$S_z = s_z\hbar, s_z = -s, -s+1, \dots +s \tag{3.15}$$

where s_z is the spin projection quantum number. Particles with integer spin (e.g., photons) are called bosons, and particles of half-integer spin (e.g., electrons) are called fermions. Based on the results of the SG experiment, electrons are found to have spin quantum number $s = 1/2$, and therefore $s_z = \pm 1/2$. We call this a "spin one-half" particle. Hence, for electrons (or any spin 1/2 particle), Eq. (3.14) gives:

$$s = \frac{1}{2} \rightarrow S = \sqrt{\frac{3}{4}}\hbar \tag{3.16}$$

and Eq. (3.15) gives:

$$s_z = \pm\frac{1}{2} \rightarrow S_z = \pm\frac{1}{2}\hbar \tag{3.17}$$

$S_z = +\hbar/2$ is called "spin up" and $S_z = -\hbar/2$ is called "spin down". Equation (3.17) are the two spin states for the electron. The SG experiment acts as a beam splitter for spin. It constitutes a measurement of the spin along an axis (along the direction of the magnetic field). We can represent the spin as a vector, as depicted in Fig. 3.11.

Fig. 3.11 Representation of spin up and spin down, and the relationship between magnetic moment and spin angular momentum, for the electron

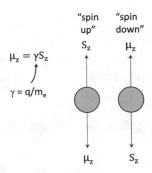

3.5 Intrinsic Magnetic Moment

Equation (3.10) also works for spin angular momentum, but with a dimensionless correction factor, g, simply called the g-factor:

$$\mu_z = \frac{q}{2m_e}L_z \rightarrow \left(g\frac{q}{2m_e}\right)S_z \tag{3.18}$$

where the orbital angular momentum, L_z, of Eq. (3.10) has been replaced with the spin angular momentum, S_z. The term in brackets in Eq. (3.18) is called the gyromagnetic ratio, γ:

$$\gamma = g\frac{q}{2m_e} \tag{3.19}$$

Hence, Eq. (3.18) becomes:

$$\mu_z = \gamma S_z \tag{3.20}$$

$g = 2.00231930436182$ for the electron, $g = +5.5856946893$ for the proton and $g = -3.82608546$ for the neutron. The electron g-factor can be measured and agrees with the theory of quantum electrodynamics! For our purposes, we can assume $g = 2$ for the electron, which yields from Eq. (3.19):

$$\gamma = \frac{q}{m_e} \quad \text{(for electron spin, } g = 2) \tag{3.21}$$

Note that magnetic moment depends on the inverse of the particle mass, so we expect the heavy protons and neutrons ($\sim 1800\ m_e$!) to have little effect on the magnetic moment of the atom as compared to the electrons, so we neglect them in the SG experiment.

According to Eqs. (3.17), (3.20) and (3.21), the intrinsic magnetic moment of the electron due to its spin is:

$$\mu_z = \pm \frac{q\hbar}{2m_e} = \pm\mu_B \tag{3.22}$$

where μ_B is called the Bohr magneton that depends only on fundamental constants. It's value is:

$$\mu_B = 9.27 \times 10^{-24} \text{ J/T} \tag{3.23}$$

3.6 Combinations of SG Apparatus

In Fig. 3.12, we show a schematic of a SG apparatus. We call this a Z apparatus because its magnetic field is oriented along z. We expect the oven to produce an equal mixture of spin up and spin down. Hence, the SG apparatus splits the two spin orientations ($s_z = \pm 1/2$) equally with 50% of atoms deflected up and 50% of atoms deflected down. The SG apparatus acts as a beam splitter or filter for the two spin states. There is nothing special about the z direction. If we rotate the Z apparatus so that its magnetic field points along some other direction, then the result will be the same. For example, if the Z apparatus is rotated so that the magnetic field is oriented along the x direction (an X apparatus), then the apparatus will split the spin states $s_x = \pm 1/2$ along the x direction. Similarly, a Y apparatus will split the spin states $s_y = \pm 1/2$ along the y direction. In fact, regardless of the orientation of the SG apparatus, we will always observe $s_n = \pm 1/2$ where n is some arbitrary direction of the magnetic field. $s_n = \pm 1/2$ forms the basis states for a spin measurement along the n axis.

Let us examine combinations of SG apparatus as in Fig. 3.13. The first is a "Z" SG apparatus, meaning the magnetic field is oriented along z. As described above, the SG apparatus splits the two spin orientations ($s_z = \pm 1/2$) with 50% of atoms deflected up and 50% of atoms deflected down, along the z direction. Next, we block the $s_z = -1/2$ output beam and let only the $s_z = +1/2$ beam go into the next apparatus. We are using the first SG apparatus as a filter. The second SG apparatus is also a Z apparatus. Since all particles entering the second Z apparatus have $s_z = +1/2$, the second apparatus lets the electrons out of the top output and nothing comes out of the bottom output. 50% of atoms from the oven will exit the two SG apparatuses.

Fig. 3.12 A SG apparatus with magnetic field oriented along the z direction

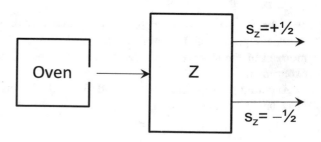

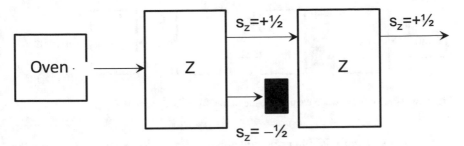

Fig. 3.13 Combination of two Z apparatuses

This experiment tells us that $s_z = +1/2$ states have no component or amplitude along $s_z = -1/2$. Hence, $s_z = +1/2$ and $s_z = -1/2$ are said to be orthogonal states.

Now suppose we rotate the second SG filter, so the magnetic field is oriented along x (Fig. 3.14). We find that half of the atoms entering the X apparatus will exit through the top $s_x = +1/2$ output, and the other half exit through the bottom $s_x = -1/2$ output. Classically, a particle with angular momentum along the z axis has no component of angular momentum along the x axis. The result of this SG experiment indicates that quantum mechanically this is not true for spins. Quantum mechanically, a state with a definite value of s_z has an amplitude along the state $s_x = +1/2$ as well as an amplitude along the state $s_x = -1/2$. We will obtain $s_x = +1/2$ or $s_x = -1/2$ randomly (50% probability) for each particle entering the X apparatus.

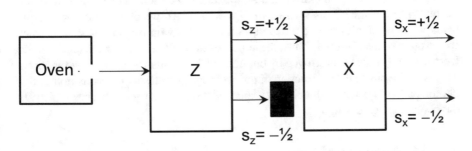

Fig. 3.14 Combination of Z and X apparatus

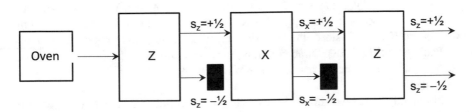

Fig. 3.15 Combination of Z, X and Z apparatus

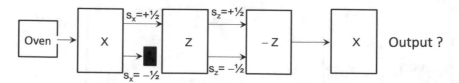

Fig. 3.16 What happens when we observe the output from the Z apparatus?

Now we add a third SG filter oriented along z (Fig. 3.15). We find that half of the atoms entering the third apparatus will exit with spin up, and half with spin down, along the z direction. This experiment shows that there is no memory of the first filter. The particles that exit the third apparatus are no longer all $s_z = +1/2$. Thus, it seems that z-oriented spins are composed of a superposition of x-oriented spins (Fig. 3.14), and x-oriented spins are composed of a superposition of z-oriented spins (Fig. 3.15).

Finally, consider the combination of SG apparatuses in Fig. 3.16. An XZ combination splits the $s_x = +1/2$ input into $s_z = \pm 1/2$. A "minus Z" apparatus, with its magnetic field oriented in the $-z$ direction, will recombine the two s_z beams and return the $s_x = +1/2$ beam. The output of the last X apparatus will be $s_x = +1/2$ only. Now what happens if you determine the path of each atom exiting the first Z apparatus?; i.e., you determine the exact state of each atom exiting the Z apparatus. For example, we might measure and determine that the state of a particular atom is $s_z = +1/2$ after the Z apparatus. This atom will next exit the $-Z$ apparatus with $s_z = +1/2$. Finally, upon exiting the X apparatus, the atom will be split into two beams with $s_x = \pm 1/2$, as we saw in Fig. 3.14. Thus, we obtain only one output beam when we don't observe the Z output, but we obtain two output beams when we do observe the Z output. Thus, the superposition of the two outputs ($s_z = \pm 1/2$) from the Z apparatus is destroyed by our observation (collapse of the wavefunction). This is analogous to the destruction of the interference pattern (superposition of two photon paths) in Young's double slit experiment when we try to observe which slit the photon passed through.

3.7   Mathematics of Spin

Wolfgang Pauli (Fig. 3.17) worked out the mathematics of spin. We cannot write an explicit wavefunction for the spin, like we did for a particle in an infinite quantum well or the H atom. Instead, the spin state is written using Dirac notation invented by Paul Dirac (Fig. 3.18). Dirac notation puts a description of the state in brackets, such as $|z_+\rangle$ or $|z_-\rangle$ for the result of a spin measurement along z. The "$|\rangle$" brackets are called a "ket".

In general, as we saw with the SG experiments, a particle can be in a superposition of spin states:

$$|\psi\rangle = \alpha|z_+\rangle + \beta|z_-\rangle \tag{3.24}$$

Fig. 3.17 Wolfgang Pauli (1900–1958); Nobel Prize in Physics in 1945. *Credit* Wikimedia Commons [8]

Fig. 3.18 Paul Dirac (1902–1984); Nobel Prize in Physics in 1933. *Credit* Wikimedia Commons [9]

where α and β are complex numbers, representing probability amplitudes. If we measure, we will obtain spin up ($|z_+\rangle$) with probability $|\alpha|^2$ or spin down ($|z_-\rangle$) with probability $|\beta|^2$. The probabilities must sum to unity:

$$|\alpha|^2 + |\beta|^2 = 1 \tag{3.25}$$

Experiments involving SG apparatuses and other interference experiments [10, 11] tell us the probability amplitudes as follows:

$$|x_+\rangle = \frac{1}{\sqrt{2}}(|z_+\rangle + |z_-\rangle) \tag{3.26}$$

$$|x_-\rangle = \frac{1}{\sqrt{2}}(|z_+\rangle - |z_-\rangle) \tag{3.27}$$

$$|y_+\rangle = \frac{1}{\sqrt{2}}(|z_+\rangle + i|z_-\rangle)) \tag{3.28}$$

$$|y_-\rangle = \frac{1}{\sqrt{2}}(|z_+\rangle - i|z_-\rangle)) \tag{3.29}$$

For example, if a spin is in the state $|x_+\rangle = \frac{1}{\sqrt{2}}(|z_+\rangle + |z_-\rangle)$, then the spin along z is completely undetermined; any measurement along z has a 50/50 chance of seeing either spin up or down, as we saw in the SG experiment of Fig. 3.14.

Exercise 3.1 How might you use the SG experiment to build a random number generator?

3.8   Vector Representation

Using Eq. (3.24), we can represent any spin state by a 2×1 column vector:

$$|\psi\rangle = \begin{pmatrix} \alpha \\ \beta \end{pmatrix} \tag{3.30}$$

where the first element of the vector is the probability amplitude for the $|z_+\rangle$ state, and the second element is the probability amplitude for the $|z_-\rangle$ state. Table 3.1 provides the vector representation for Eqs. (3.26)–(3.29).

3.9   Spin Operators

Next, as with all observables (according to one of the fundamental postulates of quantum mechanics), we should be able to write eigenvalue equations for the spin:

$$\hat{S}_x|x_+\rangle = +\frac{\hbar}{2}|x_+\rangle \tag{3.31}$$

$$\hat{S}_x|x_-\rangle = -\frac{\hbar}{2}|x_-\rangle \tag{3.32}$$

$$\hat{S}_y|y_+\rangle = +\frac{\hbar}{2}|y_+\rangle \tag{3.33}$$

Table 3.1 Dirac notation and corresponding vector representation

Dirac notation	Vector representation
$\lvert\psi\rangle = \alpha\lvert z_+\rangle + \beta\lvert z_-\rangle$	$\lvert\psi\rangle = \begin{pmatrix}\alpha\\\beta\end{pmatrix}$
$\lvert z_+\rangle$	$\lvert z_+\rangle = \begin{pmatrix}1\\0\end{pmatrix}$
$\lvert z_-\rangle$	$\lvert z_-\rangle = \begin{pmatrix}0\\1\end{pmatrix}$
$\lvert x_+\rangle = \frac{1}{\sqrt{2}}(\lvert z_+\rangle + \lvert z_-\rangle)$	$\lvert x_+\rangle = \frac{1}{\sqrt{2}}\begin{pmatrix}1\\1\end{pmatrix}$
$\lvert x_-\rangle = \frac{1}{\sqrt{2}}(\lvert z_+\rangle - \lvert z_-\rangle)$	$\lvert x_-\rangle = \frac{1}{\sqrt{2}}\begin{pmatrix}1\\-1\end{pmatrix}$
$\lvert y_+\rangle = \frac{1}{\sqrt{2}}(\lvert z_+\rangle + i\lvert z_-\rangle)$	$\lvert y_+\rangle = \frac{1}{\sqrt{2}}\begin{pmatrix}1\\i\end{pmatrix}$
$\lvert y_-\rangle = \frac{1}{\sqrt{2}}(\lvert z_+\rangle - i\lvert z_-\rangle)$	$\lvert y_-\rangle = \frac{1}{\sqrt{2}}\begin{pmatrix}1\\-i\end{pmatrix}$

$$\hat{S}_y\lvert y_-\rangle = -\frac{\hbar}{2}\lvert y_-\rangle \tag{3.34}$$

$$\hat{S}_z\lvert z_+\rangle = +\frac{\hbar}{2}\lvert z_+\rangle \tag{3.35}$$

$$\hat{S}_z\lvert z_-\rangle = -\frac{\hbar}{2}\lvert z_-\rangle \tag{3.36}$$

where $\hat{S}_x$, $\hat{S}_y$ and $\hat{S}_z$ are the "spin operators". For example, Eqs. (3.31) and (3.32) tell us that if we measure the spin along the x direction, we will obtain spin angular momentum of eigenvalue $S_x = +1/2\hbar$ from the eigenstate $\lvert x_+\rangle$, or eigenvalue $S_x = -1/2\hbar$ from the eigenstate $\lvert x_-\rangle$. The eigenvalues, $\pm\frac{\hbar}{2}$, are the result of the measurement—for example, by using a SG apparatus.

The eigenstates are represented by 2×1 column vectors, as shown in Table 3.1. For example, $\lvert x_+\rangle = \frac{1}{\sqrt{2}}\begin{pmatrix}1\\1\end{pmatrix}$ and $\lvert x_-\rangle = \frac{1}{\sqrt{2}}\begin{pmatrix}1\\-1\end{pmatrix}$. Thus, it is evident from Eqs. (3.31)–(3.36) that the spin operators must be 2×2 matrices. For example, it is easy to show by substitution in Eqs. (3.31) and (3.32) that the $\hat{S}_x$ operator is:

$$\hat{S}_x = \frac{\hbar}{2}\begin{pmatrix}0 & 1\\1 & 0\end{pmatrix} \tag{3.37}$$

because this matrix satisfies Eq. (3.31) with eigenvector $|x_+\rangle = \frac{1}{\sqrt{2}}\begin{pmatrix} 1 \\ 1 \end{pmatrix}$ and corresponding eigenvalue $+\frac{\hbar}{2}$. Similarly, $\widehat{S}_x$ satisfies Eq. (3.32) with eigenvector $|x_-\rangle = \frac{1}{\sqrt{2}}\begin{pmatrix} 1 \\ -1 \end{pmatrix}$ and corresponding eigenvalue $-\frac{\hbar}{2}$. In a similar manner:

$$\widehat{S}_y = \frac{\hbar}{2}\begin{pmatrix} 0 & -i \\ i & 0 \end{pmatrix} \tag{3.38}$$

and

$$\widehat{S}_z = \frac{\hbar}{2}\begin{pmatrix} 1 & 0 \\ 0 & -1 \end{pmatrix} \tag{3.39}$$

Exercise 3.2 Show that Eqs. (3.37)–(3.39) satisfy Eqs. (3.31)–(3.36).

3.10   Pauli Spin Matrices

The Pauli spin matrices are defined as:

$$\widehat{\sigma}_x = X = \begin{pmatrix} 0 & 1 \\ 1 & 0 \end{pmatrix} \tag{3.40}$$

$$\widehat{\sigma}_y = Y = \begin{pmatrix} 0 & -i \\ i & 0 \end{pmatrix} \tag{3.41}$$

$$\widehat{\sigma}_z = Z = \begin{pmatrix} 1 & 0 \\ 0 & -1 \end{pmatrix} \tag{3.42}$$

Physicists tend to prefer the notation $\widehat{\sigma}_x$, $\widehat{\sigma}_y$ and $\widehat{\sigma}_z$, while computer scientists prefer X, Y and Z. Using the Pauli matrices, the spin operators can be written as:

$$\widehat{S}_x = \frac{\hbar}{2}\widehat{\sigma}_x \tag{3.43}$$

$$\widehat{S}_y = \frac{\hbar}{2}\widehat{\sigma}_y \tag{3.44}$$

$$\hat{S}_z = \frac{\hbar}{2}\hat{\sigma}_z \qquad (3.45)$$

The Pauli spin matrices, plus the identity matrix, $I = \begin{pmatrix} 1 & 0 \\ 0 & 1 \end{pmatrix}$, form a basis for any complex 2×2 matrix; i.e., any 2×2 matrix with complex entries can be written as $\alpha I + \beta\hat{\sigma}_x + \delta\hat{\sigma}_y + \gamma\hat{\sigma}_z$ where α, β, δ and γ are complex numbers.

Exercise 3.3 Prove that any complex 2×2 matrix can be written as $\alpha I + \beta\hat{\sigma}_x + \delta\hat{\sigma}_y + \gamma\hat{\sigma}_z$ where α, β, δ and γ are complex numbers.

Using the rules for matrix multiplication, it is easy to prove the identities below in Exercise 3.4. This type of algebra will help us later when we examine quantum algorithms.

Exercise 3.4 Prove the following relations:

$$X^2 = Y^2 = Z^2 = -iXYZ = I$$
$$Z = -iXY$$
$$ZY = -iX$$
$$ZYX = -iI$$
$$YX = -iZ$$

3.11 Arbitrary Spin Direction

In the previous sections, we considered a spin measurement along the direction $\hat{x}$, $\hat{y}$ and $\hat{z}$ (given by the orientation of the magnetic field in the SG apparatus). Regardless of the measurement direction, the spin measurement resulted in $\pm\hbar/2$. What is the result of a spin measurement along an arbitrary direction, given by the unit vector $\hat{n}$ depicted in Fig. 3.19? We can guess that the result of a spin measurement along $\hat{n}$ will give $\pm\hbar/2$. Let us prove it.

Referring to Fig. 3.19, the unit vector $\hat{n}$ can be decomposed into components along x, y and z, and described in spherical coordinates:

$$n_x = \cos\phi \sin\theta \qquad (3.46)$$

$$n_y = \sin\phi \sin\theta \qquad (3.47)$$

Fig. 3.19 A unit vector in the direction $\hat{n}$

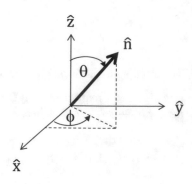

$$n_z = \cos\theta \tag{3.48}$$

The spin operator can then be expressed as:

$$\hat{S}_n = n_x\hat{S}_x + n_y\hat{S}_y + n_z\hat{S}_z \tag{3.49}$$

which results in:

$$\hat{S}_n = \frac{\hbar}{2}\begin{pmatrix} \cos\theta & e^{-i\phi}\sin\theta \\ e^{i\phi}\sin\theta & -\cos\theta \end{pmatrix} \tag{3.50}$$

Exercise 3.5 Prove Eq. (3.50) from Eq. (3.49). Show that the $\hat{S}_n$ operator has eigenvalues $\pm\frac{\hbar}{2}$ and corresponding eigenvectors $|n_+\rangle = \begin{pmatrix} \cos\frac{\theta}{2} \\ e^{i\phi}\sin\frac{\theta}{2} \end{pmatrix}$ and $|n_-\rangle = \begin{pmatrix} \sin\frac{\theta}{2} \\ -e^{i\phi}\cos\frac{\theta}{2} \end{pmatrix}$.

Not surprisingly, from Exercise 3.5, the spin measurement along an arbitrary direction results in spin angular momentum of $\pm\frac{\hbar}{2}$. The corresponding eigenvectors are:

$$|n_+\rangle = \begin{pmatrix} \cos\frac{\theta}{2} \\ e^{i\phi}\sin\frac{\theta}{2} \end{pmatrix} \tag{3.51}$$

and

$$|n_-\rangle = \begin{pmatrix} \sin\frac{\theta}{2} \\ -e^{i\phi}\cos\frac{\theta}{2} \end{pmatrix} \tag{3.52}$$

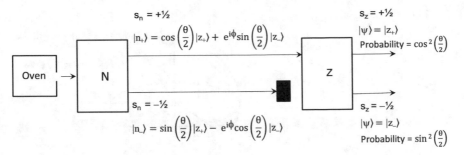

Fig. 3.20 The spin projection quantum number (s_n, s_z) for measurement (projection) along a direction $\hat{n}$ or $\hat{z}$, and the corresponding spin states

What is the meaning of these eigenvectors? According to the vector representation, the first element of each vector is the probability amplitude of the $|z_+\rangle$ state, while the second element of each vector is the probability amplitude of the $|z_-\rangle$ state. Each vector in Eqs. (3.51) and (3.52) represents a superposition of $|z_+\rangle$ and $|z_-\rangle$ states as follows:

$$|n_+\rangle = \cos\left(\frac{\theta}{2}\right)|z_+\rangle + e^{i\phi}\sin\left(\frac{\theta}{2}\right)|z_-\rangle \tag{3.53}$$

$$|n_-\rangle = \sin\left(\frac{\theta}{2}\right)|z_+\rangle - e^{i\phi}\cos\left(\frac{\theta}{2}\right)|z_-\rangle \tag{3.54}$$

What do Eqs. (3.53) and (3.54) mean physically? If we pass a beam of Ag atoms from an oven into an N apparatus (a SG apparatus with magnetic field oriented along $\hat{n}$), then the atoms will be separated into two beams, one with spin up atoms (spin projection quantum number along $\hat{n}$, $s_n = +1/2$) and the other with spin down atoms ($s_n = -1/2$), as depicted in Fig. 3.20. Each beam is a superposition of $|z_+\rangle$ and $|z_-\rangle$ states, according to Eqs. (3.53) and (3.54), as shown in Fig. 3.20. This means that if we pass the beam corresponding to $s_n = +1/2$ through a Z apparatus, then this beam will be split into the two $|z_+\rangle$ and $|z_-\rangle$ states with probability amplitudes given by Eq. (3.53). Likewise, if we pass the beam corresponding to $s_n = -1/2$ through a Z apparatus, then this beam will be split into the two $|z_+\rangle$ and $|z_-\rangle$ states with probability amplitudes given by Eq. (3.54). The probabilities after exiting the Z apparatus are given by the square of the modulus of the probability amplitudes; for the $|n_+\rangle$ state, this means $\cos^2\left(\frac{\theta}{2}\right)$ for the $|z_+\rangle$ state and $\sin^2\left(\frac{\theta}{2}\right)$ for the $|z_-\rangle$ state as shown in Fig. 3.20. Note that the probabilities sum to unity as required since $\cos^2\left(\frac{\theta}{2}\right) + \sin^2\left(\frac{\theta}{2}\right) = 1$.

Fig. 3.21 The Bloch sphere

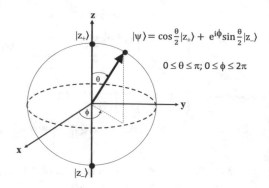

3.12   Bloch Sphere

Any spin vector, $|\psi\rangle$, can be represented by Eq. (3.53), rewritten as:

$$|\psi\rangle = \cos\frac{\theta}{2}|z_+\rangle + e^{i\phi}\sin\frac{\theta}{2}|z_-\rangle \qquad (3.55)$$

where the angles are restricted to the ranges $0 \le \theta \le \pi$ and $0 \le \phi \le 2\pi$. The "Bloch sphere" (Fig. 3.21), named after the physicist, Felix Bloch (Nobel Prize in Physics in 1952), is a useful way of representing spin states in spherical coordinates. The Bloch sphere displays the spin states as a point on the surface of the sphere, or equivalently as a vector in the sphere. The "North pole" of the Bloch sphere is the $|z_+\rangle$ state and the "South pole" is the $|z_-\rangle$ state. Any spin state can be derived from Eq. (3.55) by appropriate selection of the angles θ and ϕ. For example,

$$\theta = 0: |\psi\rangle = |z_+\rangle \qquad (3.56)$$

$$\theta = \pi: |\psi\rangle = |z_-\rangle \qquad (3.57)$$

$$\theta = \pi/2, \phi = 0: |\psi\rangle = \cos\frac{\pi}{4}|z_+\rangle + e^{i0}\sin\frac{\pi}{4}|z_-\rangle$$

$$= \frac{1}{\sqrt{2}}|z_+\rangle + \frac{1}{\sqrt{2}}|z_-\rangle$$

$$= |x_+\rangle \qquad (3.58)$$

$$\theta = \pi/2, \phi = \pi: |\psi\rangle = \cos\frac{\pi}{4}|z_+\rangle + e^{i\pi}\sin\frac{\pi}{4}|z_-\rangle$$

$$= \frac{1}{\sqrt{2}}|z_+\rangle - \frac{1}{\sqrt{2}}|z_-\rangle$$

$$= |x_-\rangle \qquad (3.59)$$

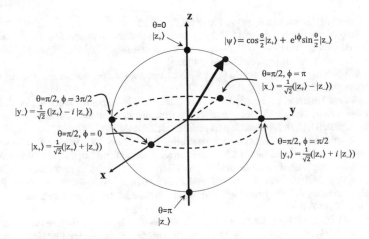

Fig. 3.22 $|x_+\rangle$, $|x_-\rangle$, $|y_+\rangle$, $|y_-\rangle$, $|z_+\rangle$ and $|z_-\rangle$ spin states in the Bloch sphere

$$\theta = \pi/2, \phi = \pi/2: |\psi\rangle = \cos\frac{\pi}{4}|z_+\rangle + e^{i\pi/2}\sin\frac{\pi}{4}|z_-\rangle$$

$$= \frac{1}{\sqrt{2}}|z_+\rangle + \frac{1}{\sqrt{2}}i|z_-\rangle$$

$$= |y_+\rangle \tag{3.60}$$

$$\theta = \pi/2, \phi = 3\pi/2 : |\psi\rangle = \cos\frac{\pi}{4}|z_+\rangle + e^{i3\pi/2}\sin\frac{\pi}{4}|z_-\rangle$$

$$= \frac{1}{\sqrt{2}}|z_+\rangle - \frac{1}{\sqrt{2}}i|z_-\rangle$$

$$= |y_-\rangle \tag{3.61}$$

In Eqs. (3.56)–(3.61), we used Eqs. (3.26)–(3.29). The above spin states are depicted in Fig. 3.22. $|x_+\rangle$ and $|x_-\rangle$ lie along the x-axis of the Bloch sphere, $|y_+\rangle$ and $|y_-\rangle$ lie along the y-axis, and $|z_+\rangle$ and $|z_-\rangle$ lie along the z-axis. Any spin state on the equator of the Bloch sphere represents an equal superposition of spin up and spin down. Opposite vectors on the Bloch sphere (e.g. $|x_+\rangle$ and $|x_-\rangle$) are orthonormal states. The Bloch sphere is a unit sphere, meaning each Bloch vector has unity magnitude; i.e., the vector is normalized.

References

1. File: Niels Bohr.jpg. (2020, October 3). *Wikimedia Commons, the free media repository*. Retrieved 13:35, December 7, 2020 from https://commons.wikimedia.org/w/index.php?title= File:Niels_Bohr.jpg&oldid=479514272
2. W. Gerlach, O. Stern, Z. Phys. **9**, 349 (1922)

3. B. Friedrich, D. Herschbach, Phys. Today **56**, 53 (2003)
4. File: Otto Stern 1950s.jpg. (2015, July 2). *Wikimedia Commons, the free media repository.* Retrieved 12:12, December 7, 2020 from https://commons.wikimedia.org/w/index.php?title= File:Otto_Stern_1950s.jpg&oldid=164764898
5. Spin Milestones, Nature, March 2008, S5-S20; https://www.nature.com/milestones/milespin/ pdf/milespin_all.pdf
6. File: George Uhlenbeck (cropped).jpg. (2017, July 24). *Wikimedia Commons, the free media repository.* Retrieved 13:23, December 7, 2020 from https://commons.wikimedia.org/w/index. php?title=File:George_Uhlenbeck_(cropped).jpg&oldid=252960944
7. File: UhlenbeckKramersGoudsmit (cropped).jpg. (2017, May 15). *Wikimedia Commons, the free media repository.* Retrieved 13:21, December 7, 2020 from https://commons.wikimedia. org/w/index.php?title=File:UhlenbeckKramersGoudsmit_(cropped).jpg&oldid=244202772
8. File: Pauli.jpg. (2016, February 7). *Wikimedia Commons, the free media repository.* Retrieved 13:27, December 7, 2020 from https://commons.wikimedia.org/w/index.php?title=File:Pauli. jpg&oldid=186866509
9. File: Paul Dirac, 1933.jpg. (2019, August 28). *Wikimedia Commons, the free media repository.* Retrieved 13:31, December 7, 2020 from https://commons.wikimedia.org/w/index.php?title= File:Paul_Dirac,_1933.jpg&oldid=363583299
10. G. Badurek et al., Phys. Lett. A **56**, 244 (1976)
11. G. Bergmann, Solid State Comm. **42**, 815 (1982)

Chapter 4
Qubits

Information in computing (classical or quantum) is represented as binary 0 and 1. Quantum computing uses quantum states to represent 0 and 1, allowing a superposition of 0 and 1 simultaneously. This superposition of 0 and 1 is called a qubit which is the fundamental unit of quantum computing.

4.1 Bits

We can represent any information using character codes where alphanumeric characters and other symbols are represented by decimal numbers. These decimal numbers, in turn, can be expressed in binary notation as a sum of increasing powers of two, $b_n*2^n \ldots b_2*2^2 + b_1*2^1 + b_0*2^0$, where $b_n, \ldots b_2, b_1, b_0$ are the binary digits called "bits" with value 0 or 1 ($b_i = \{0, 1\}$). Hence, in classical computing, we represent information using bits, represented by a string of 0's and 1's (Fig. 4.1).

For example, 2 bits can represent the numbers 0–3:

$$0 = 0^*2^1 + 0^*2^0 \rightarrow 00$$
$$1 = 0^*2^1 + 1^*2^0 \rightarrow 01$$
$$2 = 1^*2^1 + 0^*2^0 \rightarrow 10$$
$$3 = 1^*2^1 + 1^*2^0 \rightarrow 11$$

where the decimal number is 0, 1, 2 or 3 with binary representation 00, 01, 10 and 11, respectively. Similarly, 3 bits can represent the numbers 0–7:

R. LaPierre, *Introduction to Quantum Computing*, The Materials Research Society Series, https://doi.org/10.1007/978-3-030-69318-3_4

Fig. 4.1 Binary
representation of a decimal
number

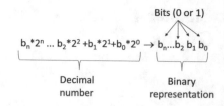

$$b_n*2^n \ldots b_2*2^2 + b_1*2^1 + b_0*2^0 \rightarrow b_n \ldots b_2\, b_1\, b_0$$

Bits (0 or 1)

Decimal number

Binary representation

$$0 = 0*2^2 + 0*2^1 + 0*2^0 \rightarrow 000$$
$$1 = 0*2^2 + 0*2^1 + 1*2^0 \rightarrow 001$$
$$2 = 0*2^2 + 1*2^1 + 0*2^0 \rightarrow 010$$
$$3 = 0*2^2 + 1*2^1 + 1*2^0 \rightarrow 011$$
$$4 = 1*2^2 + 0*2^1 + 0*2^0 \rightarrow 100$$
$$5 = 1*2^2 + 0*2^1 + 1*2^0 \rightarrow 101$$
$$6 = 1*2^2 + 1*2^1 + 0*2^0 \rightarrow 110$$
$$7 = 1*2^2 + 1*2^1 + 1*2^0 \rightarrow 111$$

As a final example, 4 bits can represent the numbers 0–15:

$$0 = 0*2^3 + 0*2^2 + 0*2^1 + 0*2^0 \rightarrow 0000$$
$$1 = 0*2^3 + 0*2^2 + 0*2^1 + 1*2^0 \rightarrow 0001$$
$$2 = 0*2^3 + 0*2^2 + 1*2^1 + 0*2^0 \rightarrow 0010$$
$$3 = 0*2^3 + 0*2^2 + 1*2^1 + 1*2^0 \rightarrow 0011$$
$$4 = 0*2^3 + 1*2^2 + 0*2^1 + 0*2^0 \rightarrow 0100$$
$$5 = 0*2^3 + 1*2^2 + 0*2^1 + 1*2^0 \rightarrow 0101$$
$$6 = 0*2^3 + 1*2^2 + 1*2^1 + 0*2^0 \rightarrow 0110$$
$$7 = 0*2^3 + 1*2^2 + 1*2^1 + 1*2^0 \rightarrow 0111$$
$$8 = 1*2^3 + 0*2^2 + 0*2^1 + 0*2^0 \rightarrow 1000$$
$$9 = 1*2^3 + 0*2^2 + 0*2^1 + 1*2^0 \rightarrow 1001$$
$$10 = 1*2^3 + 0*2^2 + 1*2^1 + 0*2^0 \rightarrow 1010$$
$$11 = 1*2^3 + 0*2^2 + 1*2^1 + 1*2^0 \rightarrow 1011$$
$$12 = 1*2^3 + 1*2^2 + 0*2^1 + 0*2^0 \rightarrow 1100$$
$$13 = 1*2^3 + 1*2^2 + 0*2^1 + 1*2^0 \rightarrow 1101$$
$$14 = 1*2^3 + 1*2^2 + 1*2^1 + 0*2^0 \rightarrow 1110$$
$$15 = 1*2^3 + 1*2^2 + 1*2^1 + 1*2^0 \rightarrow 1111$$

We can extrapolate the above examples to find, in general, that n bits can represent any number from 0 to $2^n - 1$. For example, with $n = 1$, we have $2^n - 1 = 1$, so we can represent any number from 0 to 1; with $n = 2$, we have $2^n - 1 = 3$, so we can represent any number from 0 to 3; and so on.

4.2 Qubits

In classical computing, 0 and 1 are represented by a binary state; e.g., voltage on or off; or electrical charge present or not present. In quantum computing, 0 and 1 are represented by two quantum states that differ in some quantum number. Equivalently, 0 and 1 are represented by two different eigenstates of some quantum operator. Unlike classical computing, a quantum state can exist as a superposition of the two quantum states; i.e., a superposition of 0 and 1 is possible. A "quantum bit" is called a qubit, which is the basic unit of quantum information. For example, a qubit could be a superposition of:

- Electron with spin up or spin down
- Electron in ground or excited state of an atom
- Photon in "path 1" or "path 2"
- Photon with horizontal (H) or vertical (V) polarization
- Photon with $+45°$ or $-45°$ polarization
- Photon with left circular polarization (LCP) or right circular polarization (RCP)
- Superconducting current flowing clockwise or counter-clockwise

The first part of this book is aimed at understanding quantum computers in general terms, using 0's and 1's, without any mention of specific hardware platforms (although we will frequently use electron spin or photon polarization as examples); i.e., we will first examine the software. Later, we will examine the hardware.

4.3 Dirac Ket

Dirac notation, introduced in Chap. 3, is a convenient way of representing quantum states and their superpositions. Examples of Dirac notation include:

$|z_+\rangle, |z_-\rangle$
$|\uparrow\rangle, |\downarrow\rangle$
$|\text{ground state}\rangle, |\text{excited state}\rangle$
$|H\rangle, |V\rangle$
$|+45°\rangle, |-45°\rangle$
$|LCP\rangle, |RCP\rangle$

The first two examples above are spin states, the next one is energy states, and the last three are photon polarization states. Any description of the state can go inside the brackets.

The aforementioned examples can be generalized to $|0\rangle$ and $|1\rangle$; i.e., one state represents $|0\rangle$ and the other represents $|1\rangle$. As mentioned, a qubit could be in a linear superposition of these two states:

$$|\psi\rangle = \alpha|0\rangle + \beta|1\rangle \tag{4.1}$$

where α and β are probability amplitudes (complex numbers) satisfying the normalization condition:

$$|\alpha|^2 + |\beta|^2 = 1 \tag{4.2}$$

i.e., the probabilities must sum to unity. If we measure, we will obtain $|0\rangle$ with probability $|\alpha|^2$, or $|1\rangle$ with probability $|\beta|^2$.

Exercise 4.1 Show that the following state is normalized:

$$\left(\frac{1}{2} + \frac{1}{2}i\right)|0\rangle - \frac{1}{\sqrt{2}}|1\rangle$$

As we saw in Chap. 3, a single qubit can also be represented by a column vector as follows:

$$|\psi\rangle = \alpha|0\rangle + \beta|1\rangle \rightarrow \begin{pmatrix} \alpha \\ \beta \end{pmatrix} \tag{4.3}$$

We say that a qubit is represented as a vector in a "2-dimensional complex vector space". Thus, we will see that quantum computing requires an understanding of complex linear algebra.

4.4 Computational Basis

We see from Eq. (4.3) that $|0\rangle$ can be represented as $\begin{pmatrix} 1 \\ 0 \end{pmatrix}$ and $|1\rangle$ can be represented as $\begin{pmatrix} 0 \\ 1 \end{pmatrix}$. This is known as the computational basis:

$$\text{Computational basis} \begin{cases} |0\rangle \rightarrow \begin{pmatrix} 1 \\ 0 \end{pmatrix} \\ |1\rangle \rightarrow \begin{pmatrix} 0 \\ 1 \end{pmatrix} \end{cases} \tag{4.4}$$

i.e., any 2×1 complex vector can be represented as a linear sum (superposition) of $\begin{pmatrix} 1 \\ 0 \end{pmatrix}$ and $\begin{pmatrix} 0 \\ 1 \end{pmatrix}$.

Basis states are not unique; i.e., $|0\rangle$ and $|1\rangle$ are not the only basis. For example, we could use $|+\rangle$ and $|-\rangle$ as a basis where:

Table 4.1 Vector axioms

Name	Axiom
Associativity of addition	$\mathbf{u} + (\mathbf{v} + \mathbf{w}) = (\mathbf{u} + \mathbf{v}) + \mathbf{w}$
Commutativity of addition	$\mathbf{u} + \mathbf{v} = \mathbf{v} + \mathbf{u}$
Additive identity	There exists a vector $\mathbf{0}$ such that $\mathbf{v} + \mathbf{0} = \mathbf{v}$ for all $\mathbf{v}$
Additive inverse	For every vector $\mathbf{v}$, there exists a vector $-\mathbf{v}$ such that $\mathbf{v} + (-\mathbf{v}) = \mathbf{0}$
Scalar multiplication identity	$1\mathbf{v} = \mathbf{v}$
Distributivity of vector sums	For a scalar a, $a(\mathbf{u} + \mathbf{v}) = a\mathbf{u} + a\mathbf{v}$
Distributivity of scalar sums	For scalars a and b, $(a + b)\mathbf{v} = a\mathbf{v} + b\mathbf{v}$
Associativity of scalar multiplication	$a(b\mathbf{v}) = (ab)\mathbf{v}$

$$|+\rangle = \frac{1}{\sqrt{2}}(|0\rangle + |1\rangle) \tag{4.5}$$

$$|-\rangle = \frac{1}{\sqrt{2}}(|0\rangle - |1\rangle) \tag{4.6}$$

This basis is often called the Hadamard basis, for reasons that will become clear in Chap. 7 when we examine quantum gates. A different basis is analogous to choosing a different coordinate system (e.g., Cartesian, cylindrical, or spherical coordinates) to represent a vector.

Exercise 4.2 Rewrite $|0\rangle$ and $|1\rangle$ in terms of $|+\rangle$ and $|-\rangle$. Show that $|+\rangle$ and $|-\rangle$ are orthonormal.

Like all vectors, the quantum vectors (Eq. (4.3)) obey certain vector axioms. Suppose we have vectors $\mathbf{u} = \begin{pmatrix} a \\ b \end{pmatrix}$, $\mathbf{v} = \begin{pmatrix} c \\ d \end{pmatrix}$ and $\mathbf{w} = \begin{pmatrix} e \\ f \end{pmatrix}$. The vector axioms involving $\mathbf{u}$, $\mathbf{v}$ and $\mathbf{w}$ are listed in Table 4.1. Note that we are talking here about vectors as obeying certain axioms (the vector axioms), not in the traditional sense of having magnitude and direction (Euclidean vectors). Objects that satisfy the axioms in Table 4.1 are said to occupy a "vector space".

4.5 Dirac Bra

We previously defined the Dirac ket:

$$\text{ket: } | \ \rangle \tag{4.7}$$

A related quantity, the Dirac bra, also called the "dual vector", is defined as follows:

$$\text{bra}: \langle \ | = \left((|\psi\rangle)^T \right)^* = (|\psi\rangle)^\dagger \tag{4.8}$$

where "T" instructs us to take the transpose of the vector, and "*" instructs us to take the complex conjugate of each vector element. These two actions combined are called the "conjugate transpose" and are given the symbol "$\dagger$" (pronounced "dagger"). The "conjugate transpose" is also called the "adjoint" or "Hermitian conjugate". In vector notation, we have:

$$\text{ket}: |\psi\rangle = \alpha|0\rangle + \beta|1\rangle \rightarrow \begin{pmatrix} \alpha \\ \beta \end{pmatrix} \tag{4.9}$$

$$\text{bra}: \langle\psi| = |\psi\rangle^\dagger = \alpha^*\langle 0| + \beta^*\langle 1| \rightarrow \begin{pmatrix} \alpha^* & \beta^* \end{pmatrix} \tag{4.10}$$

4.6 Inner Product

The inner product is defined as the product of a bra and a ket, $\langle\psi\|\psi\rangle$, which is usually written as $\langle\psi|\psi\rangle$ with one of the two vertical lines omitted:

$$\text{inner product}: \langle\psi\|\psi\rangle \rightarrow \langle\psi|\psi\rangle \tag{4.11}$$

The inner product provides the origin of the name "bra" and "ket": putting the bra and ket together, we get a "bra-ket" or "bracket" ($\langle\ \rangle$). Using the vector representation, we get:

$$\langle\psi|\psi\rangle = \begin{pmatrix} \alpha^* & \beta^* \end{pmatrix} \begin{pmatrix} \alpha \\ \beta \end{pmatrix} = \alpha^*\alpha + \beta^*\beta = |\alpha|^2 + |\beta|^2 \tag{4.12}$$

Hence, the inner product is a scalar, and is equivalent to the dot product for Euclidean vectors. Vector spaces that include an inner product are called "Hilbert" spaces.

Comparing Eq. (4.12) with (4.2), we see that if a state $|\psi\rangle$ is normalized, then $\langle\psi|\psi\rangle = 1$:

$$\text{Normalization}: \langle\psi|\psi\rangle = 1 \tag{4.13}$$

If two vectors (ψ_1, ψ_2) are orthogonal, then $\langle\psi_1|\psi_2\rangle = 0$:

$$\text{Orthogonality}: \langle\psi_1|\psi_2\rangle = 0 \tag{4.14}$$

If two vectors are normalized and orthogonal, we say they are orthonormal:

$$\text{Orthonormal: } \langle \psi_i | \psi_j \rangle = \delta_{ij} \tag{4.15}$$

where δ_{ij} is the Kronecker function previously defined in Eq. (2.31).

Exercise 4.3 Prove the following:

(a) $\langle \psi_1 | \psi_2 \rangle^* = \langle \psi_2 | \psi_1 \rangle$
(b) $\langle \psi_1 | \psi_2 \rangle$ can be a complex number, but $\langle \psi_1 | \psi_1 \rangle$ is real and positive
(c) $\langle c \psi_1 | \psi_2 \rangle = c^* \langle \psi_1 | \psi_2 \rangle$
(d) $\langle \psi_1 | c \psi_2 \rangle = c \langle \psi_1 | \psi_2 \rangle$

Equation (4.15) is the Dirac notation for Eq. (2.30). It also expresses the orthonormal condition for spin states where an explicit form for the wavefunction does not exist. Recalling the vector representations for the spin states in Table 3.1, we can prove the following:

$$\langle x_+ | x_- \rangle = 0$$
$$\langle x_+ | x_+ \rangle = 1$$
$$\langle x_- | x_- \rangle = 1$$
$$\langle y_+ | y_- \rangle = 0$$
$$\langle y_+ | y_+ \rangle = 1$$
$$\langle y_- | y_- \rangle = 1$$
$$\langle z_+ | z_- \rangle = 0$$
$$\langle z_+ | z_+ \rangle = 1$$
$$\langle z_- | z_- \rangle = 1$$

Hence, $|x_+\rangle$ and $|x_-\rangle$ are orthonormal, $|y_+\rangle$ and $|y_-\rangle$ are orthonormal, and $|z_+\rangle$ and $|z_-\rangle$ are orthonormal.

Exercise 4.4 Prove the orthonormal relations listed above for the spin states.

For the computational basis, we have:

$$\text{kets: } |0\rangle \rightarrow \begin{pmatrix} 1 \\ 0 \end{pmatrix}, |1\rangle \rightarrow \begin{pmatrix} 0 \\ 1 \end{pmatrix} \tag{4.16}$$

$$\text{bras: } \langle 0| \rightarrow (1 \ 0), \langle 1| \rightarrow (0 \ 1) \tag{4.17}$$

Then the orthonormality condition becomes:

$$\langle 0|0\rangle = (1\ 0)\begin{pmatrix} 1 \\ 0 \end{pmatrix} = 1 \tag{4.18}$$

$$\langle 1|1\rangle = (0\ 1)\begin{pmatrix} 0 \\ 1 \end{pmatrix} = 1 \tag{4.19}$$

$$\langle 0|1\rangle = (1\ 0)\begin{pmatrix} 0 \\ 1 \end{pmatrix} = 0 \tag{4.20}$$

$$\langle 1|0\rangle = (0\ 1)\begin{pmatrix} 1 \\ 0 \end{pmatrix} = 0 \tag{4.21}$$

where Eqs. (4.18) and (4.19) is the normalization condition for the computational basis, and Eqs. (4.20) and (4.21) is the orthogonality condition. More generally, Eqs. (4.18, 4.19, 4.20, 4.21) can be combined into a single orthonormality condition:

$$\langle i|j\rangle = \delta_{ij}, \ i, j = \{0, 1\} \tag{4.22}$$

Suppose we have a state $|\psi\rangle = \alpha|z^+\rangle + \beta|z^-\rangle$. What does $\langle z^+|\psi\rangle$ mean? We can easily calculate $\langle z^+|\psi\rangle = \langle z^+|(\alpha|z^+\rangle + \beta|z^-\rangle) = \alpha\langle z^+|z^+\rangle + \beta\langle z^+|z^-\rangle = \alpha^*1 + \beta^*0 = \alpha$. Hence, $\langle z^+|\psi\rangle$ is the *probability amplitude* of measuring the particle in state $|z^+\rangle$ given an initial state $|\psi\rangle$; i.e., how much of the state $|z^+\rangle$ is to be found in state $|\psi\rangle$. Therefore, the inner product quantifies "how much" of one vector is contained in another; this is a generalization of the dot product for Euclidean vectors. $|\langle z^+|\psi\rangle|^2 = |\alpha|^2$ is the *probability* of measuring the particle in the state $|z^+\rangle$ given that the state was initially in the state $|\psi\rangle$.

4.7 Outer Product

Consider two vectors: $|\psi\rangle$ with its vector representation $\begin{pmatrix} \alpha \\ \beta \end{pmatrix}$ and $|\phi\rangle$ with its vector representation $\begin{pmatrix} \delta \\ \epsilon \end{pmatrix}$. The "outer product" is defined as:

$$|\psi\rangle\langle\phi| = \begin{pmatrix} \alpha \\ \beta \end{pmatrix} \otimes (\delta^*\ \epsilon^*) = \begin{pmatrix} \alpha\delta^* & \alpha\epsilon^* \\ \beta\delta^* & \beta\epsilon^* \end{pmatrix} \tag{4.23}$$

where "$\otimes$" instructs us to apply the outer product between the two vectors. The outer product is a ket followed by a bra, which is the reverse order compared to the inner product. Note that we can't apply the usual rules for matrix multiplication here: it makes no sense to multiply a 2×1 matrix $\begin{pmatrix} \alpha \\ \beta \end{pmatrix}$ by a 1×2 matrix $(\delta^* \epsilon^*)$. The "$\otimes$" notation indicates that this is a special type of matrix multiplication, called the tensor product, resulting in the matrix $\begin{pmatrix} \alpha\delta^* & \alpha\epsilon^* \\ \beta\delta^* & \beta\epsilon^* \end{pmatrix}$. The resulting matrix is an operator, similar to the Pauli matrices we saw earlier.

We can also see that outer products are operators as follows. Suppose we have an outer product $|u\rangle\langle v|$ acting on a vector $|w\rangle$. Using the vector axioms, we can write this as:

$$(|u\rangle\langle v|)|w\rangle = |u\rangle(\langle v|w\rangle) = (\langle v|w\rangle)|u\rangle \tag{4.24}$$

where $\langle v|w\rangle$ is a scalar. Hence, $|u\rangle\langle v|$ acting on a vector $|w\rangle$ produces another vector $|u\rangle$ multiplied by a scalar $\langle v|w\rangle$. This is the definition of an operator. Hence, the outer product is simply a way of representing an operator in Dirac notation. We'll deal more with the tensor product in Chap. 9.

Exercise 4.5 Show that each Pauli operator can be expressed in terms of an outer product as follows:

$$\hat{\sigma}_x = \begin{bmatrix} 0 & 1 \\ 1 & 0 \end{bmatrix} = |1\rangle\langle 0| + |0\rangle\langle 1|$$

$$\hat{\sigma}_y = \begin{bmatrix} 0 & -i \\ i & 0 \end{bmatrix} = i|1\rangle\langle 0| - i|0\rangle\langle 1|$$

$$\hat{\sigma}_z = \begin{bmatrix} 1 & 0 \\ 0 & -1 \end{bmatrix} = |0\rangle\langle 0| - |1\rangle\langle 1|$$

4.8 Projection Operators

An outer product $|u\rangle\langle u|$ is known as a projection operator. They "project" a vector $|\psi\rangle$ onto the vector $|u\rangle$. Suppose $|\psi\rangle = \alpha|0\rangle + \beta|1\rangle$. Then the operator $|0\rangle\langle 0|$ applied to $|\psi\rangle$ results in:

$$|0\rangle\langle 0|\psi\rangle = |0\rangle\langle 0|(\alpha|0\rangle + \beta|1\rangle) = |0\rangle(\alpha\langle 0|0\rangle + \beta\langle 0|1\rangle) = \alpha|0\rangle \tag{4.25}$$

where we have used the vector axioms and the orthonormality condition, $\langle 0|0 \rangle = 1$ and $\langle 0|1 \rangle = 0$. Hence, $|0\rangle\langle 0|$ projects the vector $|\psi\rangle = \alpha|0\rangle + \beta|1\rangle$ onto the vector $|0\rangle$, resulting in the vector $\alpha|0\rangle$. Similarly,

$$|1\rangle\langle 1|\psi\rangle = |1\rangle\langle 1|(\alpha|0\rangle + \beta|1\rangle) = |1\rangle(\alpha\langle 1|0\rangle + \beta\langle 1|1\rangle) = \beta|1\rangle \tag{4.26}$$

Hence, using Eqs. (4.25) and (4.26), we can express $|\psi\rangle = \alpha|0\rangle + \beta|1\rangle$ as:

$$|\psi\rangle = \alpha|0\rangle + \beta|1\rangle = |0\rangle\langle 0|\psi\rangle + |1\rangle\langle 1|\psi\rangle \tag{4.27}$$

or, since $\langle 0|\psi\rangle$ and $\langle 1|\psi\rangle$ are scalars, we are free to move them in front of the $|0\rangle$ and $|1\rangle$ vectors:

$$|\psi\rangle = \langle 0|\psi\rangle|0\rangle + \langle 1|\psi\rangle|1\rangle \tag{4.28}$$

Exercise 4.6 Show that $|0\rangle\langle 0| + |1\rangle\langle 1| = I$, the identity matrix. This is called the "resolution of the identity" or the "completeness relation".

4.9 Bloch Sphere

As we saw in Chap. 3, we can represent any single qubit as a vector or point on the "Bloch sphere" (Fig. 4.2). Thus, instead of $|\psi\rangle = \alpha|0\rangle + \beta|1\rangle$, we can write:

$$|\psi\rangle = \cos\frac{\theta}{2}|0\rangle + e^{i\phi}\sin\frac{\theta}{2}|1\rangle \tag{4.29}$$

4.10 Two Qubits

For two qubits, we can write:

$$|\psi\rangle = \underbrace{(\alpha_1|0\rangle_1 + \beta_1|1\rangle_1)}_{\text{qubit 1}} \underbrace{(\alpha_2|0\rangle_2 + \beta_2|1\rangle_2)}_{\text{qubit 2}} \tag{4.30}$$

where the subscript indicates the qubit number. Expanding this two-qubit state gives:

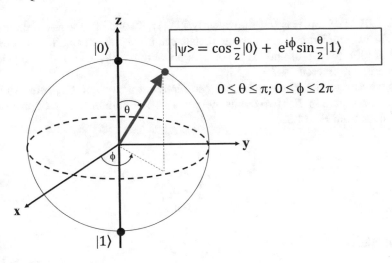

Fig. 4.2 Bloch sphere with $|0\rangle$ and $|1\rangle$ at the poles

$$|\psi\rangle = \alpha_1\alpha_2|0\rangle_1|0\rangle_2 + \alpha_1\beta_2|0\rangle_1|1\rangle_2 + \beta_1\alpha_2|1\rangle_1|0\rangle_2 + \beta_1\beta_2|1\rangle_1|1\rangle_2 \qquad (4.31)$$

or

$$|\psi\rangle = \alpha_{00}|00\rangle + \alpha_{01}|01\rangle + \alpha_{10}|10\rangle + \alpha_{11}|11\rangle \qquad (4.32)$$

where we have simplified the expression by abandoning the particle subscripts and compacting the ket notation, so that, for example, $|0\rangle_1|0\rangle_2$ is written as $|00\rangle$. In the latter notation, it is understood that the first number in the ket is the state of the first particle and the second number is the state of the second particle. For brevity, we also define the complex coefficients (probability amplitudes) as $\alpha_{00} = \alpha_1\alpha_2$, $\alpha_{01} = \alpha_1\beta_2$, $\alpha_{10} = \beta_1\alpha_2$, and $\alpha_{11} = \beta_1\beta_2$. The normalization condition requires:

$$|\alpha_{00}|^2 + |\alpha_{01}|^2 + |\alpha_{10}|^2 + |\alpha_{11}|^2 = 1 \qquad (4.33)$$

i.e., the probabilities must sum to unity.

Alternatively, 00, 01, 10 and 11 are binary for 0, 1, 2 and 3, respectively. Thus, we can rewrite Eqs. (4.32) and (4.33) as:

$$|\psi\rangle = \alpha_0|0\rangle + \alpha_1|1\rangle + \alpha_2|2\rangle + \alpha_3|3\rangle \qquad (4.34)$$

$$|\alpha_0|^2 + |\alpha_1|^2 + |\alpha_2|^2 + |\alpha_3|^2 = 1 \qquad (4.35)$$

where we have changed the subscripts on the probability amplitudes accordingly.

$|00\rangle, |01\rangle, |10\rangle$ and $|11\rangle$ (or, equivalently, $|0\rangle, |1\rangle, |2\rangle$ and $|3\rangle$) form an orthonormal basis for two qubits; i.e., any two-qubit state can be expressed as a linear superposition of $|00\rangle, |01\rangle, |10\rangle$ and $|11\rangle$. You will prove that $|00\rangle, |01\rangle, |10\rangle$ and $|11\rangle$ are orthonormal in Exercise 4.7.

As we saw with a single qubit, we can represent a two-qubit system using the vector representation. For two qubits, we write a 4×1 column vector of probability amplitudes from Eq. (4.32):

$$|\psi\rangle = \begin{pmatrix} \alpha_{00} \\ \alpha_{01} \\ \alpha_{10} \\ \alpha_{11} \end{pmatrix} \tag{4.36}$$

or, from Eq. (4.34):

$$|\psi\rangle = \begin{pmatrix} \alpha_0 \\ \alpha_1 \\ \alpha_2 \\ \alpha_3 \end{pmatrix} \tag{4.37}$$

Exercise 4.7 How would you write the vectors for $|00\rangle, |01\rangle, |10\rangle$ and $|11\rangle$? Show that these four states are orthonormal.

For example, suppose $|\psi\rangle = \left(\frac{1}{2} + \frac{i}{2}\right)|00\rangle + \frac{1}{2}|01\rangle - \frac{i}{2}|11\rangle$. We could write this in the vector representation as:

$$|\psi\rangle = \begin{pmatrix} \frac{1}{2} + \frac{i}{2} \\ \frac{1}{2} \\ 0 \\ -\frac{i}{2} \end{pmatrix}$$

Upon measurement, what is the probability of obtaining $|00\rangle, |01\rangle, |10\rangle$ or $|11\rangle$? The probabilities are given by the square of the modulus (absolute value) of the probability amplitudes:

$$P_{00} = \left|\frac{1}{2} + \frac{i}{2}\right|^2 = \frac{1}{2}$$

$$P_{01} = \left|\frac{1}{2}\right|^2 = \frac{1}{4}$$

$$P_{10} = 0$$

$$P_{11} = \left| -\frac{i}{2} \right|^2 = \frac{1}{4}$$

Note that the probabilities sum to unity as required.

4.11 Partial Measurement Rule

Suppose we have a state:

$$|\psi\rangle = \alpha_{00}|00\rangle + \alpha_{01}|01\rangle + \alpha_{10}|10\rangle + \alpha_{11}|11\rangle$$

If $|\psi\rangle$ is normalized, then:

$$|\alpha_{00}|^2 + |\alpha_{01}|^2 + |\alpha_{10}|^2 + |\alpha_{11}|^2 = 1$$

What is the probability of measuring $|0\rangle$ in the left-most qubit? The probability is:

$$P_0 = |\alpha_{00}|^2 + |\alpha_{01}|^2$$

Upon measurement, the state will collapse to a new state:

$$|\psi'\rangle = \alpha_{00}|00\rangle + \alpha_{01}|01\rangle$$

This state is no longer normalized. We must "renormalize" the state:

$$|\psi'\rangle = (\alpha_{00}|00\rangle + \alpha_{01}|01\rangle)/\left(|\alpha_{00}|^2 + |\alpha_{01}|^2\right)^{1/2}$$

This renormalization procedure is known as the "partial measurement rule".
For example, suppose:

$$|\psi\rangle = \left(\frac{1}{2} + \frac{i}{2} \right)|00\rangle + \frac{1}{2}|01\rangle - \frac{i}{2}|11\rangle$$

The probability of measuring $|0\rangle$ in the left-most qubit is:

$$P_0 = \left| \frac{1}{2} + \frac{i}{2} \right|^2 + \left| \frac{1}{2} \right|^2 = \frac{1}{2} + \frac{1}{4} = \frac{3}{4}$$

The resulting state after measurement is:

$$|\psi'\rangle = \left(\frac{1}{2} + \frac{i}{2}\right)|00\rangle + \frac{1}{2}|01\rangle$$

This state is not normalized. After renormalization, the state becomes:

$$|\psi'\rangle = \left[\left(\frac{1}{2} + \frac{i}{2}\right)|00\rangle + \frac{1}{2}|01\rangle\right]/\sqrt{3/4}$$

Exercise 4.8 Show that the above state is normalized.

4.12 Three Qubits

The previous arguments can be extended to three qubits:

$$\begin{aligned}
|\psi\rangle &= \alpha_{000}|000\rangle + \alpha_{001}|001\rangle + \alpha_{010}|010\rangle + \alpha_{100}|100\rangle \\
&+ \alpha_{011}|011\rangle + \alpha_{101}|101\rangle + \alpha_{110}|110\rangle + \alpha_{111}|111\rangle
\end{aligned} \quad (4.38)$$

or

$$|\psi\rangle = \alpha_0|0\rangle + \alpha_1|1\rangle + \alpha_2|2\rangle + \alpha_3|3\rangle + \alpha_4|4\rangle + \alpha_5|5\rangle + \alpha_6|6\rangle + \alpha_7|7\rangle \quad (4.39)$$

$|0\rangle, |1\rangle, \ldots |7\rangle$ form an orthonormal basis set; i.e., any three-qubit system can be expressed as a superposition of $|0\rangle, |1\rangle, \ldots |7\rangle$. Using the vector representation:

$$|\psi\rangle = \begin{pmatrix} \alpha_0 \\ \alpha_1 \\ \vdots \\ \alpha_7 \end{pmatrix} \quad (4.40)$$

The normalization condition is:

$$|\alpha_0|^2 + |\alpha_1|^2 + |\alpha_2|^2 + |\alpha_3|^2 + |\alpha_4|^2 + |\alpha_5|^2 + |\alpha_6|^2 + |\alpha_7|^2 = 1 \quad (4.41)$$

4.13 Multiple Qubits

When we consider more than a single qubit, we see that the number of terms in the superposition quickly grow. In fact, the number of terms grows exponentially with n as 2^n. Hence, when dealing with multiple qubits, it becomes useful to introduce a more compact notation using the summation operator:

$$|\psi\rangle = \sum_{x\in\{0,1\}^n} \alpha_x|x\rangle, \alpha_x \in C \tag{4.42}$$

where $x\in\{0,1\}^n$ means the summation runs over all possible binary strings x of length n (i.e., for n qubits). For example, if $n = 2$, then $x = 00, 01, 10$, and 11, reproducing Eq. (4.32). $\alpha_x \in C$ means that α_x is an element of the complex numbers. The normalization condition, Eq. (4.33), becomes:

$$\sum_{x\in\{0,1\}^n} |\alpha_x|^2 = 1 \tag{4.43}$$

Equivalently, using decimal notation:

$$|\psi\rangle = \sum_{x=0}^{2^n-1} \alpha_x|x\rangle, \quad \alpha_x \in C \tag{4.44}$$

where x is a decimal integer $(0, 1, 2, \dots 2^n - 1)$, as opposed to the binary string in Eq. (4.42). Recall that an n-qubit system can represent any integer from 0 to $2^n - 1$, which defines the range for x. For example, if $n = 2$, then $x = 0, 1, 2$ and 3, reproducing Eq. (4.34). The normalization condition, Eq. (4.35), becomes:

$$\sum_{x=0}^{2^n-1} |\alpha_x|^2 = 1, \alpha_x \in C \tag{4.45}$$

Also, we can represent a multiple qubit system using the vector representation:

$$|\psi\rangle = \begin{pmatrix} \alpha_0 \\ \alpha_1 \\ \vdots \\ \alpha_{2^n-1} \end{pmatrix} \tag{4.46}$$

For example, how would you write $|\psi\rangle = \frac{1}{\sqrt{2}}|0000\rangle + \frac{1}{\sqrt{2}}|1111\rangle$ as a vector? The vector representation of $|0000\rangle$ is a 16×1 column vector with the first entry containing 1 and the remaining 15 entries containing 0. $|1111\rangle$ is the 16×1 column

vector containing 0 in the first 15 entries and 1 in the last entry. Thus, $|\psi\rangle$ is a 16 ×
1 column vector containing $\frac{1}{\sqrt{2}}$ as the first entry, followed by 14 zeros, and then $\frac{1}{\sqrt{2}}$
in the last entry:

$$|\psi\rangle = \frac{1}{\sqrt{2}} \begin{pmatrix} 1 \\ 0 \\ \vdots \\ 0 \\ 0 \end{pmatrix} + \frac{1}{\sqrt{2}} \begin{pmatrix} 0 \\ 0 \\ \vdots \\ 0 \\ 1 \end{pmatrix} = \begin{pmatrix} \frac{1}{\sqrt{2}} \\ 0 \\ \vdots \\ 0 \\ \frac{1}{\sqrt{2}} \end{pmatrix}$$

We see that Dirac notation becomes very convenient and compact as the number of
qubits becomes large and the quantum vector is sparse (meaning most of the entries
are zero).

The dimension of the vector grows exponentially: 2-dimensional vector for 1
qubit, 4-dimensional vector for 2 qubits, 8-dimensional vector for 3 qubits, 16-
dimensional vector for 4 qubits, etc. In general, we need a 2^n-dimensional vector
for n qubits; i.e., an n-qubit system is a superposition of 2^n possible states. For 300
qubits, we need to keep track of $2^{300} \sim 10^{90}$ probability amplitudes in the superposition. This is greater than the number of particles in the observable universe. Where
does nature store all this information? Nobody knows. It is one of the mysteries of
quantum mechanics.

Quantum systems are very different than classical systems. For each new particle
added to a classical system, we only increase the number of required variables polynomially (e.g., the position and momentum of each particle). In quantum systems, we
increase the number of variables exponentially with each new particle (qubit) added
to the system because quantum systems allow superpositions of states. Quantum
computers might be more powerful than classical computers by taking advantage of
these superpositions. As we saw in Fig. 1.16, quantum computing relies on quantum
superpositions (also called "quantum parallelism") of binary states (qubits). For
example, with 4 qubits, we can input 16 states at once rather than each of the 4
states sequentially. Each additional qubit doubles the computing power. It is this
superposition of states that gives an exponential speedup for some computations.

Chapter 5
Entanglement

There exist multiple qubit states that cannot be expressed as a product of individual qubit states. These states are called entangled states, which is another powerful resource used in quantum computing.

5.1 Composite States

Let us revisit the two-qubit system examined in Chap. 4. For two qubits, we can write:

$$|\Psi\rangle = \underbrace{(\alpha_1|0\rangle_1 + \beta_1|1\rangle_1)}_{\text{qubit 1}} \underbrace{(\alpha_2|0\rangle_2 + \beta_2|1\rangle_2)}_{\text{qubit 2}} \tag{5.1}$$

where the subscript on the ket indicates the qubit number. Expanding this two-qubit state gives:

$$\begin{aligned}|\psi\rangle &= \alpha_1\alpha_2|0\rangle_1|0\rangle_2 + \alpha_1\beta_2|0\rangle_1|1\rangle_2 \\ &+ \beta_1\alpha_2|1\rangle_1|0\rangle_2 + \beta_1\beta_2|1\rangle_1|1\rangle_2\end{aligned} \tag{5.2}$$

For example,

$$\underbrace{\left(\frac{1}{\sqrt{2}}|0\rangle + \frac{1}{\sqrt{2}}|1\rangle\right)}_{\text{qubit 1}} \underbrace{\left(\frac{1}{2}|0\rangle - \frac{\sqrt{3}}{2}|1\rangle\right)}_{\text{qubit 2}}$$

$$= \frac{1}{2\sqrt{2}}|00\rangle - \frac{\sqrt{3}}{2\sqrt{2}}|01\rangle + \frac{1}{2\sqrt{2}}|10\rangle - \frac{\sqrt{3}}{2\sqrt{2}}|11\rangle \tag{5.3}$$

© The Author(s), under exclusive license to Springer Nature Switzerland AG 2021
R. LaPierre, *Introduction to Quantum Computing*, The Materials Research Society Series,
https://doi.org/10.1007/978-3-030-69318-3_5

where the subscripts have been omitted. In this example, measurement of the first (left-most) qubit gives $|0\rangle$ with probability $\left|\frac{1}{2\sqrt{2}}\right|^2 + \left|-\frac{\sqrt{3}}{2\sqrt{2}}\right|^2 = \frac{1}{2}$, and $|1\rangle$ with probability $\left|\frac{1}{2\sqrt{2}}\right|^2 + \left|-\frac{\sqrt{3}}{2\sqrt{2}}\right|^2 = \frac{1}{2}$. This measurement does not affect the state of the second qubit.

5.2 Separable States

Suppose you are given a composite state of two qubits (for example, Eq. (5.3)) and asked for the state of the individual qubits. To answer this question, we would work backwards to factor the state as follows:

$$
\begin{aligned}
&\frac{1}{2\sqrt{2}}|00\rangle - \frac{\sqrt{3}}{2\sqrt{2}}|01\rangle + \frac{1}{2\sqrt{2}}|10\rangle - \frac{\sqrt{3}}{2\sqrt{2}}|11\rangle \\
&= \underbrace{\left(\frac{1}{\sqrt{2}}|0\rangle + \frac{1}{\sqrt{2}}|1\rangle\right)}_{\text{qubit 1}} \underbrace{\left(\frac{1}{2}|0\rangle - \frac{\sqrt{3}}{2}|1\rangle\right)}_{\text{qubit 2}}
\end{aligned}
\tag{5.4}
$$

In this example, it is easy to work backwards to obtain the state of the two individual qubits. This is known as a separable state.

5.3 Entanglement

There are some states for which this factoring is impossible; i.e., you cannot write the composite state as a product of individual qubit states:

$$
|\psi\rangle \neq |\psi\rangle_1 |\psi\rangle_2
\tag{5.5}
$$

These are known as entangled states. In entangled states, you cannot talk about the state of qubits individually—they are somehow intertwined. For example, consider the state:

$$
|\psi\rangle = \frac{1}{\sqrt{2}}|00\rangle + \frac{1}{\sqrt{2}}|11\rangle
\tag{5.6}
$$

This state cannot be factored into the product of two individual qubit states (try it).

Let us determine if $|\psi\rangle$ in Eq. (5.6) can be written as the product of two qubit states:

$$|\psi\rangle = \frac{1}{\sqrt{2}}|00\rangle + \frac{1}{\sqrt{2}}|11\rangle \overset{?}{\rightarrow} (\alpha_1|0\rangle + \beta_1|1\rangle)(\alpha_2|0\rangle + \beta_2|1\rangle) \qquad (5.7)$$

Expanding Eq. (5.7), we obtain:

$$\frac{1}{\sqrt{2}}|00\rangle + \frac{1}{\sqrt{2}}|11\rangle = \alpha_1\alpha_2|00\rangle + \alpha_1\beta_2|01\rangle + \beta_1\alpha_2|10\rangle + \beta_1\beta_2|11\rangle \qquad (5.8)$$

To satisfy Eq. (5.8), we must have:

$$\alpha_1\alpha_2 = \frac{1}{\sqrt{2}} \qquad (5.9)$$

$$\alpha_1\beta_2 = 0 \qquad (5.10)$$

$$\beta_1\alpha_2 = 0 \qquad (5.11)$$

$$\beta_1\beta_2 = \frac{1}{\sqrt{2}} \qquad (5.12)$$

However, it is impossible to satisfy Eqs. (5.9–5.12) simultaneously. If Eq. (5.10) is satisfied, then either $\alpha_1 = 0$ or $\beta_2 = 0$. If $\alpha_1 = 0$, then Eq. (5.9) cannot be satisfied. If $\beta_2 = 0$, then Eq. (5.12) cannot be satisfied. Similarly, if Eq. (5.11) is satisfied, then either $\beta_1 = 0$ or $\alpha_2 = 0$. If $\beta_1 = 0$, then Eq. (5.12) cannot be satisfied. If $\alpha_2 = 0$, then Eq. (5.9) cannot be satisfied. We conclude that Eq. (5.7) cannot be satisfied; i.e., we cannot write $|\psi\rangle = \frac{1}{\sqrt{2}}|00\rangle + \frac{1}{\sqrt{2}}|11\rangle$ as a product of two states. It is an entangled state. You can't say "sometimes the state is in $|00\rangle$ and sometimes $|11\rangle$". The state is in both $|00\rangle$ and $|11\rangle$ simultaneously (it is a superposition). If we measure qubit 1 and find $|0\rangle$, then qubit 2 will instantaneously be found in the state $|0\rangle$. If we measure qubit 1 in state $|1\rangle$, then qubit 2 will instantaneously be found in state $|1\rangle$. We say that the two states are "correlated". Upon measurement, you will obtain either the state $|00\rangle$ or $|11\rangle$, each with probability 1/2. After measurement, the two qubits, $|00\rangle = |0\rangle_1|0\rangle_2$ or $|11\rangle = |1\rangle_1|1\rangle_2$, are no longer entangled (the states of the two qubits become separable).

5.4 Bell States

There are four entangled two-qubit states that are commonly encountered, known as the Bell states:

$$|\Phi^+\rangle = \frac{1}{\sqrt{2}}(|00\rangle + |11\rangle) \qquad (5.13)$$

$$|\Psi^+\rangle = \frac{1}{\sqrt{2}}(|01\rangle + |10\rangle) \tag{5.14}$$

$$|\Phi^-\rangle = \frac{1}{\sqrt{2}}(|00\rangle - |11\rangle) \tag{5.15}$$

$$|\Psi^-\rangle = \frac{1}{\sqrt{2}}(|01\rangle - |10\rangle) \tag{5.16}$$

These are also known as EPR pairs, named after Albert Einstein, Boris Podolsky and Nathan Rosen.

Let us examine some examples. In the case of photons, an entangled state could exist between horizontally and vertically polarized light:

$$|\psi\rangle = \frac{1}{\sqrt{2}}|H\rangle_A|V\rangle_B + \frac{1}{\sqrt{2}}|V\rangle_A|H\rangle_B = \frac{1}{\sqrt{2}}(|HV\rangle + |VH\rangle) \tag{5.17}$$

Equation (5.17) is an example of Eq. (5.14) where horizontally polarized light (H) represents 0 and vertically polarized light represents 1. Here, we have labelled the first qubit as "A" and the second qubit as "B". If we measure photon A to have H polarization, then photon B will definitely have V polarization. If we measure photon A to have V polarization, then photon B will definitely have H polarization.

In the case of electron spins, an entangled state could exist between spin up and spin down:

$$|\psi\rangle = \frac{1}{\sqrt{2}}(|\uparrow\rangle_A|\downarrow\rangle_B + |\downarrow\rangle_A|\uparrow\rangle_B) = \frac{1}{\sqrt{2}}(|\uparrow\downarrow\rangle + |\downarrow\uparrow\rangle) \tag{5.18}$$

Equation (5.18) is another example of Eq. (5.14) where spin up represents 0 and spin down represents 1. If we measure electron A to have spin up, then electron B will definitely have spin down. If we measure electron A to have spin down, then electron B will definitely have spin up.

In the previous examples, we could prepare the entangled state between particle A and B, and then separate the two particles by large distances—in fact, they could be on opposite sides of the universe. Let's focus on the example of electron spins. After separating the electrons, an individual (let's call her Alice) could perform a measurement on electron A (using a Stern-Gerlach apparatus) and another individual (let's call him Bob) could perform a subsequent spin measurement on particle B. If Alice measures spin up, Bob will measure spin down. If Alice measures spin down, Bob will measure spin up. The measurements are correlated between Alice and Bob. This occurs no matter the distance between Alice and Bob. There appears to be instantaneous action at a distance, or "non-locality" in entanglement, which Einstein called "spooky action at a distance". Could Alice and Bob use entanglement to communicate instantaneously across vast distances? The theory of special relativity

Fig. 5.1 Does entanglement
violate special relativity?

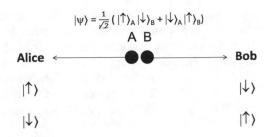

$$|\psi\rangle = \frac{1}{\sqrt{2}}(|\uparrow\rangle_A|\downarrow\rangle_B + |\downarrow\rangle_A|\uparrow\rangle_B)$$

states that information cannot travel faster than the speed of light. Does entanglement violate special relativity?

In fact, entanglement does not violate special relativity because no information is being transmitted (Fig. 5.1). Upon measurement, Alice will collapse the entangled wavefunction, $|\psi\rangle = \frac{1}{\sqrt{2}}(|\uparrow\downarrow\rangle + |\downarrow\uparrow\rangle)$, to a separable state—either $|\uparrow\downarrow\rangle$ with probability 1/2, or $|\downarrow\uparrow\rangle$ with probability 1/2; i.e., Alice obtains either spin up or spin down with 50% probability (i.e., random), and Bob obtains the opposite spin. Alice cannot control which of these two states she obtains. Subsequent measurement by Bob will result in the opposite spin state to Alice, but his measurement will likewise appear to him to be completely random—no information is sent. Special relativity remains intact.

5.5 Quantum Eraser

Entanglement allows us to perform some interesting experiments. In Chap. 1, we saw how the interference pattern disappeared in the double slit experiment if we made a measurement to determine which slit each photon passed through. The quantum eraser experiment uses entanglement to perform such a measurement. Figure 5.2 shows two entangled photon pairs, with each photon in a pair emitted at right angles to each other (one of the pairs is indicated by the solid lines, and the other pair is represented by the dashed lines). One photon, called the "idler" photon, passes through slit 1 or 2. The other photon, called the "signal" photon, is emitted towards

Fig. 5.2 Quantum eraser
experiment

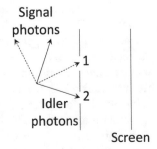

a photodetector. If two separate detectors are used for each of the signal photons, then we could determine which slit the corresponding idler photon passed through. Thus, the interference pattern on the screen disappears. On the other hand, if a single detector is used for both signal photons, then we erase any information about which slit the idler photons passed through; i.e., we cannot know if the detector signal is associated with an idler photon emitted towards slit 1 or 2. Thus, the interference pattern is observed on the screen. This is called the "quantum eraser" experiment.

5.6   Quantum Metrology

Quantum entanglement is evident in nature, ranging from the covalent bond of the hydrogen molecule to a possible role in bird navigation [1]. Quantum science, especially entanglement, is being explored in quantum metrology to improve the performance of clocks, magnetometers, gravimeters, accelerometers, and optical microscopy. For example, quantum imaging exploits the correlation between entangled photons to overcome the Rayleigh diffraction [2] and shot noise limitations [3] in optical microscopy. Quantum illumination, also called "quantum radar", proposes to use entanglement in detection and ranging systems [4]. We will see that entanglement is also used in quantum communication, teleportation, and quantum computing. Superposition, interference and entanglement are new types of resources that do not exist in classical computing that can be used in quantum computing.

5.7   EPR Paradox and Hidden Variables

In a famous 1935 paper, Albert Einstein, Boris Podolsky and Nathan Rosen (known as EPR) sought to demonstrate by the "EPR paradox" that quantum mechanics was incomplete [5]. EPR were concerned by the instantaneous action at a distance, or "non-locality", implied by entanglement. Quantum mechanics also seems to violate "realism". "Realism" means that particles have definite properties that are independent of any measurement.

Suppose we toss a coin. In principle, it is possible to know whether it will land heads or tails if we keep track of a lot of information about the system (called "degrees of freedom"), such as the forces applied during the toss, the air currents, the height of the toss, etc. However, all these physical properties are impossible to calculate in practice, so the most we can do is ascribe a probability distribution for the toss outcome resulting in $P_{heads} = 1/2$ and $P_{tails} = 1/2$. This outcome results from averaging over the many degrees of freedom that we don't have access to. This principle also forms the basis for statistical thermodynamics.

Einstein and many others believed that quantum mechanics was like this; i.e., they proposed that the probabilities in quantum mechanics are deterministic (versus probabilistic) and have some underlying causes that are "hidden"; i.e., that we cannot

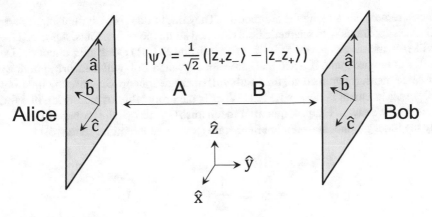

Fig. 5.3 Apparatus for a Bell test

access (analogous to the unknown forces during the coin toss). These underlying causes were called "hidden variables". If we knew the hidden variables, we would be able to calculate a definite measurement outcome, rather than just probabilities.

Many quantum pioneers, exemplified by Einstein, believed in "local realism" where the state of particles is defined when they are created. However, the "hidden variables" only allow us to determine the probability of these states. Einstein famously said: "God does not play dice with the universe". Also, with regards to realism, Einstein said "Do you believe the moon exists only when you look at it?".

Others, exemplified by Bohr, believed in the possibility of entanglement. They believed that no definitive statements about a physical system may be made until a measurement is made. Particle properties do not exist until we measure them. It turns out the Bohr was correct; but how do we prove it?

5.8 The Bell Test

In 1964, the physicist, John Bell, proposed a test for quantum mechanics by measuring the spin states along three different directions ($\hat{a}$, $\hat{b}$, $\hat{c}$) for many entangled pairs of electron spins (Fig. 5.3) [6]. In the following sections, we will examine the experiment that he proposed.

5.9 Measurements Along One Direction

Suppose two electrons are initially created with an entangled spin state of $|\psi_i\rangle = \frac{1}{\sqrt{2}}(|z_+z_-\rangle - |z_-z_+\rangle)$, as shown in Fig. 5.3. The two electrons are separated—one is sent to Alice and the other is sent to Bob. Suppose Alice and Bob choose the same measurement basis along the $\hat{z}$ direction. They might do this by both using a Stern-Gerlach apparatus with magnetic field oriented along the z-axis. Thus, Alice and Bob will each measure either spin up ($|z_+\rangle$) or spin down ($|z_-\rangle$) along the $\hat{z}$ direction. Due to the entangled state, if Alice measures spin up, then Bob will measure spin down. If Alice measures spin down, then Bob will measure spin up. Suppose the final state after measurement is $|\psi_f\rangle = |z_+z_-\rangle$. What is the probability of measuring this final state? As we saw in Chap. 4, quantum mechanics tells us that the probability is given by the inner product between the final state ($|\psi_f\rangle$) and the initial state ($|\psi_i\rangle$):

$$P(z_+z_-) = |\langle\psi_f|\psi_i\rangle|^2$$

$$= \left| \langle z_+z_-| \frac{1}{\sqrt{2}}(|z_+z_-\rangle - |z_-z_+\rangle) \right|^2$$

$$= \frac{1}{2}|\langle z_+z_-|z_+z_-\rangle - \langle z_+z_-|z_-z_+\rangle|^2 \tag{5.19}$$

In Eq. (5.19), we have encountered something new. What does $\langle z_+z_-|z_+z_-\rangle$ or $\langle z_+z_-|z_-z_+\rangle$ mean? Remember that $\langle z_+z_-|z_+z_-\rangle = ({}_1\langle z_+|{}_2\langle z_-|)(|z_+\rangle_1|z_-\rangle_2)$ where we have added the particle number to each bra and ket. The inner product is defined between the bra and ket of the *same particle*. Thus, collecting the bra and ket of the same particle gives:

$$\langle z_+z_-|z_+z_-\rangle = ({}_1\langle z_+|{}_2\langle z_-|)(|z_+\rangle_1|z_-\rangle_2)$$
$$= ({}_1\langle z_+|z_+\rangle_1)({}_2\langle z_-|z_-\rangle_2) = (1)(1) = 1 \tag{5.20}$$

where we have used the orthonormality condition. Similarly,

$$\langle z_+z_-|z_-z_+\rangle = ({}_1\langle z_+|{}_2\langle z_-|)(|z_-\rangle_1|z_+\rangle_2) = ({}_1\langle z_+|z_-\rangle_1)({}_2\langle z_-|z_+\rangle_2) = (0)\,(0) = 0 \tag{5.21}$$

where we have again used the orthonormality condition. Thus,

$$P(z_+z_-) = \frac{1}{2}|\langle z_+|z_+\rangle\langle z_-|z_-\rangle - \langle z_+z_-|z_-z_+\rangle|^2$$

$$= \frac{1}{2}|1 - 0|^2$$

$$= \frac{1}{2} \tag{5.22}$$

Exercise 5.1 Show that $P(z_-z_+) = 1/2$

EPR would say that the particles are not entangled. Instead, EPR would propose that the measurement outcome is reproduced with the particle distribution in Table 5.1; i.e., half the A/B particle pairs (first row of Table 5.1) are produced with spin up/down and the other half of the particle pairs (second row of Table 5.1) are produced

Table 5.1 Proposed EPR particle distribution for measurement along one direction	Distribution	Particle A (Alice)	Particle B (Bob)
	1/2	z_+	z_-
	1/2	z_-	z_+

with spin down/up. This is not a superposition as in the entangled state. Rather, it is a classical probability distribution. Hidden variables preclude us from predicting the exact spin state of each pair. With this particle distribution, we cannot distinguish between the predictions of EPR and quantum mechanics.

5.10 Measurements Along Two Directions

Suppose instead that Alice and Bob can measure along two directions, say $\hat{z}$ and $\hat{x}$. If Alice chooses to measure along $\hat{z}$ and Bob chooses to measure along $\hat{x}$, then quantum mechanics predicts that the probability of both measuring spin up is:

$$
\begin{aligned}
P(z_+x_+) &= \left| \langle z_+x_+ | \frac{1}{\sqrt{2}} (|z_+z_-\rangle - |z_-z_+\rangle) \right|^2 \\
&= \frac{1}{2} |\langle z_+x_+|z_+z_-\rangle - \langle z_+x_+|z_-z_+\rangle|^2 \\
&= \frac{1}{2} |\langle z_+|z_+\rangle\langle x_+|z_-\rangle - \langle z_+|z_-\rangle\langle x_+|z_+\rangle|^2 \\
&= \frac{1}{2} |(1)\langle x_+|z_-\rangle - (0)\langle x_+|z_+\rangle|^2 \\
&= \frac{1}{2} |\langle x_+|z_-\rangle|^2
\end{aligned}
\tag{5.23}
$$

Using the fact that $\langle x_+| = \frac{1}{\sqrt{2}} (\langle z_+| + \langle z_-|)$ from Eq. (3.26), we get:

$$
\begin{aligned}
P(z_+x_+) &= \frac{1}{2} \left| \frac{1}{\sqrt{2}} (\langle z_+| + \langle z_-|)|z_-\rangle \right|^2 \\
&= \frac{1}{2} \left| \frac{1}{\sqrt{2}} (\langle z_+|z_-\rangle + \langle z_-|z_-\rangle) \right|^2 \\
&= \frac{1}{2} \left| \frac{1}{\sqrt{2}} (0 + 1) \right|^2 = \frac{1}{4}
\end{aligned}
$$

Exercise 5.2 Prove $P(z_+x_-) = \frac{1}{4}$, $P(z_-x_+) = \frac{1}{4}$, and $P(z_-x_-) = \frac{1}{4}$.

Table 5.2 Proposed EPR particle distribution for measurement along two directions

Distribution	Particle A (Alice)	Particle B (Bob)
$\frac{1}{4}$	x_+, z_+	x_-, z_-
$\frac{1}{4}$	x_+, z_-	x_-, z_+
$\frac{1}{4}$	x_-, z_+	x_+, z_-
$\frac{1}{4}$	x_-, z_-	x_+, z_+

EPR would say that the particles are not entangled; instead, the particles are produced with the classical probability distribution in Table 5.2. For example, x_+, z_1 in the first row of Table 5.2 means Alice would obtain a spin of $+ \hbar/2$ when measuring along $\hat{x}$ or $+ \hbar/2$ when measuring along $\hat{z}$ for particle A; while Bob would measure the opposite spin for particle B. $\frac{1}{4}$ of the particle pairs produced are of this type. Once again, we cannot distinguish between the predictions of EPR and quantum mechanics.

5.11 Measurements Along Three Directions

Finally, let us measure along three directions $\left(\hat{a}, \hat{b}, \hat{c}\right)$ where we will find a contradiction between EPR and quantum mechanics. Suppose N particles are distributed as in Table 5.3. For example, a_+, b_-, c_+ (third row of Table 5.3) means Alice would obtain either a spin of $+ \hbar/2$ when measuring along $\hat{a}$, $-\hbar/2$ when measuring along $\hat{b}$, or $+ \hbar/2$ when measuring along $\hat{c}$ for particle A. Note that Bob would obtain the opposite spin for each corresponding particle B; i.e., he would obtain a_-, b_+, c_-. There are N_3 of these instances. Instead of 1/8 probability for each row in Table 5.3, we will generalize and use $N_1, N_2, ..., N_8$ for each of the possibilities. Note that the total number of A/B particle pairs is $N = N_1 + N_2 + ... + N_8$. Also note that if we represent $+$ by 0 and $-$ by 1, then the eight possibilities listed in Table 5.3 for particle A represent the 8 possible basis states 000, 001, 010, 011, 100, 101, 110, and 111.

The probability distribution of Table 5.3 predicts the following:

Table 5.3 Proposed EPR particle distribution for measurement along three directions

Distribution	Particle A (Alice)	Particle B (Bob)
N_1	a_+, b_+, c_+	a_-, b_-, c_-
N_2	a_+, b_+, c_-	a_-, b_-, c_+
N_3	a_+, b_-, c_+	a_-, b_+, c_-
N_4	a_+, b_-, c_-	a_-, b_+, c_+
N_5	a_-, b_+, c_+	a_+, b_-, c_-
N_6	a_-, b_+, c_-	a_+, b_-, c_+
N_7	a_-, b_-, c_+	a_+, b_+, c_-
N_8	a_-, b_-, c_-	a_+, b_+, c_+

$$P(a_+b_+) = \frac{N_3 + N_4}{N} \tag{5.24}$$

$$P(b_+c_+) = \frac{N_2 + N_6}{N} \tag{5.25}$$

$$P(a_+c_+) = \frac{N_2 + N_4}{N} \tag{5.26}$$

where we have divided by the total number of particles, N, to produce a probability. As before, $P(a_+b_+)$ in Eq. (5.24) is the probability that Alice will measure spin up along $\hat{a}$, while Bob measures spin up along $\hat{b}$, and similar for Eqs. (5.25) and (5.26). Next, the following inequality clearly holds:

$$N_3 + N_4 + N_2 + N_6 \geq N_2 + N_4 \tag{5.27}$$

Dividing Eq. (5.27) by N and substituting Eqs. (5.24–5.26) gives:

$$P(a_+b_+) + P(b_+c_+) \geq P(a_+c_+) \tag{5.28}$$

Equation (5.28) is called a Bell inequality. The Bell inequality represents what would be expected classically. In fact, Eq. (5.28) applies to any three properties of a system. For example, a_+ could represent the number of students in a classroom with blue eyes, while a_- are the number without blue eyes; b_+ could represent the number with black hair, while b_- are the number without black hair; and c_+ could represent the number of students with height $\geq$ 5'8", while c_- is the number with height < 5'8". You can understand Eq. (5.28) by considering a Venn diagram.

Do the predictions of quantum mechanics satisfy the Bell inequality, Eq. (5.28)? For the 3 directions, let us choose $\hat{a}$ as the $\hat{z}$ direction, $\hat{b}$ as the 45° direction (between the $\hat{x}$ and $\hat{z}$ axis), and $\hat{c}$ as the $\hat{x}$ direction. To evaluate Bell's inequality, it is useful to express all the states in terms of the $\hat{z}$ direction. The Bloch sphere can help us with this task. The $\hat{a}$ direction is easy:

$$|a_+\rangle = |z_+\rangle \tag{5.29}$$

$$|a_-\rangle = |z_-\rangle \tag{5.30}$$

These two states are the poles of the Bloch sphere (Fig. 5.4).
Next, using the Bloch vector, we can express the states along the $\hat{b}$ direction:

$$|\psi\rangle = \cos\left(\frac{\theta}{2}\right)|z_+\rangle + e^{i\phi}\sin\left(\frac{\theta}{2}\right)|z_-\rangle \tag{5.31}$$

Fig. 5.4 $|a_+\rangle$ and $|a_-\rangle$ states shown in the Bloch sphere

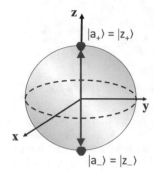

$$\theta = \pi/4, \ \phi = 0: \quad |b_+\rangle = \cos\frac{\pi}{8}|z_+\rangle + \sin\frac{\pi}{8}|z_-\rangle \tag{5.32}$$

$$\theta = 3\pi/4, \ \phi = \pi: |b_-\rangle = \cos\frac{3\pi}{8}|z_+\rangle - \sin\frac{3\pi}{8}|z_-\rangle \tag{5.33}$$

as illustrated in the Bloch sphere of Fig. 5.5.

Finally, for the $\hat{c}$ direction, we have:

$$\theta = \pi/2, \ \phi = 0 : |c_+\rangle = |x_+\rangle = \frac{1}{\sqrt{2}}(|z_+\rangle + |z_-\rangle) \tag{5.34}$$

$$\theta = \pi/2, \ \phi = \pi : |c_-\rangle = |x_-\rangle = \frac{1}{\sqrt{2}}(|z_+\rangle - |z_-\rangle) \tag{5.35}$$

as illustrated in the Bloch sphere of Fig. 5.6.

With these definitions, we can now calculate the first term in Bell's inequality:

$$P(a_+b_+) = \left| \langle a_+b_+| \frac{1}{\sqrt{2}}(|z_+z_-\rangle - |z_-z_+\rangle) \right|^2$$

$$= \frac{1}{2}|\langle a_+b_+|z_+z_-\rangle - \langle a_+b_+|z_-z_+\rangle|^2$$

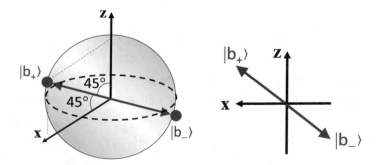

Fig. 5.5 $|b_+\rangle$ and $|b_-\rangle$ states shown in the Bloch sphere, and in the x–z plane

Fig. 5.6 $|c_+\rangle$ and $|c_-\rangle$ states shown in the Bloch sphere, along the x-axis

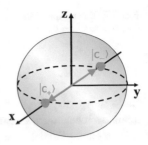

Substituting Eqs. (5.29) and (5.30), we obtain:

$$P(a_+b_+) = \frac{1}{2}|\langle z_+b_+|z_+z_-\rangle - \langle z_+b_+|z_-z_+\rangle|^2$$

$$= \frac{1}{2}|\langle z_+|z_+\rangle\langle b_+|z_-\rangle - \langle z_+|z_-\rangle\langle b_+|z_+\rangle|^2$$

$$= \frac{1}{2}|\langle b_+|z_-\rangle|^2$$

Substituting Eq. (5.32) gives:

$$P(a_+b_+) = \frac{1}{2}\left|(\cos\frac{\pi}{8}\langle z_+| + \sin\frac{\pi}{8}\langle z_-|)|z_-\rangle\right|^2$$

$$= \frac{1}{2}\sin^2\frac{\pi}{8}$$

or

$$P(a_+b_+) \sim 0.073 \tag{5.36}$$

Exercise 5.3 Show that the other terms in Bell's inequality are:

$$P(b_+c_+) = \frac{1}{2}\sin^2\frac{\pi}{8} \sim 0.073 \tag{5.37}$$

$$P(a_+c_+) = \frac{1}{4} = 0.25 \tag{5.38}$$

Using Eqs. (5.36–5.38), we can evaluate Bell's inequality:

$$P(a_+b_+) + P(b_+c_+) \geq P(a_+c_+)$$
$$0.073 + 0.073 \geq 0.25 \tag{5.39}$$
$$0.146 \geq 0.25 \quad \text{(false!)}$$

Fig. 5.7 Illustration of the
case where $\theta_{ab} = \theta_{bc} = \theta$

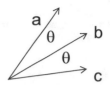

Bell's inequality is what we expect classically, but we see that it is violated in quantum mechanics! Because measurements on one particle affect the other particle (due to entanglement), the correlations do not obey Bell's classical inequality. Here, we have a discrepancy between quantum mechanics and classical physics due to entanglement.

Exercise 5.4 Show the above Bell inequality is satisfied if the initial state is not entangled.

Exercise 5.5 For the entangled state of Fig. 5.3, show in general that $P = 1/2$ $\sin^2(\theta/2)$ where θ is the angle between Bob's and Alice's spin measurement direction.

In general, for three directions Bell's inequality becomes:

$$\frac{1}{2} \sin^2(\theta_{ab}/2) + \frac{1}{2} \sin^2(\theta_{bc}/2) \geq \frac{1}{2} \sin^2(\theta_{ac}/2) \qquad (5.40)$$

If we simplify the geometry and assume $\theta_{ab} = \theta_{bc} = \theta$ (Fig. 5.7), then the Bell inequality in Eq. (5.28) becomes:

$$\sin^2(\theta/2) \geq \frac{1}{2} \sin^2\theta \qquad (5.41)$$

In this case, Bell's inequality is violated if $\theta < 90°$. For the $\hat{a}$, $\hat{b}$ and $\hat{c}$ directions chosen in the previous analysis, we had $\theta = 45°$, which satisfies this condition.

5.12 CHSH Inequality

There are other versions of the Bell inequality, depending on other possible combinations of probabilities. For example, a commonly used version was derived by Clauser, Horne, Shimony and Holt (CHSH) in 1969 [7]. Suppose a and a' are the results from two measurement directions for Alice, and b and b' are the results from two measurement directions for Bob. Suppose the observables a, a', b, b' are assigned values $\pm$ 1. If a and $a' = \pm 1$, it follows that either $a + a' = 0$, in which case $a - a' = \pm 2$. Otherwise, $a - a' = 0$, in which case $a + a' = \pm 2$. Therefore, we define a quantity S:

$$S = (a' + a)b + (a' - a)b' = a'b + ab + a'b' - ab' = \pm 2 \qquad (5.42)$$

Quantities such as "ab" represent a coincidence measurement, i.e., a and b are obtained in separate measurements by Alice and Bob and the product ab is determined. After repeated measurements, the average value of S will depend on the probability distribution of measuring $+ 2$ or $- 2$. Therefore, $|\langle S \rangle| \leq 2$ where $\langle \ \rangle$ represents an average over many coincidence measurements. Thus,

$$|\langle S \rangle| = |\langle a'b \rangle + \langle ab \rangle + \langle a'b' \rangle - \langle ab' \rangle| \leq 2 \qquad (5.43)$$

Equation (5.43) is the CHSH inequality. Different versions of the CHSH inequality exist, depending on how Eq. (5.42) is expressed. The signs in Eq. (5.43) are not so important as long as one of them is different than the others.

Equation (5.43) is a classical prediction. What does quantum mechanics predict? Suppose the CHSH inequality is evaluated using electron spin. The experimental arrangement is identical to Fig. 5.3. If spin measurements are made, then $+1$ represents spin parallel to the measurement axis (spin up) and -1 represents spin anti-parallel to the measurement axis (spin down). For example, suppose Alice chooses to measure spin along the $a \equiv \hat{z}$ and $a' \equiv \hat{x}$ directions, while Bob chooses to measure along the $b \equiv x + \hat{z}$ and $b' \equiv \hat{x} - \hat{z}$ directions (Fig. 5.8). Later, it will be easy to show (Exercise 9.8) that, for electron spin, $\langle ab \rangle = - \cos\theta$ where θ is the difference between the a and b measurement directions. Thus, $\langle ab \rangle = - \cos(45°) = -\frac{1}{\sqrt{2}}$. Similarly, $\langle a'b \rangle = -\frac{1}{\sqrt{2}}$, $\langle a'b' \rangle = -\frac{1}{\sqrt{2}}$, and $\langle ab' \rangle = \frac{1}{\sqrt{2}}$, which gives $|\langle S \rangle| = 2\sqrt{2}$; i.e., greater than the classical prediction of 2! It can be shown that the upper bound of $|\langle S \rangle|$ is $2\sqrt{2}$ in the case when a, b, a' and b' are separated by successive 45° angles as shown in Fig. 5.8 [8].

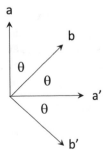

Fig. 5.8 Measurement along a, a', b and b' directions for CHSH inequality. For spin measurements, the CHSH inequality is saturated when a, b, a' and b' are separated by successive $\theta = 45°$ angles

Fig. 5.9 Alain Aspect.
Credit Wikimedia commons
[10]

5.13   Testing Bell's Inequality

Instead of using particle spin, experimental tests of Bell's theorem typically use entangled photon polarization states like $|\psi\rangle = \frac{1}{\sqrt{2}}(|V\rangle_A|H\rangle_B - |H\rangle_A|V\rangle_B)$ and test a violation of the CHSH inequality. The physicist, Alain Aspect (Fig. 5.9), famously performed an experiment in 1981, using entangled photons (rather than electron spin as proposed by John Bell), showing violation of the CHSH inequality as a result of quantum mechanics [9]. Polarization-entangled photon pairs can be produced by a process called "spontaneous parametric down conversion" in "non-linear crystals" (e.g., lithium niobate), or by using electronic transitions in quantum dots (more on quantum dots in Chap. 20). In addition, photodetectors capable of detecting single photons are needed, such as avalanche photodiodes (APDs). Polarizers are used to split different polarization states to separate detectors. We can choose random polarizations for each measurement using, for example, a Pockels cell. The path length between the source and Alice's detector is shorter than that for Bob, so Alice performs her measurement before Bob. The experiment is repeated many times with single photons, and Alice's measurement results are correlated with Bob's to verify a violation of Bell's inequality or the CHSH inequality (Fig. 5.10).

5.14 Loopholes

There are so-called "loopholes" in the Bell test that might reproduce the results of the Bell experiment without the need to introduce quantum mechanics. For example:

1. Alice and Bob may be too close together, so the first photon qubit measurement by Alice may be able to communicate its result somehow to influence the result of the second qubit measurement by Bob. This is called the locality or causality loophole.
2. The qubits may not always be entangled (too few entangled particles from the source), or there may be imperfections in the detectors such that they fail to detect some photons (detector efficiency is less than 100%), or they detect photons even when the light source is off (so-called dark current). This is called the detection loophole.
3. The basis choices may not be random. This is called the freedom of choice loophole.

The first loophole is closed by a sufficient distance between Alice and Bob, so Bob performs his measurement before a possible signal could travel from Alice to Bob (even if traveling at the speed of light) [11]. Also, the efficiency of sources and detectors have improved, and methods of generating random basis selections have been implemented [12, 13] to close the other loopholes. For example, 100,000 gamers were used to generate random input bits for random selection of the basis choice in a Bell test to close the "freedom of choice" loophole [12]. In another implementation of the Bell test, photons from distant quasars (7.78 Gy old) were used to generate random input bits for random selection of the basis choice in a Bell test [13].

Violation of Bell's inequality shows that quantum mechanics violates locality, realism or both (local realism). Bohr was correct. However, the "spooky action at a distance" still seems mysterious. As we will see in the following chapters, entangled states also provide a means of secure communication and is used in quantum computing.

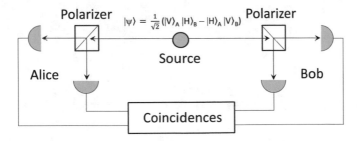

Fig. 5.10 Optical implementation of the Bell test

References

1. H.G. Hiscock et al., PNAS **113**, 4634 (2016)
2. E. Toninelli et al., Optica **6**, 347 (2019)
3. G. Brida, M. Genovese, I.R. Berchera, Nat Photonics **4**, 227 (2010)
4. J.H. Shapiro, IEEE Aerosp Electron Syst Mag **35**, 8 (2020)
5. A. Einstein, B. Podolsky, N. Rosen, Phys Rev **47**, 777 (1935)
6. J.S. Bell, Physics Physique Fizika **1**, 195 (1964)
7. J.F. Clauser, M.A. Horne, A. Shimony, R.A. Holt, Phys Rev Lett **23**, 880 (1969)
8. Lecture notes of John Preskill. Chapter 4. https://theory.caltech.edu/~preskill/ph219/index. html#lecture
9. Aspect A, Grangier P, Roger G (1981) Phys Rev Lett 47:460
10. Attribution: The Royal Society. This file is licensed under the Creative Commons Attribution-Share Alike 3.0 Unported license (https://creativecommons.org/licenses/by-sa/3.0/deed.en). File: Alain-Aspect-ForMemRS.jpg. (2020, September 17). *Wikimedia Commons, the free media repository*. Retrieved 14:41, December 7, 2020 from https://commons.wikimedia.org/ w/index.php?title=File:Alain-Aspect-ForMemRS.jpg&oldid=462669099
11. G. Weihs et al., Phys Rev Lett **81**, 5039 (1998)
12. C. Abellán et al., Nature **557**, 212 (2018)
13. D. Rauch, Phys Rev Lett **121**, 080403 (2018)

Chapter 6
Quantum Key Distribution

Quantum key distribution (QKD) provides a means of secure communication between two parties. QKD exploits the principle that you cannot eavesdrop on a quantum communication channel without producing a detectable disturbance.

6.1 Cryptography

Cryptography refers to the secure transmission of messages or data between two parties using an encryption method. A third party who intercepts the message cannot decipher it without a "secret key". A method developed by Rivest, Shamir and Adleman (called RSA) is the current classical encryption method, which is based on the difficulty of finding the prime factors of large integers. RSA is the basis for the global financial system, internet transactions, and online banking. However, RSA can be theoretically cracked by Shor's quantum algorithm (covered in Chap. 13). Therefore, a more secure encryption method is required. Quantum key distribution (QKD) can provide a solution to this problem. How does it work?

First, we need a binary representation for the information to be transmitted. For example, a 5-bit representation of the alphabet might look like Table 6.1. Next, we need to encrypt the message that we want to send. We can do this using binary addition, also called "addition modulo 2". Addition modulo 2, symbolized by $\oplus$, is the result of adding the two numbers, dividing by 2 and taking the remainder. Thus, $0 \oplus 0 = 0$, $1 \oplus 0 = 1$, $0 \oplus 1 = 1$, and $1 \oplus 1 = 0$. These rules of binary addition are summarized in Table 6.2.

We encrypt the message using binary addition (addition modulo 2) of the message and a random key (a completely random string of 0's and 1's):

$$E = M \oplus K \tag{6.1}$$

© The Author(s), under exclusive license to Springer Nature Switzerland AG 2021
R. LaPierre, *Introduction to Quantum Computing*, The Materials Research Society Series,
https://doi.org/10.1007/978-3-030-69318-3_6

Table 6.1 Binary representation of the alphabet

A	0	0	0	0	0
B	0	0	0	0	1
C	0	0	0	1	0
D	0	0	0	1	1
E	0	0	1	0	0
F	0	0	1	0	1
G	0	0	1	1	0
H	0	0	1	1	1
I	0	1	0	0	0
J	0	1	0	0	1
K	0	1	0	1	0
L	0	1	0	1	1
M	0	1	1	0	0
N	0	1	1	0	1
O	0	1	1	1	0
P	0	1	1	1	1
Q	1	0	0	0	0
R	1	0	0	0	1
S	1	0	0	1	0
T	1	0	0	1	1
U	1	0	1	0	0
V	1	0	1	0	1
W	1	0	1	1	0
X	1	0	1	1	1
Y	1	1	0	0	0
Z	1	1	0	0	1

Table 6.2 Binary addition table

$0 \oplus 0 = 0$
$0 \oplus 1 = 1$
$1 \oplus 0 = 1$
$1 \oplus 1 = 0$

where E is the encrypted message, M is the message, and K is the key. The random key scrambles the message (it becomes random). The message is recovered by applying the key again:

$$M = E \oplus K \tag{6.2}$$

Table 6.3 One-time pad for encryption/decryption

Word	P					A					S					S				
Binary Word	0	1	1	1	1	0	0	0	0	0	1	0	0	1	0	1	0	0	1	0

$\oplus$

Key (random)	0	1	1	0	1	0	0	0	1	1	1	1	0	1	1	0	0	1	0	0
Encrypted Message	0	0	0	1	0	0	0	0	1	1	0	1	0	0	1	1	0	1	1	0

$\oplus$

Key (same as above)	0	1	1	0	1	0	0	0	1	1	1	1	0	1	1	0	0	1	0	0
Binary Word	0	1	1	1	1	0	0	0	0	0	1	0	0	1	0	1	0	0	1	0
Word	P					A					S					S				

This works because, in modulo 2 arithmetic, $K + K = 0$ for any string K, where "0" means a string of all zeros. Therefore, $M = E \oplus K = (M \oplus K) \oplus K = M \oplus (K \oplus K) = M$. For example, Table 6.3 shows the encryption/decryption of the word "pass".

6.2 One-Time Pad

The above method of encrypting/decrypting a message is called a "one-time pad", proposed by Gilbert Vernam of AT&T in 1926 [1]. In 1949, Claude Shannon showed, using information theory, that the one-time pad method is completely secure [2]. It is the only provably secure cryptosystem known today, providing the secret key is truly random, the key is as long as the message, and the key is used only once. However, the one-time pad requires the communicators (Alice and Bob) to share the secret key. This is the problem of "key distribution". This is the vulnerable part of the encryption where the secret key might be intercepted by a third party.

6.3 Polarization Basis

Quantum key distribution (QKD) uses qubits rather than bits for key distribution. QKD typically uses the polarization of single photons to represent 0's and 1's and

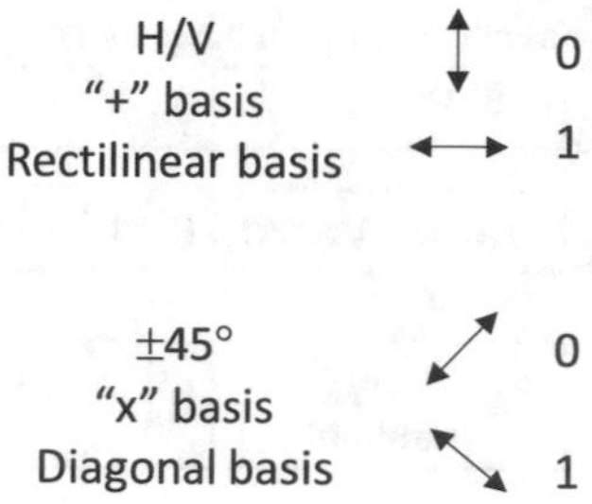

Fig. 6.1 Rectilinear and
diagonal basis used for QKD

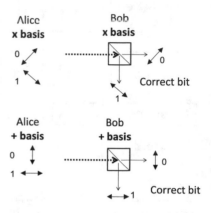

Fig. 6.2 If Alice's and
Bob's polarization basis
match, then Bob's bit
measurement (0 or 1) will
match what Alice sent

create a key (one-time pad). For example, vertical polarization (V) can represent "0" and horizontal polarization (H) can represent "1". This is referred to as the "+" basis or "rectilinear" basis. Also, +45° polarization can represent "0" and −45° polarization can represent "1". This is referred to as the "×" basis or "diagonal" basis. These are shown in Fig. 6.1.

Alice and Bob can generate or detect different light polarizations of single photons using a polarizing beam splitter or "filter". For example, in Fig. 6.2, Alice sends a photon, representing 0 or 1, using the diagonal or rectilinear basis. Bob uses a polarizing beam splitter to send the different polarization states (0 or 1) to different photodetectors, so Bob can detect the polarization state based on what detector "clicks". If Alice's and Bob's polarization basis match, then Bob's bit measurement (0 or 1) will match what Alice sent, as shown in Fig. 6.2. Otherwise, if the polarization basis of Alice and Bob do not match (Fig. 6.3), then Bob will obtain 0 or 1 randomly. This occurs because a photon with +45° or −45° polarization state can be written as a superposition of H and V polarization states with equal probability amplitude of $\frac{1}{\sqrt{2}}$ or probability 1/2. Similarly, a photon with H or V polarization state can be written as a superposition of +45° and −45° polarization states with equal probability amplitude of $\frac{1}{\sqrt{2}}$ or probability 1/2. Thus, QKD depends on the complementarity of observables—not all observables of a quantum system can be simultaneously determined. For example, if you send H/V polarization, then the ±

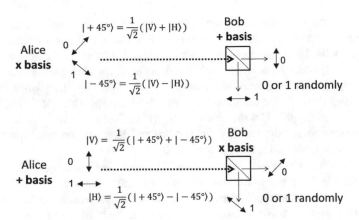

Fig. 6.3 If Alice's and Bob's polarization basis do not match, then Bob obtains 0 or 1 randomly

$45°$ polarization is random. If you send $\pm 45°$ polarization, then the H/V polarization is random. Using these principles, we can develop a QKD scheme.

6.4 BB84

One possible QKD scheme is BB84, developed in 1984 by Charles Bennett of IBM and Gilles Brassard of the University of Montreal [3]. BB84 is illustrated in Table 6.4. QKD relies on a non-orthogonal basis (e.g., "+" and "×" basis) to generate a secret key. First, Alice sends individual photons to Bob using a quantum communication channel (e.g., an optical fiber or free space) capable of preserving quantum information (e.g., the polarization state of individual photons). Alice chooses a random polarization basis ("+" or "×") and sends 0 or 1 randomly. Bob measures each photon, using a random detection polarization basis ("+" or "×" polarizer). When

Table 6.4 BB84 protocol

Alice's transmission basis	×	+	×	+	×	×	×	×	+
Alice's bit sequence	0	0	1	0	1	0	1	1	1
Alice's polarization	╱	│	╲	│	╲	╱	╲	╲	─
Bob's detection basis	+	+	+	+	×	+	+	×	+
Bob's bit measurement	0 or 1	0	0 or 1	0	1	0 or 1	0 or 1	1	1
Basis comparison	✗	✓	✗	✓	✓	✗	✗	✓	✓
Sifted key		0		0	1			1	1

Bob guesses the right basis (the same one Alice used), then he will measure the correct bit (0 or 1) as in Fig. 6.2. However, if Bob chooses the incorrect basis, then he is equally likely to get the right or wrong bit (he obtains 0 or 1 randomly) as in Fig. 6.3. Next, Alice shares with Bob the basis used for each measurement, using a public classical communication channel (e.g., the internet). They only keep each measurement (0 or 1) for the measurements that used the same basis (50% of the time on average, signified by the check mark "$\sqrt{}$" in Table 6.4), and discard the other measurement results (signified by "χ"). Alice and Bob use the resulting measurements (a shared, identical, secret string of 0's and 1's) to generate the secure key. This key is called the "sifted" key, which will be about half the length of the original key. The key is a secret key known only to Alice and Bob. Alice and Bob can then use the secret key to encrypt a message and communicate over a public channel.

If a third party (usually referred to as Eve, for "eavesdropper") intercepts the photon from Alice and measures its polarization, she will need to choose a polarization basis which will only match Alice's polarization basis 50% of the time on average. In turn, Bob will also choose a polarization basis and he gets the correct measurements from Eve only 50% of the time. Thus, only 25% of Bob's bit string will be correct if Eve intercepted Alice's key. Hence, Eve's presence would be revealed if Alice and Bob compare some test bits. The test bits are discarded and do not form part of the sifted key.

Exercise 6.1 Why do we need single photons for QKD to be secure? Why not use "classical light" composed of many photons to represent 0 and 1?

6.5 No Cloning Theorem

Could Eve intercept the message and make a copy? Suppose the quantum state is $\alpha|H\rangle + \beta|V\rangle$. A measurement collapses the state to one basis state with a particular amplitude (α or β). We cannot copy a quantum state, because we can't simultaneously measure both probability amplitudes (α and β). In QKD, this means that Eve cannot make a copy of Alice's photon and send it to Bob (otherwise, she could decrypt her copy later, when Alice and Bob compare their random basis selections). This also means that many techniques of classical information theory (such as protecting information by making redundant copies, or having a "fanout" from a single bit) are impossible in quantum computing. This is referred to as the "no cloning theorem" and will be discussed further in Chap. 7.

6.6 Technology

QKD requires single photon sources and detectors. QKD can be implemented over optical fiber or in free space. For optical fibers, single photon sources and detectors operating near a wavelength of 1550 nm are required, where fiber has low absorption loss. For free space, single photon sources and detectors are required in the visible or near-infrared (NIR) range where efficient detectors exist, or in the mid-wavelength infrared (MWIR) or long-wavelength infrared (LWIR) where the atmosphere is more transparent. Single photon sources with high repetition rate, high coupling efficiency into optical fiber, and no cooling are desired. Faster and more sensitive detectors with lower dark noise are desired. "Quantum repeaters" are also required for long distances (hundreds of kilometers) in optical fiber. A review of single photon sources and detectors is available in Ref. [4].

6.7 Other QKD Schemes

In 1991, Arthur Ekert at the University of Oxford proposed the E91 QKD protocol using entangled photons, which eliminates the one-time pad [5]. Alice and Bob receive one photon each of an entangled pair of photons from a source located either at Alice's or Bob's position, or somewhere else. For each photon they receive, Alice and Bob choose a measurement basis and write down the result of the measurement. Afterwards, they discuss their choice of basis and retrieve a sifted key. Alice and Bob can calculate whether Bell's inequality is violated or not. If it is not violated, then Eve must have destroyed the entanglement by eavesdropping. These implementations are referred to as entanglement-based protocols, Ekert protocols, or EPR protocols.

6.8 Quantum Random Number Generator

QKD depends on our ability to create a completely random key and generate a random basis selection. This can be done using a random number generator. The problem is that classical methods of generating random numbers are not truly random—they rely on an algorithm and are thus "pseudo-random". Quantum physics can be used to generate a truly random string of 0's and 1's. If we pass photons polarized at an angle θ through a linear polarizer oriented along the vertical, then a fraction $\cos^2\theta$ of photons pass through and $1-\cos^2\theta = \sin^2\theta$ of photons are absorbed. But if photons are all identical, why are some absorbed and some are not? In quantum mechanics, we can only predict probabilities. Each photon state can be considered as a superposition of horizontal (H) and vertical (V) polarization:

$$|\text{photon}, \theta\rangle = \cos\theta|\text{photon}, V\rangle + e^{i\phi}\sin\theta|\text{photon}, H\rangle \qquad (6.3)$$

We recognize Eq. (6.3) as the general Bloch vector. Thus, each photon exits the polarizer with probability $\cos^2\theta$ in the vertical polarization state and probability $\sin^2\theta$ in the horizontal polarization state. We can use this to build a truly random number generator for QKD.

6.9 Quantum Money

Similar ideas to QKD could be used to prevent money from being counterfeited. This is known as quantum money, developed by Stephen Wiesner around 1970. Suppose a banknote is embedded with qubits, where each qubit is either in the computational basis ($|0\rangle, |1\rangle$) or the Hadamard basis ($|+\rangle = \frac{1}{\sqrt{2}}(|0\rangle + |1\rangle)$ and $|-\rangle = \frac{1}{\sqrt{2}}(|0\rangle - |1\rangle)$). For example, the banknote might be embedded with the qubit sequence $|0\rangle|+\rangle|1\rangle|1\rangle|-\rangle|0\rangle|+\rangle$. The bank keeps track of each qubit and basis sequence using the serial number of each banknote. Upon measuring the qubit sequence, a person will only reproduce the correct sequence if they know the basis used for each qubit. Otherwise, a measurement will give the correct result only 50% of the time. Hence, it is not possible to copy the banknote without knowing the basis used for each qubit in the serial number. A person wishing to verify the authenticity of a banknote can only do so by contacting the bank and obtaining the basis used for each qubit. Measuring with the correct basis will then yield the correct sequence, which can be verified by the bank for that serial number. Of course, implementing qubits into banknotes is a challenging problem!

6.10 Outlook

The first quantum key exchange took place in the lab in 1989 by Charles Bennett and John Smolin. An entertaining report is available in Ref. [6]. On Apr 21, 2004, the physicist, Anton Zeilinger, at the University of Vienna received a US$3500 donation for his research from a bank in Vienna via secure quantum communication through a 1450 m long optical fiber cable [7]. Today, quantum key exchange is possible over hundreds of kilometers using fiber optics or free space. A global quantum-encrypted network ("quantum internet") is expected over the next decade using optical fiber or satellites. ID Quantique, MagiQ Technologies and Crypto4A Technologies are examples of companies involved in the commercialization of QKD.

References

1. G. Vernam, Trans AIEE **45**, 295 (1926)
2. C.E. Shannon, Bell Syst Tech J **28**, 656 (1949)

3. C.H. Bennett, G. Brassard, Quantum cryptography: public key distribution and coin tossing. in International Conference on Computers, Systems and Signal Processing, Bangalore, India, 10–12 December (1984), pp. 175–179
4. M.D. Eisaman et al., Rev Sci Instrum **82**, (2011)
5. A.K. Ekert, Phys Rev Lett **67**, 661 (1991)
6. J.A. Smolin, IBM J Res Dev **48**, 47 (2004)
7. E. Schillinger, Nature **428**, 883 (2004)

Chapter 7
Quantum Gates

In classical and quantum computing, we implement gate operations for information processing. Quantum gates, like classical gates, transform some input state into an output state. However, unlike classical gates, quantum gates allow a superposition of states and must satisfy certain criteria such as reversibility.

7.1 Classical Gates

Before introducing quantum gates, we should talk about gates used in classical computing. The gate operations that implement Boolean logic include the NOT, AND, OR, XOR, NAND and NOR operators. Figure 7.1 shows the circuit symbol for these gates and their truth table where the input (A, B) and output states are tabulated.

7.2 Quantum Gates

In quantum computing, we implement some operator $\widehat{U}$ that transforms an initial input state, $|\psi_i\rangle$, to a final output state, $|\psi_f\rangle$:

$$\text{input:} |\psi_i\rangle \xrightarrow{\widehat{U}} \text{output:} |\psi_f\rangle \tag{7.1}$$

or

$$\hat{U} |\psi_i\rangle = |\psi_f\rangle \tag{7.2}$$

In the vector representation, $|\psi_i\rangle$ and $|\psi_f\rangle$ are $N \times 1$ column vectors (columns of probability amplitudes) where $N = 2^n$ and n is the number of qubits. Thus, the $\widehat{U}$

© The Author(s), under exclusive license to Springer Nature Switzerland AG 2021
R. LaPierre, *Introduction to Quantum Computing*, The Materials Research Society Series,
https://doi.org/10.1007/978-3-030-69318-3_7

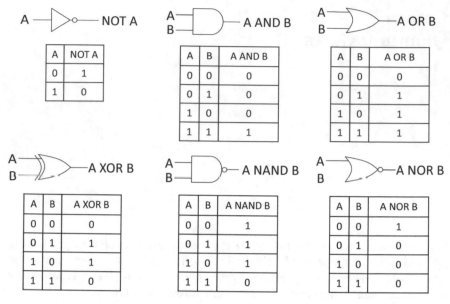

Fig. 7.1 Classical logic gates and their truth tables

operator is represented by an $N \times N$ matrix, U (note that the subscripts start at 0):

$$
\begin{pmatrix}
U_{00} & U_{01} & \cdots & U_{(0)(N-1)} \\
U_{10} & U_{11} & \cdots & U_{(1)(N-1)} \\
\vdots & \vdots & \ddots & \vdots \\
U_{(N-1)(0)} & U_{(N-1)(1)} & \cdots & U_{(N-1)(N-1)}
\end{pmatrix}
\begin{pmatrix}
\alpha_0 \\
\vdots \\
\alpha_i \\
\vdots \\
\alpha_{N-1}
\end{pmatrix}
=
\begin{pmatrix}
\beta_o \\
\vdots \\
\beta_i \\
\vdots \\
\beta_{N-1}
\end{pmatrix}
\qquad (7.3)
$$

For example, an operator for a single qubit ($N = 2^n = 2^1 = 2$) is a 2×2 matrix. The $\widehat{U}$ operator is also called a quantum gate or a transformation matrix.

7.3 Circuit or Gate Model of Quantum Computing

The circuit or gate model of quantum computing is depicted in Fig. 7.2. The inputs on the left are known as the "register". Each box represents a quantum operator or gate ($\widehat{U}$) that accepts one or more qubits as input and transforms it to produce an output of qubit(s) in a possibly different state. Quantum gates can act on a single qubit or multiple qubits. Later, we will prove that each gate must have the same number of inputs as outputs. Thus, a single-qubit gate accepts a single-qubit input and produces a single-qubit output; a two-qubit gate accepts a two-qubit input and

Fig. 7.2 Gate or circuit
model of quantum
computing

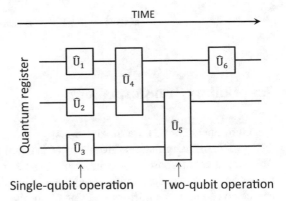

Fig. 7.3 Order of gate
operations

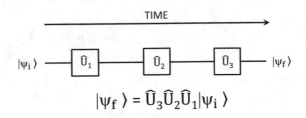

produces a two-qubit output; or, in general, an n-qubit gate accepts an n-qubit input
and produces an n-qubit output.

   Remember that the qubits are the states of physical particles such as spin residing
on an electron, or the polarization of photons. They are not really "inputs" in the
classical sense. The quantum circuit represents time running from left to right. The
"wires" represent the sequence of operations; they are not wires in the conventional
sense. For example, the wires of the circuit diagram could correspond to a single
electron or photon, and each gate corresponds to a change in the spin of that electron
or polarization of the photon. Note that the output states may be entangled, meaning
we may not be able to write them as a product of individual qubit states.

   Note that time in the circuit goes from left to right, but the corresponding unitary
operators (matrices) must be multiplied together from right to left, so that the earliest
gates are applied to the state first (Fig. 7.3).

7.4   Linear Transformations

The $\hat{U}$ operators (represented as matrices) are called "linear transformations". They
transform one vector into another vector. They are "linear" because they obey the
vector axioms:

$$\hat{U}(a|v\rangle) = a\,\hat{U}|v\rangle \qquad\qquad (7.4)$$

$$\hat{U}(|v_1\rangle + |v_2\rangle) = \hat{U}|v_1\rangle + \hat{U}|v_2\rangle \tag{7.5}$$

7.5 Unitary Transformations

Quantum operators $(\widehat{U})$ are unitary transformations. If a matrix U is unitary, then $U^{-1} = U^\dagger$, meaning that the inverse of U is equal to its conjugate transpose $(U^\dagger)$. The conjugate transpose $(\dagger)$ was introduced in Chap. 4, and means we transpose the matrix and then take the complex conjugate of each entry. The transpose of a matrix can be considered as a mirror reflection about the diagonal of the matrix; i.e., each entry, U_{ij} becomes U_{ji}. For example, for a single qubit:

$$U = \begin{pmatrix} a & c \\ b & d \end{pmatrix} \tag{7.6}$$

and

$$U^\dagger = \begin{pmatrix} a^* & b^* \\ c^* & d^* \end{pmatrix} \tag{7.7}$$

Let us examine the action of U on $|0\rangle$ and $|1\rangle$:

$$\hat{U}|0\rangle = \begin{pmatrix} a & c \\ b & d \end{pmatrix}\begin{pmatrix} 1 \\ 0 \end{pmatrix} = \begin{pmatrix} a \\ b \end{pmatrix} = a|0\rangle + b|1\rangle \tag{7.8}$$

$$\hat{U}|1\rangle = \begin{pmatrix} a & c \\ b & d \end{pmatrix}\begin{pmatrix} 0 \\ 1 \end{pmatrix} = \begin{pmatrix} c \\ d \end{pmatrix} = c|0\rangle + d|1\rangle \tag{7.9}$$

Now $U^{-1} = U^\dagger$ implies:

$$UU^\dagger = U^\dagger U = I \tag{7.10}$$

where $I = \begin{pmatrix} 1 & 0 \\ 0 & 1 \end{pmatrix}$ is the 2×2 identity matrix. Starting with $U^\dagger U = I$ $(UU^\dagger = I$ gives the same result), we get:

$$U^\dagger U = \begin{pmatrix} a^* & b^* \\ c^* & d^* \end{pmatrix}\begin{pmatrix} a & c \\ b & d \end{pmatrix} = \begin{pmatrix} a^*a + b^*b & a^*c + b^*d \\ c^*a + d^*b & c^*c + d^*d \end{pmatrix} = \begin{pmatrix} 1 & 0 \\ 0 & 1 \end{pmatrix} \tag{7.11}$$

The diagonal elements of $U^\dagger U$ give:

$$a^*a + b^*b = |a|^2 + |b|^2 = 1 \tag{7.12}$$

$$c^*c + d^*d = |c|^2 + |d|^2 = 1 \tag{7.13}$$

Equations (7.12) and (7.13) tell us that the output of $\hat{U}|0\rangle$ in Eq. (7.8) and $\hat{U}|1\rangle$ in Eq. (7.9) are normalized. The off-diagonal elements of $U^\dagger U$ give:

$$a^*c + b^*d = 0 \tag{7.14}$$

$$c^*a + d^*b = 0 \tag{7.15}$$

Equations (7.14) and (7.15) state that the output of $\hat{U}|0\rangle$ in Eq. (7.8) and $\hat{U}|1\rangle$ in Eq. (7.9) are orthogonal. Thus, Eqs. (7.12)–(7.15) tell us that $\hat{U}$ preserves the orthonormality required for quantum states.

A more elegant and general way of showing that $\hat{U}$ preserves orthonormality is the following. Suppose there exists an orthonormal basis:

$$\langle \psi_i | \psi_j \rangle = \delta_{ij} \tag{7.16}$$

Suppose $\widehat{U}$ is applied to the basis vectors, $|\psi_i\rangle$, producing new vectors, $|\phi_i\rangle$:

$$\hat{U}|\psi_i\rangle = |\phi_i\rangle \tag{7.17}$$

The dual vectors, $\langle \phi_i |$, are:

$$\langle \phi_i | = \left(\hat{U}|\psi_i\rangle \right)^\dagger = \langle \psi_i | \hat{U}^\dagger \tag{7.18}$$

Thus, the inner product is:

$$\langle \phi_i | \phi_j \rangle = \langle \psi_i | \hat{U}^\dagger \hat{U} | \psi_j \rangle = \langle \psi_i | I | \psi_j \rangle = \langle \psi_i | \psi_j \rangle = \delta_{ij} \tag{7.19}$$

Thus, the $|\phi_i\rangle$ vectors are also orthonormal. Therefore, unitary transformations preserve inner products; i.e., they preserve the length and angle between vectors in the Bloch sphere. As such, transformations performed on a single qubit may be visualized as rotations of any vector in the Bloch sphere.

7.6 Reversibility

Since $U^{-1} = U^\dagger$, it follows that all unitary matrices have an inverse. Unitary transformations are therefore always reversible. Thus, only reversible gates can be

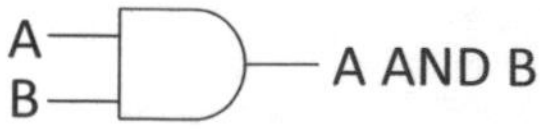

Fig. 7.4 The classical AND
gate is not reversible

A ───┐
     ⟩── A AND B
B ───┘

A	B	A AND B
0	0	0
0	1	0
1	0	0
1	1	1

implemented in quantum computing, and all quantum computers are reversible. For example, the classical NOT gate is reversible; knowing the output state, we know what the input state was. In contrast, the classical AND, OR and NAND gates are not reversible; knowing the output state, we cannot know what the input state was. For example, if the output of a classical AND gate is 0, then we cannot tell if the input was 0 and 0, 0 and 1, or 1 and 0 (Fig. 7.4). Any logic gate that has fewer output than input lines inevitably discards information, because we cannot deduce the input from the output. Hence, the requirement for reversibility means that the number of inputs in a quantum circuit must equal the number of outputs.

Today's classical computers are not reversible; they dissipate heat, as we know from experience. The irreversibility and dissipation of energy in classical computing is related to the erasure of information. Rolf Landauer discovered that erasure of a bit must produce entropy of at least $k\ln2$ and expend energy of $kT\ln2$ at temperature T where k is the Boltzmann constant, but there is no necessary entropy cost to copy a bit or measure one. This is because erasure is an irreversible process and must inevitably increase entropy (like all irreversible processes). In the 1970s, Charles Bennett showed that any computation can, in principle, be done reversibly (i.e., without erasing information). For example, any classical circuit can be entirely simulated using reversible "Toffoli" or "Fredkin" gates (these gates are discussed later). Therefore, it is possible in principle to take the output state and run the computation backwards to recover the input state, returning the computer to its initial state without dissipating energy. Since any classical circuit can be made reversible, this means that any classical circuit can be simulated by a quantum circuit. The only operator that is not reversible is the operation of measurement. Upon measurement, the quantum state collapses to one of the basis states. Subsequent measurements will yield the same result.

7.7 Hermitian Operators

An operator is said to be Hermitian or self-adjoint is it is equal to its complex conjugate, $\widehat{U} = \widehat{U}^\dagger$. In quantum mechanics, all observable properties must be represented by Hermitian operators. For example, let us consider a general 2×2 matrix operator and its conjugate transpose. If $\widehat{U} = \widehat{U}^\dagger$, then:

$$\begin{pmatrix} a & c \\ b & d \end{pmatrix} = \begin{pmatrix} a^* & b^* \\ c^* & d^* \end{pmatrix} \tag{7.20}$$

Equating the matrix elements gives $a = a^*$, $d = d^*$, $b = c^*$ and $c = b^*$. The criteria $a = a^*$ and $d = d^*$ mean that the diagonal entries, a and d, are real numbers. The other two criteria indicate that the off-diagonal entries are the complex conjugate of each other. For example, consider the following matrix, U, corresponding to some measurable observable:

$$\begin{pmatrix} a & 0 \\ 0 & d \end{pmatrix} \begin{pmatrix} 1 \\ 0 \end{pmatrix} = a \begin{pmatrix} 1 \\ 0 \end{pmatrix} \tag{7.21}$$

and

$$\begin{pmatrix} a & 0 \\ 0 & d \end{pmatrix} \begin{pmatrix} 0 \\ 1 \end{pmatrix} = d \begin{pmatrix} 0 \\ 1 \end{pmatrix} \tag{7.22}$$

Hence, the diagonal entries, a and d, are eigenvalues of the operator U with eigenstates $\begin{pmatrix} 1 \\ 0 \end{pmatrix}$ and $\begin{pmatrix} 0 \\ 1 \end{pmatrix}$, respectively. Since the eigenvalues, a and d, correspond to some observable (measurable quantity) of the operator U, they must be real numbers. This criterion is met by demanding that U be Hermitian.

A more satisfying and general proof that all quantum operators are Hermitian is the following. Suppose λ is the eigenvalue of some operator U:

$$U|\psi\rangle = \lambda|\psi\rangle \tag{7.23}$$

Applying a bra, $\langle\psi|$, to both sides gives the expectation value of the operator:

$$\langle\psi|U|\psi\rangle = \langle\psi|(U|\psi\rangle) = \langle\psi|(\lambda|\psi\rangle) = \lambda\langle\psi|\psi\rangle = \lambda \tag{7.24}$$

since $\langle\psi|\psi\rangle = 1$ if $|\psi\rangle$ is normalized. Taking the adjoint of both sides of Eq. (7.23) gives us:

$$\langle\psi|U^\dagger = \lambda^*\langle\psi| \tag{7.25}$$

Applying a ket to both sides gives us:

$$\langle\psi|U^\dagger|\psi\rangle = \lambda^*\langle\psi|\psi\rangle = \lambda^* \tag{7.26}$$

A measurable observable must be a real number, so $\lambda = \lambda^*$. Hence, comparing Eqs. (7.24) and (7.26), we must have $U = U^\dagger$; i.e., U is Hermitian. Examples of Hermitian operators include the Hamiltonian, $\widehat{H}$, corresponding to a measurement of

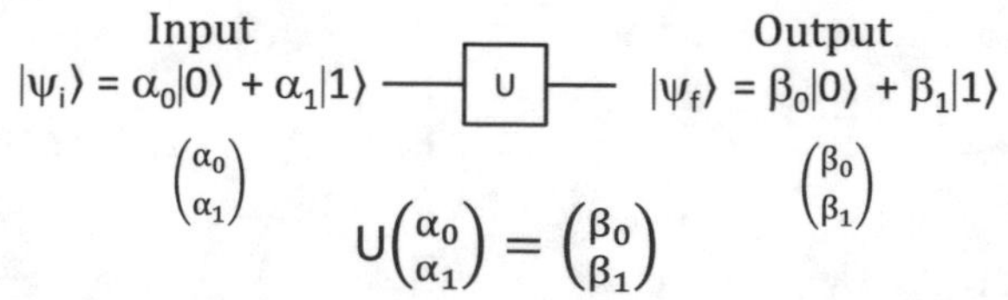

Fig. 7.5 A single-qubit gate

Fig. 7.5 A single-qubit gate

Input

$|\psi_i\rangle = \alpha_0|0\rangle + \alpha_1|1\rangle$ ──□ U □── $|\psi_f\rangle = \beta_0|0\rangle + \beta_1|1\rangle$

$\begin{pmatrix} \alpha_0 \\ \alpha_1 \end{pmatrix}$

Output

$\begin{pmatrix} \beta_0 \\ \beta_1 \end{pmatrix}$

$$U\begin{pmatrix} \alpha_0 \\ \alpha_1 \end{pmatrix} = \begin{pmatrix} \beta_0 \\ \beta_1 \end{pmatrix}$$

the energy, and the spin operators, $\widehat{S}_i$ ($i = x, y, z$), corresponding to a measurement of spin. Note that not all quantum gates correspond to a measurable observable. Hence, while all quantum gates must be unitary, not all quantum gates are Hermitian. For example, the phase gate considered below is not Hermitian. We will now describe some commonly used quantum gates. We will first consider single-qubit operations, followed by multi-qubit operations.

7.8 Single-Qubit (Unary) Gates

A generic single-qubit (or unary) gate is shown in Fig. 7.5. A unitary transformation, $\widehat{U}$, represented by the matrix, $U = \begin{pmatrix} a & c \\ b & d \end{pmatrix}$, transforms an input qubit with superposition $|\psi_i\rangle = \alpha_0|0\rangle + \alpha_1|1\rangle$, into the output qubit with superposition $|\psi_f\rangle = \beta_0|0\rangle + \beta_1|1\rangle$:

$$U\begin{pmatrix} \alpha_0 \\ \alpha_1 \end{pmatrix} = \begin{pmatrix} \beta_0 \\ \beta_1 \end{pmatrix} \tag{7.27}$$

All single-qubit transformations are represented by a 2×2 matrix. Since there are an infinite number of 2×2 matrices, there are an infinite number of single-qubit transformations. In practice, there are only a few important transformations as described in the following sections.

7.9 NOT (X) Gate

The NOT gate or X gate, also known as a bit flip gate, flips $|0\rangle$ and $|1\rangle$:

$$\begin{aligned} X|0\rangle &= |1\rangle \\ X|1\rangle &= |0\rangle \end{aligned} \tag{7.28}$$

The matrix representation and Dirac notation for the X gate is:

$$X = \begin{pmatrix} 0 & 1 \\ 1 & 0 \end{pmatrix} = |1\rangle\langle 0| + |0\rangle\langle 1| \tag{7.29}$$

The circuit representation for the X gate is:

or

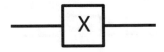

The X gate can also be represented as $X|j\rangle = |j \oplus 1\rangle$ where $\oplus$ denotes binary addition (addition modulo 2, or the XOR operation; Table 6.2).

In Eq. (7.29), X is also shown in Dirac outer product notation that was introduced in Chap. 4. You can check that this outer product notation is correct by applying it to $|0\rangle$ and $|1\rangle$. Thus, $X|0\rangle = (|1\rangle\langle 0| + |0\rangle\langle 1|) |0\rangle = |1\rangle\langle 0|0\rangle + |0\rangle\langle 1|0\rangle = |1\rangle (1) + |0\rangle (0) = |1\rangle$. In a similar fashion, $X|1\rangle = |0\rangle$. Alternatively, remembering the rule for outer products (tensor products) gives $|1\rangle\langle 0| + |0\rangle\langle 1| = \begin{pmatrix} 0 \\ 1 \end{pmatrix} \otimes (10) + \begin{pmatrix} 1 \\ 0 \end{pmatrix} \otimes (01) =$

$$\begin{pmatrix} 0 & 0 \\ 1 & 0 \end{pmatrix} + \begin{pmatrix} 0 & 1 \\ 0 & 0 \end{pmatrix} = \begin{pmatrix} 0 & 1 \\ 1 & 0 \end{pmatrix}.$$

Exercise 7.1 Show that X is a unitary transformation.

Exercise 7.2 Suppose $|\psi\rangle = \alpha|0\rangle + \beta|1\rangle$. Using the matrix and Dirac notation, determine NOT$|\psi\rangle$.

Exercise 7.2 shows that the X gate exchanges the probability amplitudes for $|0\rangle$ and $|1\rangle$. The X gate is equivalent to the classical NOT gate only when applied to the basis states, $|0\rangle$ and $|1\rangle$, without superposition.

7.10 Y Gate

The Y gate swaps the amplitudes of $|0\rangle$ and $|1\rangle$, multiplies each amplitude by i, and negates the amplitude of $|1\rangle$. It is a phase shift and bit flip gate:

$$Y|0\rangle = i|1\rangle$$
$$Y|1\rangle = -i|0\rangle \tag{7.30}$$

The matrix representation and Dirac notation for the Y gate is:

$$Y = \begin{pmatrix} 0 & -i \\ i & 0 \end{pmatrix} = i|1\rangle\langle0| - i|0\rangle\langle1| \tag{7.31}$$

The circuit representation for Y is:

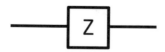

Exercise 7.3 Show that the outer product notation for Y gives the correct output. Show that the matrix for Y gives the correct output. Show that Y is unitary.

7.11   Z Gate

The Z gate negates the amplitude of $|1\rangle$. It is a phase shift gate:

$$Z|0\rangle = |0\rangle$$
$$Z|1\rangle = -|1\rangle \tag{7.32}$$

The matrix representation and Dirac notation for the Z gate is:

$$Z = \begin{pmatrix} 1 & 0 \\ 0 & -1 \end{pmatrix} = |0\rangle\langle0| - |1\rangle\langle1| \tag{7.33}$$

The circuit representation for Z is:

The Z gate can also be represented as $Z|j\rangle = (-1)^j|j\rangle$.

7.12   Pauli Gates

The three X, Y, Z gates are identical to the Pauli matrices $(\sigma_x, \sigma_y, \sigma_z)$ that we introduced in Chap. 3 when we discussed spin. They are named after the physicist, Wolfgang Pauli. Physicists usually prefer to use the $\sigma_x, \sigma_y, \sigma_z$ notation, while computer scientists prefer X, Y, Z. We will use both notations interchangeably.

7.13   Hadamard Gate

The Hadamard gate, H, named after the mathematician Jacques Hadamard, allows us to create superpositions. H applied to $|0\rangle$ or $|1\rangle$ gives the following output:

$$H|0\rangle = \tfrac{1}{\sqrt{2}}(|0\rangle + |1\rangle) = |+\rangle$$
$$H|1\rangle = \tfrac{1}{\sqrt{2}}(|0\rangle - |1\rangle) = |-\rangle$$

(7.34)

Thus, H allows us to prepare superposition states from the computational basis states. The matrix representation and Dirac notation for the H gate is:

$$H = \frac{1}{\sqrt{2}}\begin{pmatrix} 1 & 1 \\ 1 & -1 \end{pmatrix} = |0\rangle\langle+| + |1\rangle\langle-|$$

(7.35)

The circuit representation for H is:

Do not confuse the Hadamard transform with the Hamiltonian (also denoted by H).

Exercise 7.5 Show that the outer product notation for H gives the correct output. Show that the matrix for H gives the correct output. Show that H is unitary.

Note that, in terms of spin, we previously denoted $|0\rangle$ as $|z_+\rangle$, and $|1\rangle$ as $|z_-\rangle$. Therefore, $H|z_+\rangle = \frac{1}{\sqrt{2}}(|z_+\rangle + |z_-\rangle) = |x_+\rangle$, according to Table 3.1. Similarly, $H|z_-\rangle = \frac{1}{\sqrt{2}}(|z_+\rangle - |z_-\rangle) = |x_-\rangle$. Thus, H converts the z-basis into the x-basis.

Exercise 7.6 shows that H is its own inverse. If you apply H twice to $|0\rangle$ or $|1\rangle$, then $|0\rangle$ or $|1\rangle$ is recovered. Later, we will see that H can be considered as a quantum Fourier transform.

Exercise 7.6 Show that if we apply H to the superposition $\frac{1}{\sqrt{2}}(|0\rangle + |1\rangle)$, then we obtain $|0\rangle$. Show that if we apply H to the superposition $\frac{1}{\sqrt{2}}(|0\rangle - |1\rangle)$, then we obtain $|1\rangle$.

Exercise 7.7 Show that the output of the Hadamard gate can be represented as:

$$H|x\rangle = \frac{|0\rangle + (-1)^x|1\rangle}{\sqrt{2}}$$

Exercise 7.8 Show that $H = \frac{1}{\sqrt{2}}(X + Z)$.

7.14   Identity Operator

The identity operator, I, does nothing to a qubit. It is represented by the matrix:

$$I = \begin{pmatrix} 1 & 0 \\ 0 & 1 \end{pmatrix} = |0\rangle\langle 0| + |1\rangle\langle 1| \tag{7.36}$$

Exercise 7.9 Show that the identity matrix can also be written as $|+\rangle\langle+| + |-\rangle\langle-|$, where $|+\rangle = \frac{1}{\sqrt{2}}(|0\rangle + |1\rangle)$ and $|-\rangle = \frac{1}{\sqrt{2}}(|0\rangle - |1\rangle)$. This is another form of the "resolution of the identity" that was mentioned in Exercise 4.6.

Exercise 7.10 Show the following:

$$\boxed{H}\boxed{X}\boxed{H} = \boxed{Z}$$

$$\boxed{H}\boxed{Z}\boxed{H} = \boxed{X}$$

$$\boxed{X}\boxed{X} = \boxed{Y}\boxed{Y} = \boxed{Z}\boxed{Z} = \boxed{I}$$

$$\boxed{Z}\boxed{Y}\boxed{X} = \boxed{i\,I}$$

7.15 Phase Gate

The phase gate rotates the phase of the $|1\rangle$ state by an angle ϕ about the z-axis in the Bloch sphere:

$$R_\phi|0\rangle = |0\rangle$$
$$R_\phi|1\rangle = e^{i\phi}|1\rangle$$

(7.37)

The matrix representation and Dirac notation for the phase gate is:

$$R_\phi = \begin{pmatrix} 1 & 0 \\ 0 & e^{i\phi} \end{pmatrix} = |0\rangle\langle0| + e^{i\phi}|1\rangle\langle1|$$

(7.38)

The circuit representation is:

Note that $Z = R_\pi$ is a special case of the rotation gate since $e^{i\pi} = -1$.

Other special cases of R_ϕ are the S and T gates:

$$S = R_{\pi/2} = \begin{bmatrix} 1 & 0 \\ 0 & e^{i\pi/2} \end{bmatrix} = \begin{bmatrix} 1 & 0 \\ 0 & i \end{bmatrix} \tag{7.39}$$

$$T = R_{\pi/4} = \begin{bmatrix} 1 & 0 \\ 0 & e^{i\pi/4} \end{bmatrix} \tag{7.40}$$

Confusingly, the S gate is also sometimes called the $\pi/4$ gate, since S can be rewritten more symmetrically as $S = \begin{bmatrix} 1 & 0 \\ 0 & e^{i\pi/2} \end{bmatrix} = e^{i\pi/4} \begin{bmatrix} e^{-i\pi/4} & 0 \\ 0 & e^{i\pi/4} \end{bmatrix}$. A phase factor like $e^{i\pi/4}$ in front of the matrix is called a global phase factor. Although we cannot ignore relative phase differences between $|0\rangle$ and $|1\rangle$, we can ignore global phase factors because they are applied to the entire qubit superposition (such as $|\psi\rangle = e^{i\theta}(\alpha|0\rangle + \beta|1\rangle)$) and will not affect probabilities since $|e^{i\theta}|^2 = 1$. Hence, the S gate can be expressed as:

$$S = \begin{bmatrix} e^{-i\pi/4} & 0 \\ 0 & e^{i\pi/4} \end{bmatrix} \tag{7.41}$$

within a global phase factor. Similarly, T is called the $\pi/8$ gate since:

$$T = \begin{bmatrix} e^{-i\pi/8} & 0 \\ 0 & e^{i\pi/8} \end{bmatrix} \tag{7.42}$$

within a global phase factor.

7.16 Two-Qubit (Binary) Gates

Let us now consider two-qubit gates. A generic two-qubit (or binary) gate is shown in Fig. 7.6. A unitary transformation, $\widehat{U}$, represented by a matrix, U, transforms a two-qubit input with superposition $|\psi_i\rangle = \alpha_{00}|00\rangle + \alpha_{01}|01\rangle + \alpha_{10}|10\rangle + \alpha_{11}|11\rangle$, into the two-qubit output with a different superposition $|\psi_f\rangle = \beta_{00}|00\rangle + \beta_{01}|01\rangle + \beta_{10}|10\rangle + \beta_{11}|11\rangle$. U must be a 4×4 matrix:

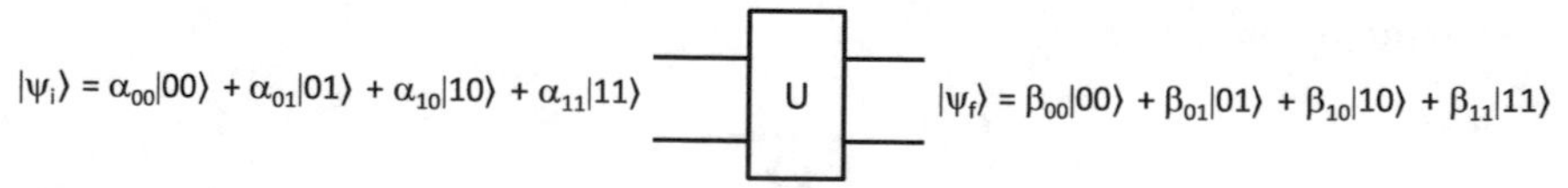

$|\psi_i\rangle = \alpha_{00}|00\rangle + \alpha_{01}|01\rangle + \alpha_{10}|10\rangle + \alpha_{11}|11\rangle$ U $|\psi_f\rangle = \beta_{00}|00\rangle + \beta_{01}|01\rangle + \beta_{10}|10\rangle + \beta_{11}|11\rangle$

Fig. 7.6 A two-qubit gate

$$U \begin{pmatrix} \alpha_{00} \\ \alpha_{01} \\ \alpha_{10} \\ \alpha_{11} \end{pmatrix} = \begin{pmatrix} \beta_{00} \\ \beta_{01} \\ \beta_{10} \\ \beta_{11} \end{pmatrix} \qquad (7.43)$$

We will consider some specific multi-qubit gates in the following sections.

7.17 Controlled NOT (CNOT) Gate

The controlled NOT (CNOT) gate flips the second qubit (called the target qubit) only if the first qubit (called the control qubit) is in the $|1\rangle$ state. The circuit representation for the CNOT gate and its transformation is shown in Fig. 7.7.

The input to the CNOT gate is $|c\rangle \, |t\rangle$ where $|c\rangle$ represents the control qubit and $|t\rangle$ represents the target qubit. The output of the CNOT gate is $|c\rangle \, |c \oplus t\rangle$ where $c \oplus t$ is the binary addition of c and t. If $c = 0$, then $c \oplus t = 0 \oplus t = t$ and nothing happens. If $c = 1$, then $c \oplus t = 1 \oplus t$ and the output is flipped, since $1 \oplus 0 = 1$ and $1 \oplus 1 = 0$. The matrix representation and Dirac notation for the CNOT gate is:

$$\text{CNOT} = \begin{pmatrix} 1 & 0 & 0 & 0 \\ 0 & 1 & 0 & 0 \\ 0 & 0 & 0 & 1 \\ 0 & 0 & 1 & 0 \end{pmatrix} = |0\rangle\langle 0| \otimes I + |1\rangle\langle 1| \otimes X \qquad (7.44)$$

In Eq. (7.44), $\otimes$ refers to the tensor product introduced in Chap. 4 and discussed further in Chap. 9. Using the Dirac and tensor notation, we calculate the output of the CNOT gate below for each of the possible inputs: $|0\rangle|0\rangle$, $|0\rangle|1\rangle$, $|1\rangle \, |0\rangle$ and $|1\rangle|1\rangle$. In the Dirac notation for the CNOT gate, $|0\rangle\langle 0|$ and $|1\rangle\langle 1|$ operate on the first (control) qubit, while I (the identity operator) and X operate on the second (target) qubit. In the first example below, the arrows indicate the qubit to which each operator applies.

For $|00\rangle$ input:

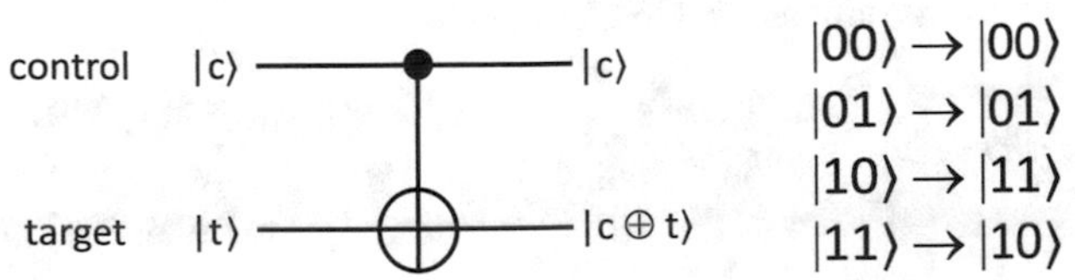

$$|00\rangle \rightarrow |00\rangle$$
$$|01\rangle \rightarrow |01\rangle$$
$$|10\rangle \rightarrow |11\rangle$$
$$|11\rangle \rightarrow |10\rangle$$

$$\alpha_{00}|00\rangle + \alpha_{01}|01\rangle + \alpha_{10}|10\rangle + \alpha_{11}|11\rangle \xrightarrow{\text{CNOT}} \alpha_{00}|00\rangle + \alpha_{01}|01\rangle + \alpha_{10}|11\rangle + \alpha_{11}|10\rangle$$

Fig. 7.7 Circuit representation and transformation of the CNOT gate

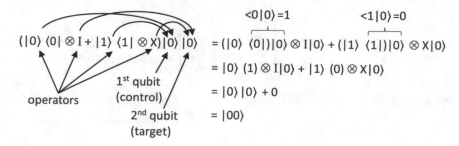

Similarly, for $|01\rangle$ input:

$$(|0\rangle\langle 0| \otimes I + |1\rangle\langle 1| \otimes X)|0\rangle|1\rangle = (|0\rangle\langle 0|)|0\rangle \otimes I|1\rangle + (|1\rangle\langle 1|)|0\rangle \otimes X|1\rangle$$
$$= |0\rangle(1) \otimes I|1\rangle + |1\rangle(0) \otimes X|1\rangle$$
$$= |0\rangle|1\rangle + 0$$
$$= |01\rangle$$

For $|10\rangle$ input:

$$(|0\rangle\langle 0| \otimes I + |1\rangle\langle 1| \otimes X)|1\rangle|0\rangle = (|0\rangle\langle 0|)|1\rangle \otimes I|0\rangle + (|1\rangle\langle 1|)|1\rangle \otimes X|0\rangle$$
$$= |0\rangle(0) \otimes I|0\rangle + |1\rangle(1) \otimes X|0\rangle$$
$$= 0 + |1\rangle|1\rangle$$
$$= |11\rangle$$

For $|11\rangle$ input:

$$(|0\rangle\langle 0| \otimes I + |1\rangle\langle 1| \otimes X)|1\rangle|1\rangle = (|0\rangle\langle 0|)|1\rangle \otimes I|1\rangle + (|1\rangle\langle 1|)|1\rangle \otimes X|1\rangle$$
$$= |0\rangle(0) \otimes I|1\rangle + |1\rangle(1) \otimes X|1\rangle$$
$$= 0 + |1\rangle|0\rangle$$
$$= |10\rangle$$

Comparing the above with the expected output of a CNOT gate, we see that $|0\rangle\langle 0| \otimes I + |1\rangle\langle 1| \otimes X$ reproduces the CNOT gate.

Exercise 7.11 Show that CNOT is unitary.

Exercise 7.12 What is the CNOT output for a control qubit of $|1\rangle$ and a target qubit of $|t\rangle = \alpha|0\rangle + \beta|1\rangle$?

Exercise 7.13 Show the following:

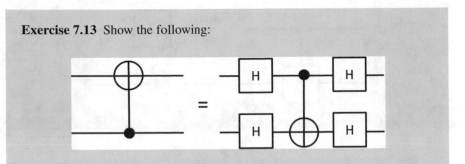

An input of $|x0\rangle$ to the CNOT gate (where $x = 0$ or 1) will give $|x\rangle|x \oplus 0\rangle = |xx\rangle$; this looks like cloning, which is forbidden. However, this only works for the classical states, $|x\rangle = |0\rangle$ or $|1\rangle$, but not a general state (qubit) of $\alpha|0\rangle + \beta|1\rangle$ (try it). We can use CNOT to copy classical bits but not general qubits.

7.18 Other Control Gates

In general, you can implement a control gate on any single-qubit unitary operator $\widehat{U}$, as shown in Fig. 7.8, by inserting the U gate into $|0\rangle\langle 0| \otimes I + |1\rangle\langle 1| \otimes U$. For example, the control-Z (also called CZ or CPHASE) gate is another common two-qubit gate. The transformation matrix for the CPHASE gate is:

$$\text{CPHASE} = \begin{pmatrix} 1 & 0 & 0 & 0 \\ 0 & 0 & 1 & 0 \\ 0 & 1 & 0 & 0 \\ 0 & 0 & 0 & -1 \end{pmatrix} = |0\rangle\langle 0| \otimes I + |1\rangle\langle 1| \otimes Z \qquad (7.45)$$

Exercise 7.14 Show that the CPHASE is symmetric; i.e., the transformation matrix is identical whether the control gate is at the top or bottom:

Fig. 7.8 Circuit representation of a generic control gate, implementing the transformation U

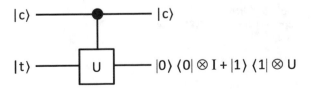

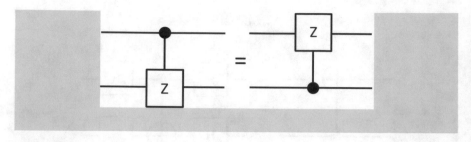

Exercise 7.15 Show the following:

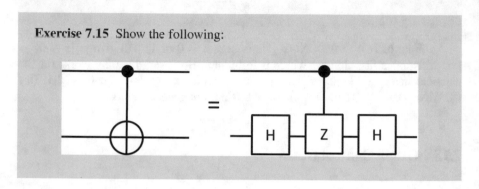

7.19 SWAP Gate

The SWAP gate swaps two qubits as shown in Fig. 7.9. Hence,

$$\text{SWAP}|x_1\rangle|x_2\rangle = |x_2\rangle|x_1\rangle \tag{7.46}$$

The matrix representation and Dirac notation for the SWAP gate is:

$$\text{SWAP} = \begin{pmatrix} 1 & 0 & 0 & 0 \\ 0 & 0 & 1 & 0 \\ 0 & 1 & 0 & 0 \\ 0 & 0 & 0 & 1 \end{pmatrix} = |00\rangle\langle00| + |10\rangle\langle01| + |11\rangle\langle11| + |01\rangle\langle10|$$

$$\tag{7.47}$$

Fig. 7.9 Circuit
representation of the SWAP
gate

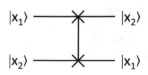

Exercise 7.16 Check that the outer product notation gives the SWAP matrix.

Exercise 7.17 Show that a two-qubit SWAP gate can be made from 3 CNOT gates as follows:

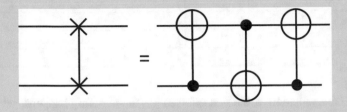

7.20 Fredkin Gate

We can also consider three-qubit gates. The Fredkin gate (also called the controlled SWAP or CSWAP gate) is a three-qubit gate that performs a controlled SWAP as shown by the circuit representation in Fig. 7.10 and the truth table of Table 7.1.

The matrix representation and Dirac notation for the Fredkin gate is:

$$
\text{CSWAP} =
\begin{pmatrix}
1 & 0 & 0 & 0 & 0 & 0 & 0 & 0 \\
0 & 1 & 0 & 0 & 0 & 0 & 0 & 0 \\
0 & 0 & 1 & 0 & 0 & 0 & 0 & 0 \\
0 & 0 & 0 & 1 & 0 & 0 & 0 & 0 \\
0 & 0 & 0 & 0 & 1 & 0 & 0 & 0 \\
0 & 0 & 0 & 0 & 0 & 0 & 1 & 0 \\
0 & 0 & 0 & 0 & 0 & 1 & 0 & 0 \\
0 & 0 & 0 & 0 & 0 & 0 & 0 & 1
\end{pmatrix}
= |0\rangle\langle 0| \otimes I \otimes I + |1\rangle\langle 1| \otimes \text{SWAP} \qquad (7.48)
$$

Fig. 7.10 Circuit representation of Fredkin gate

$$
\begin{aligned}
|c\rangle &\longrightarrow\!\!\bullet\!\!\longrightarrow |c\rangle \\
|I_1\rangle &\longrightarrow\!\!\times\!\!\longrightarrow |O_1\rangle \\
|I_2\rangle &\longrightarrow\!\!\times\!\!\longrightarrow |O_2\rangle
\end{aligned}
$$

Table 7.1 Truth table for the Fredkin gate

INPUT			OUTPUT		
c	I_1	I_2	c	O_1	O_2
0	0	0	0	0	0
0	0	1	0	0	1
0	1	0	0	1	0
0	1	1	0	1	1
1	0	0	1	0	0
1	0	1	1	1	0
1	1	0	1	0	1
1	1	1	1	1	1

From the Dirac notation, if the first qubit is $|0\rangle$, then nothing happens; while if the first qubit is $|1\rangle$, then the SWAP gate is implemented on qubits 2 and 3.

Exercise 7.18 Using only Dirac notation, show that $|0\rangle\langle0| \otimes I \otimes I + |1\rangle\langle1| \otimes$ SWAP gives the correct output for the CSWAP gate.

7.21   Toffoli Gate

Adding another control gate to CNOT, we get the three-qubit controlled-controlled-NOT (CCNOT) gate, also called the Toffoli gate. This applies the NOT gate to the third qubit (target) only if both of the first two qubits ($|c_1\rangle$, $|c_2\rangle$) are $|1\rangle$, as shown in Table 7.2; i.e., $|c_1\rangle$, $|c_2\rangle$ and $|t\rangle$ inputs are transformed to $|c_1\rangle$, $|c_2\rangle$ and $|t \oplus c_1 \cdot c_2\rangle$.

Table 7.2 Truth table for the Toffoli gate

INPUT			OUTPUT		
c_1	c_2	t	c_1	c_2	$t \oplus c_1 c_2$
0	0	0	0	0	0
0	0	1	0	0	1
0	1	0	0	1	0
0	1	1	0	1	1
1	0	0	1	0	0
1	0	1	1	0	1
1	1	0	1	1	1
1	1	1	1	1	0

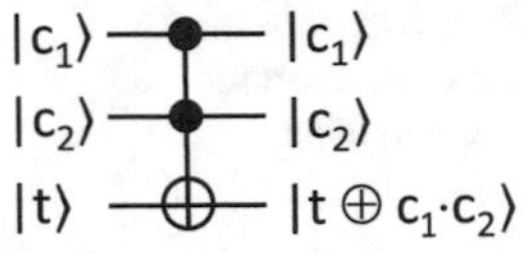

Fig. 7.11 Circuit
representation of the Toffoli
gate

$$|c_1\rangle \text{———}\bullet\text{———} |c_1\rangle$$
$$|c_2\rangle \text{———}\bullet\text{———} |c_2\rangle$$
$$|t\rangle \text{———}\oplus\text{———} |t \oplus c_1 \cdot c_2\rangle$$

The Toffoli gate was originally devised by Tommaso Toffoli as a reversible classical logic gate. Depending on the input, the gate can perform many different logical operations. For example, Toffoli can implement AND (fix the 3rd input to 0), NOT (fix the 1st and 2nd input to 1) and NAND (fix the 3rd input to 1). It is known that all classical operations can be performed using AND and NOT gates, or the NAND gate alone. Since the Toffoli gate can simulate a NAND gate, the Toffoli gate is universal for classical computing; i.e., any classical computer can be built up of Toffoli gates alone. The Toffoli gate is a valid transformation not only in classical circuits, but also in quantum circuits, implying that quantum circuits can simulate any classical circuit.

The circuit representation of the Toffoli gate is shown in Fig. 7.11. Note that the Toffoli gate is reversible as required for quantum transformations. If you apply the Toffoli gate twice, you get $|c_1\rangle |c_2\rangle |t \oplus c_1.c_2\rangle$ as the output after the first application of the Toffoli gate, and $|c_1\rangle |c_2\rangle |t \oplus c_1.c_2 \oplus c_1.c_2\rangle$ after the second application of the Toffoli gate. Note that $t \oplus c_1.c_2 \oplus c_1.c_2 = t$ since $c_1.c_2 \oplus c_1.c_2 = 0$, which returns the system to the original input state. Therefore, the Toffoli gate is its own inverse.

The matrix representation for the Toffoli gate is:

$$\text{Toffoli} = \begin{pmatrix} 1 & 0 & 0 & 0 & 0 & 0 & 0 & 0 \\ 0 & 1 & 0 & 0 & 0 & 0 & 0 & 0 \\ 0 & 0 & 1 & 0 & 0 & 0 & 0 & 0 \\ 0 & 0 & 0 & 1 & 0 & 0 & 0 & 0 \\ 0 & 0 & 0 & 0 & 1 & 0 & 0 & 0 \\ 0 & 0 & 0 & 0 & 0 & 1 & 0 & 0 \\ 0 & 0 & 0 & 0 & 0 & 0 & 0 & 1 \\ 0 & 0 & 0 & 0 & 0 & 0 & 1 & 0 \end{pmatrix} \tag{7.49}$$

It is easily shown that this matrix is unitary and can therefore be implemented as a quantum transformation.

7.22 Measurement

The circuit representation for measurement is shown in Fig. 7.12. The single line on the left-hand side represents a qubit where a superposition is allowed ($|\psi\rangle = \alpha|0\rangle + \beta|1\rangle$). The two lines on the right-hand side represents a classical bit (0 or 1 with no superposition). The possible measurement results of the system are the basis

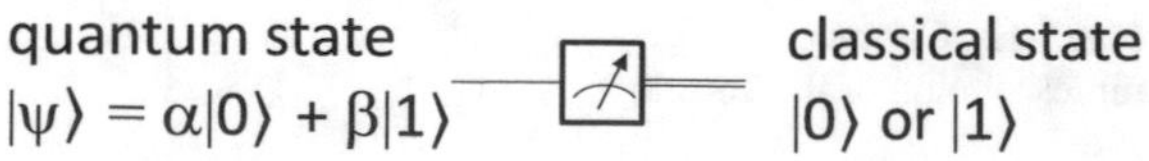

Fig. 7.12 The circuit representation for measurement

quantum state

$|\psi\rangle = \alpha|0\rangle + \beta|1\rangle$ —⊟— classical state

$|0\rangle$ or $|1\rangle$

states. Measurement takes a quantum state and collapses it to one of the basis states. We measure $|0\rangle$ with probability $|\alpha|^2$, or $|1\rangle$ with probability $|\beta|^2$. The collapse is considered to be instantaneous. If two different qubits are entangled, measurement of one qubit affects the state of the other qubit by collapsing its state too. Measurement is irreversible. If we repeat the measurement, we get the same answer. Everything in quantum mechanics is deterministic (e.g., can be described by a unitary matrix), except for the process of measurement.

Exercise 7.19 Draw a quantum circuit that implements a random number generator.

Hence, quantum mechanics presents a strange dichotomy between the world of measurement and quantum processes. The evolution of quantum processes is deterministic and described by unitary transformations, which describe how the probability amplitudes change over time. However, the process of measurement is nondeterministic and involves collapse of the state into one of the classical basis states with a corresponding probability. This is the Copenhagen interpretation of quantum mechanics. What constitutes a measurement? Is consciousness required to collapse a wavefunction? Can a cat collapse a wavefunction? Can't the measurement system be described as part of a larger quantum system? These issues have sparked endless philosophical debate, ever since the inception of quantum mechanics over a hundred years ago. Other interpretations of quantum mechanics and new physics have been proposed as an alternative to the Copenhagen interpretation, such as the "many-worlds" interpretation or "dynamical collapse" theories. However, most physicists still ascribe to the Copenhagen interpretation, and sweep these philosophical issues under the rug.

Exercise 7.20 Show that the gates X, Y, Z, H, and CNOT are Hermitian. Show that the phase gate is not Hermitian.

7.23 Bell State Circuit

Our first quantum circuit is shown in Fig. 7.13. We can analyze this circuit by multiplying the input qubit vectors by the matrix for each operator. For the first qubit, the

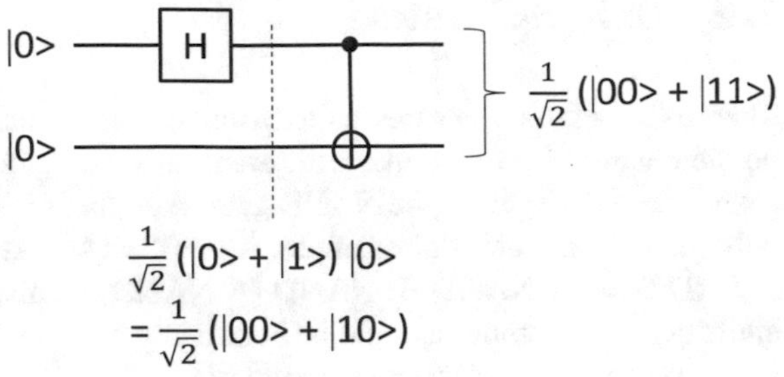

Fig. 7.13 The Bell state circuit

$$\frac{1}{\sqrt{2}} (|00> + |11>)$$

$$\frac{1}{\sqrt{2}} (|0> + |1>) |0>$$
$$= \frac{1}{\sqrt{2}} (|00> + |10>)$$

Hadamard gate gives $H|0\rangle = \frac{1}{\sqrt{2}}(|0\rangle + |1\rangle)$. Combining this with the second qubit gives $\frac{1}{\sqrt{2}}(|0\rangle + |1\rangle)|0\rangle = \frac{1}{\sqrt{2}}(|00\rangle + |10\rangle)$. This state is indicated by the dashed line in Fig. 7.13. After the CNOT gate, we get $\text{CNOT}\left[\frac{1}{\sqrt{2}}(|00\rangle + |10\rangle)\right] = \frac{1}{\sqrt{2}}(|00\rangle + |11\rangle)$, which is an entangled state. This circuit is called a Bell state circuit because it creates one of the Bell entangled states. To create entangled states, we need two-qubit gates like Fig. 7.13.

> **Exercise 7.21** For the Bell state circuit, what is the output if the input qubits are $|01\rangle$, $|10\rangle$, or $|11\rangle$?

7.24 Greenberger-Horne-Zeilinger (GHZ) State

The Bell state circuit produces bipartite entanglement; i.e., entanglement between two particles. Figure 7.14 creates an entangled state of three qubits called a Greenberger-Horne-Zeilinger (GHZ) state. An entangled state of three or more particles is called multipartite entanglement.

> **Exercise 7.22** Show that the circuit in Fig. 7.14 produces the indicated output.

Fig. 7.14 Greenberger-Horne-Zeilinger (GHZ) circuit

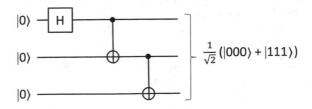

$$\frac{1}{\sqrt{2}}(|000\rangle + |111\rangle)$$

7.25 Universal Gates

There exists a family of gates, called a universal set, that are sufficient to form all other possible gates. There are many different universal gate sets. One possible universal set for classical gates is the NAND gate. All other classical gates can be built using only the NAND gate. For example, A or B = (A NAND A) NAND (B NAND B), A AND B = (A NAND B) NAND (A NAND B), and NOT A = A NAND A. An entire classical computer could be built using only the NAND gate. The Toffoli gate is also universal for classical computing.

Universal sets for quantum computers consist of a single-qubit rotation $U(\theta, \phi)$ in the Bloch sphere together with a two-qubit operation such as the controlled-NOT (CNOT) gate. As we saw with the Bell circuit, two-qubit operations are needed to achieve entanglement. With single-qubit rotation and the CNOT gate we can perform every possible unitary transformation on n qubits, allowing for the implementation of any quantum algorithm. The Solovay-Kitaev theorem [1, 2] states that a small subset of gates can approximate to arbitrary precision any unitary transformation.

7.26 No Cloning Theorem

Now that we have introduced quantum gates, we can revisit the "no cloning theorem" introduced in Chap. 6. This theorem states that it is impossible to make a copy of an unknown quantum state. It was first proven in 1982 by Wooters and Zurek [3]. In QKD, this means that Eve cannot make a copy of Alice's photon and send it on to Bob (otherwise, she could decrypt her copy later, when Alice and Bob compare their random basis selections).

Suppose we want to clone an unknown quantum state, $|\psi\rangle = \alpha|0\rangle + \beta|1\rangle$. We use some operator, U_{copy}, as shown in Fig. 7.15.

$$U_{\text{copy}}|\psi\rangle_1|0\rangle_2 = |\psi\rangle_1|\psi\rangle_2 \tag{7.50}$$

The left-hand side of Eq. (7.50) gives:

$$U_{\text{copy}}|\psi\rangle_1|0\rangle_2 = U_{\text{copy}}(\alpha|0\rangle + \beta|1\rangle)_1|0\rangle_2$$
$$= U_{\text{copy}}(\alpha|0\rangle_1|0\rangle_2 + \beta|1\rangle_1|0\rangle_2)$$

Fig. 7.15 Quantum circuit to demonstrate the no-cloning theorem

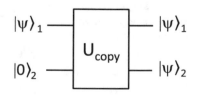

$$= \alpha|0\rangle_1|0\rangle_2 + \beta|1\rangle_1|1\rangle_2 \qquad (7.51)$$

Note that the result is an entangled state. The right-hand side of Eq. (7.50) gives:

$$\begin{aligned}|\psi\rangle_1|\psi\rangle_2 &= (\alpha|0\rangle + \beta|1\rangle)_1(\alpha|0\rangle + \beta|1\rangle)_2 \\ &= \alpha^2|0\rangle_1|0\rangle_2 + \alpha\beta|0\rangle_1|1\rangle_2 + \beta\alpha|1\rangle_1|0\rangle_2 + \beta^2|1\rangle_1|1\rangle_2 \qquad (7.52)\end{aligned}$$

Equation (7.51) does not equal (7.52) unless $\alpha = 1$ and $\beta = 0$ or vice versa; i.e., only $|0\rangle$ or $|1\rangle$ states can be copied (classical states), but not a general quantum state with superposition. An alternative viewpoint is that the matrix corresponding to U_{copy} is not unitary, and therefore cannot be implemented.

References

1. Russian Mathematical Surveys **52**, 1191 (1997)
2. https://arxiv.org/abs/quant-ph/0505030
3. W.K. Wootters, W.H. Zurek, Nature **299**, 802 (1982)

Chapter 8
Teleportation

Teleportation uses entanglement to move a quantum state from one quantum system to another, without having to send the physical system. We might use teleportation to transport states over long distances in a quantum circuit.

8.1 Teleporting Quantum States

Classical information is easily copied and transmitted as a sequence of digital 0's and 1's as electrical, optical, radio or microwave signals. Can we do the same with quantum information? How do we transfer a quantum state from Alice to Bob? Do we have to send Bob the entire physical system, or is there a way we can just send the state?

If our quantum system is in a state $|\psi\rangle = |01001\rangle$, then we can transmit this; this state contains only classical information (no superposition). But suppose the quantum state is a superposition state such as $|\psi\rangle = \frac{1}{\sqrt{2}} (|01001\rangle + |10101\rangle)$. In general, we cannot know the state because we cannot measure the coefficients of the state vector (the quantum probability amplitudes) due to the "measurement problem"; i.e., collapse into one state upon measurement. What about copying the state of Alice to a second quantum system and giving that to Bob ("quantum memory stick"). This will not work either—we cannot copy the state because of the no-cloning theorem discussed in Chap. 7.

8.2 Teleportation Circuit

Teleportation uses entanglement to move a quantum state from one quantum system to another, without having to send the physical system. Figure 8.1 shows the teleportation circuit that can transport a general quantum state (superposition), $|\psi\rangle = \alpha|0\rangle$

© The Author(s), under exclusive license to Springer Nature Switzerland AG 2021
R. LaPierre, *Introduction to Quantum Computing*, The Materials Research Society Series,
https://doi.org/10.1007/978-3-030-69318-3_8

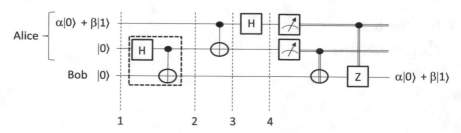

Fig. 8.1 The teleportation circuit

$+ \beta|1\rangle$, from Alice to Bob. The register (step 1 of Fig. 8.1) contains three qubits. The first qubit is the quantum state to be transported that resides with Alice, while qubits 2 and 3 are both set to $|0\rangle$. Hence, the state at step 1 of Fig. 8.1 is:

$$|\psi_1\rangle = (\alpha|0\rangle + \beta|1\rangle)_1|0\rangle_2|0\rangle_3 \tag{8.1}$$

where we have labelled the qubits with a subscript.

An entangled state is created by sending qubits 2 and 3 into a Bell state circuit (dashed box in Fig. 8.1). The total state at step 2 of Fig. 8.1 after the Bell state circuit is:

$$|\psi_2\rangle = (\alpha|0\rangle + \beta|1\rangle)[\tfrac{1}{\sqrt{2}}(|00\rangle + |11\rangle)] = \tfrac{1}{\sqrt{2}}(\alpha|000\rangle + \alpha|011\rangle + \beta|100\rangle + \beta|111\rangle)$$

$$\underbrace{}_{\substack{\text{state to be}\\ \text{teleported}}}\underbrace{\phantom{[\tfrac{1}{\sqrt{2}}(|00\rangle+|11\rangle)]}}_{\substack{\text{entangled}\\ \text{state}}} \tag{8.2}$$

After entanglement, the two qubits (qubits 2 and 3) are separated—one stays with Alice and the other goes to Bob. Hence, teleportation relies on a previous sharing of an entangled state.

Alice performs a CNOT operation that affects the target qubit 2 based on the control qubit 1. The resulting state at step 3 is:

$$|\psi_3\rangle = \frac{1}{\sqrt{2}}(\alpha|000\rangle + \alpha|011\rangle + \beta|110\rangle + \beta|101\rangle) \tag{8.3}$$

Next, Alice implements a Hadamard gate. The resulting state at step 4 is:

$$|\psi_4\rangle = \frac{1}{2}[\alpha(|0\rangle + |1\rangle)|00\rangle + \alpha(|0\rangle + |1\rangle)|11\rangle + \beta(|0\rangle - |1\rangle)|10\rangle + \beta(|0\rangle - |1\rangle)|01\rangle] \tag{8.4}$$

At this point, it is useful to rearrange Eq. (8.4) so that the first two qubits are the four possible basis states:

$$|\psi_4\rangle = \frac{1}{2}[|00\rangle(\alpha|0\rangle + \beta|1\rangle) + |01\rangle(\alpha|1\rangle + \beta|0\rangle) + |10\rangle(\alpha|0\rangle - \beta|1\rangle)$$
$$+ |11\rangle(\alpha|1\rangle + \beta|0\rangle)] \tag{8.5}$$

Now the third qubit term (Bob's) looks (or almost looks) like the state we want to transport.

Alice performs measurements on her qubits and sends the classical result (00, 01, 10, or 11) to Bob by a classical channel (denoted by the double lines in Fig. 8.1). Bob uses this information as the control for a CNOT and a control-Z gate with the following result:

$$\text{If Bob receives } |00\rangle \; : \; \alpha|0\rangle + \beta|1\rangle \xrightarrow{\text{CNOT}} \alpha|0\rangle + \beta|1\rangle \xrightarrow{Z} \alpha|0\rangle + \beta|1\rangle$$

$$\text{If Bob receives } |01\rangle \; : \; \alpha|1\rangle + \beta|0\rangle \xrightarrow{\text{CNOT}} \alpha|0\rangle + \beta|1\rangle \xrightarrow{Z} \alpha|0\rangle + \beta|1\rangle$$

$$\text{If Bob receives } |10\rangle \; : \; \alpha|0\rangle - \beta|1\rangle \xrightarrow{\text{CNOT}} \alpha|0\rangle - \beta|1\rangle \xrightarrow{Z} \alpha|0\rangle + \beta|1\rangle$$

$$\text{If Bob receives } |11\rangle \; : \; \alpha|1\rangle - \beta|0\rangle \xrightarrow{\text{CNOT}} \alpha|0\rangle - \beta|1\rangle \xrightarrow{Z} \alpha|0\rangle + \beta|1\rangle$$

Hence, the state is successfully transported.

Teleportation does not violate special relativity since Alice needs to send the results of her measurement (00, 01, 10, or 11) to Bob by a classical channel. Teleportation does not violate the no cloning theorem since Alice's quantum state is destroyed in the measurement process (collapses to 00, 01, 10, or 11).

Quantum teleportation was first proposed by Bennett et al. [1] and was first demonstrated experimentally by Bouwmeester et al. [2] using photons. Teleportation has been demonstrated in many other quantum systems (e.g., trapped ion quantum computing in Chap. 21). We might use teleportation to transport states over long distances in a quantum circuit. It may also be used as an "optical repeater" in quantum key distribution (QKD). An optical communications repeater is a device used in fiber-optic communication systems to regenerate an optical signal. Such repeaters are used to extend the reach of optical communication links by overcoming loss due to attenuation of the optical fiber.

Exercise 8.1 How might we use teleportation to design a quantum optical repeater? Sketch the quantum circuit of your repeater.

8.3 Superdense Coding

Superdense coding is another application of entanglement that allows us to send two classical bits of information using only a single qubit of quantum information. The quantum circuit that accomplishes this is shown in Fig. 8.2. In a sense, this is the

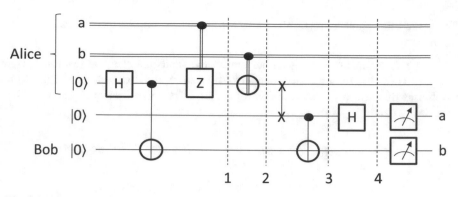

Fig. 8.2 Superdense coding circuit

Table 8.1 State at each step of superdense coding circuit

ab	State of qubit 3 and 5 after step 1	State of qubit 3 and 5 after step 2	State of qubit 4 and 5 after step 3	State of qubit 4 and 5 after step 4
00	$\frac{1}{\sqrt{2}}(\lvert 00\rangle + \lvert 11\rangle)$	$\frac{1}{\sqrt{2}}(\lvert 00\rangle + \lvert 11\rangle)$	$\frac{1}{\sqrt{2}}(\lvert 0\rangle + \lvert 1\rangle)\lvert 0\rangle$	$\lvert 00\rangle$
01	$\frac{1}{\sqrt{2}}(\lvert 00\rangle + \lvert 11\rangle)$	$\frac{1}{\sqrt{2}}(\lvert 10\rangle + \lvert 01\rangle)$	$\frac{1}{\sqrt{2}}(\lvert 1\rangle + \lvert 0\rangle)\lvert 1\rangle$	$\lvert 01\rangle$
10	$\frac{1}{\sqrt{2}}(\lvert 00\rangle - \lvert 11\rangle)$	$\frac{1}{\sqrt{2}}(\lvert 00\rangle - \lvert 11\rangle)$	$\frac{1}{\sqrt{2}}(\lvert 0\rangle - \lvert 1\rangle)\lvert 0\rangle$	$\lvert 10\rangle$
11	$\frac{1}{\sqrt{2}}(\lvert 00\rangle - \lvert 11\rangle)$	$\frac{1}{\sqrt{2}}(\lvert 10\rangle - \lvert 01\rangle)$	$\frac{1}{\sqrt{2}}(\lvert 1\rangle - \lvert 0\rangle)\lvert 1\rangle$	$-\lvert 11\rangle$

reverse of teleportation. Suppose Alice wants to send the two classical bits, a and b, to Bob. First, an entangled state is created using the first Hadamard gate and CNOT gate of Fig. 8.2. One qubit of the entangled state goes to Alice, and the other qubit goes to Bob. Referring to Table 8.1, Alice performs a controlled Z gate (step 1) and a CNOT gate (step 2) on her qubit, using the classical bits, a and b, as the controls. After performing these operations on her qubit, she sends her qubit to Bob, represented by the SWAP operation. Using Alice's qubit as a control, Bob then performs a CNOT gate (step 3), followed by a Hadamard gate (step 4). The result of a measurement on these two qubits will result in a and b. Hence, a and b are transmitted to Bob.

Superdense coding was experimentally demonstrated in 1995 [3]. It seems that one qubit of information is equal to two classical bits of information. Holevo's theorem [4] formalizes this statement. If Alice and Bob share no prior entanglement, then it is necessary for them to exchange at least n qubits (or bits) for Alice to communicate n classical bits of information to Bob. If they do share prior entanglement, then they must exchange at least $n/2$ qubits.

References

1. C.H. Bennett et al., Phys. Rev. Lett. **70**, 1895 (1993)
2. D. Bouwmeester et al., Nature **390**, 575 (1997)
3. K. Mattle et al., Phys. Rev. Lett. **76**, 4656 (1996)
4. A.S. Holevo, Problems of Information Transmission **9**, 177 (1973)

Chapter 9
Tensor Products

Tensor products provide a mathematical tool for dealing with multiple qubits. They assist in formulating the vector representation of multiple qubits, and the analysis of quantum circuits.

9.1 Tensor Product of Qubits

Tensors were introduced in Chap. 4. Here, we will formalize the idea of tensors. Tensor products provide a mathematical tool for dealing with multiple qubits. Suppose, for example, we have two single-qubit vectors:

$$|v\rangle \doteq \begin{pmatrix} \alpha \\ \beta \end{pmatrix} \tag{9.1}$$

$$|w\rangle = \begin{pmatrix} \delta \\ \gamma \end{pmatrix} \tag{9.2}$$

where α, β, δ, and γ are complex probability amplitudes. This means that the probability of qubit $|v\rangle$ being in the state $|0\rangle$ or $|1\rangle$ is $|\alpha|^2$ and $|\beta|^2$, respectively. Similarly, the probability of qubit $|w\rangle$ being in the state $|0\rangle$ or $|1\rangle$ is $|\delta|^2$ and $|\gamma|^2$, respectively.

Now consider the two-qubit system:

$$|v\rangle |w\rangle \tag{9.3}$$

Using Eqs. (9.1) and (9.2), we can write the two-qubit system as:

$$|v\rangle |w\rangle = (\alpha|0\rangle + \beta|1\rangle)(\delta|0\rangle + \gamma|1\rangle) \tag{9.4}$$

Expanding Eq. (9.4) gives:

$$|v\rangle|w\rangle = \alpha\delta|00\rangle + \alpha\gamma|01\rangle + \beta\delta|10\rangle + \beta\gamma|11\rangle \tag{9.5}$$

Equation (9.5) can be written in the vector representation as:

$$|v\rangle|w\rangle = \begin{pmatrix} \alpha\delta \\ \alpha\gamma \\ \beta\delta \\ \beta\gamma \end{pmatrix} \tag{9.6}$$

The column vector in Eq. (9.6) can also be obtained using the tensor product, defined as:

$$|v\rangle \otimes |w\rangle = \begin{pmatrix} \alpha\begin{pmatrix} \delta \\ \gamma \end{pmatrix} \\ \beta\begin{pmatrix} \delta \\ \gamma \end{pmatrix} \end{pmatrix} = \begin{pmatrix} \alpha\delta \\ \alpha\gamma \\ \beta\delta \\ \beta\gamma \end{pmatrix} \tag{9.7}$$

We read " $\otimes$ " as "tensor". Equation (9.7) tells us that the probability of $|v\rangle|w\rangle$ being $|0\rangle|0\rangle$, $|0\rangle|1\rangle$, $|1\rangle|0\rangle$ and $|1\rangle|1\rangle$ is $|\alpha\delta|^2$, $|\alpha\gamma|^2$, $|\beta\delta|^2$ and $|\beta\gamma|^2$, respectively. Hence, although the expansion of Eq. (9.4) into Eq. (9.5) seems straightforward, we were actually implementing a tensor product. $|v\rangle \otimes |w\rangle$, $|v\rangle_1|w\rangle_2$, $|v\rangle|w\rangle$, $|vw\rangle$ and $|v, w\rangle$ are all equivalent notations.

Using the tensor product, the basis states for a two-qubit system are:

$$|0\rangle = |00\rangle = |0\rangle \otimes |0\rangle = \begin{pmatrix} 1 \\ 0 \end{pmatrix} \otimes \begin{pmatrix} 1 \\ 0 \end{pmatrix} = \begin{pmatrix} 1 \\ 0 \\ 0 \\ 0 \end{pmatrix} \tag{9.8}$$

$$|1\rangle = |01\rangle = |0\rangle \otimes |1\rangle = \begin{pmatrix} 1 \\ 0 \end{pmatrix} \otimes \begin{pmatrix} 0 \\ 1 \end{pmatrix} = \begin{pmatrix} 0 \\ 1 \\ 0 \\ 0 \end{pmatrix} \tag{9.9}$$

$$|2\rangle = |10\rangle = |1\rangle \otimes |0\rangle = \begin{pmatrix} 0 \\ 1 \end{pmatrix} \otimes \begin{pmatrix} 1 \\ 0 \end{pmatrix} = \begin{pmatrix} 0 \\ 0 \\ 1 \\ 0 \end{pmatrix} \tag{9.10}$$

$$|3\rangle = |11\rangle = |1\rangle \otimes |1\rangle = \begin{pmatrix} 0 \\ 1 \end{pmatrix} \otimes \begin{pmatrix} 0 \\ 1 \end{pmatrix} = \begin{pmatrix} 0 \\ 0 \\ 0 \\ 1 \end{pmatrix} \tag{9.11}$$

Note that $|x\rangle$ denotes a 1 in the $x + 1$ row of the column vector (because we start counting at 0). For example, let us examine a three-qubit system:

$$|5\rangle = |101\rangle = |1\rangle|0\rangle|1\rangle = |1\rangle \otimes |0\rangle \otimes |1\rangle = \begin{pmatrix} 0 \\ 1 \end{pmatrix} \otimes \begin{pmatrix} 1 \\ 0 \end{pmatrix} \otimes \begin{pmatrix} 0 \\ 1 \end{pmatrix} = (0\,0\,0\,0\,0\,1\,0\,0)^{\mathrm{T}}$$

$$(9.12)$$

Note that we use the transpose (T) in the last term of Eq. (9.12) to indicate a column vector. 101 is binary for the number 5. Thus, 101 represents a 1 in the 6th row of the column vector.

9.2 Tensor Product of Operators

Suppose we have two single-qubit operations, U_1 and U_2. How do we represent this as a single two-qubit operation? We use the tensor product:

$$U = U_1 \otimes U_2 = \begin{pmatrix} a & c \\ b & d \end{pmatrix} \otimes \begin{pmatrix} e & g \\ f & h \end{pmatrix}$$

$$(9.13)$$

The tensor product in this instance is defined as follows:

$$\begin{pmatrix} a & c \\ b & d \end{pmatrix} \otimes \begin{pmatrix} e & g \\ f & h \end{pmatrix} = \begin{pmatrix} a\begin{pmatrix} e & g \\ f & h \end{pmatrix} & c\begin{pmatrix} e & g \\ f & h \end{pmatrix} \\ b\begin{pmatrix} e & g \\ f & h \end{pmatrix} & d\begin{pmatrix} e & g \\ f & h \end{pmatrix} \end{pmatrix} = \begin{pmatrix} ae & ag & ce & cg \\ af & ah & cf & ch \\ be & bg & de & dg \\ bf & bh & df & dh \end{pmatrix}$$

$$(9.14)$$

Hence, the circuit representations in Fig. 9.1 are equivalent.

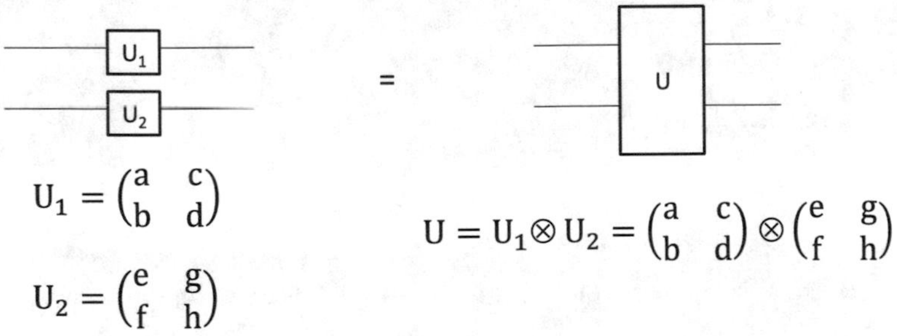

$$U_1 = \begin{pmatrix} a & c \\ b & d \end{pmatrix}$$

$$U_2 = \begin{pmatrix} e & g \\ f & h \end{pmatrix}$$

$$U = U_1 \otimes U_2 = \begin{pmatrix} a & c \\ b & d \end{pmatrix} \otimes \begin{pmatrix} e & g \\ f & h \end{pmatrix}$$

Fig. 9.1 Tensor product for a two-qubit operator

For example, suppose we have an input state of $|00\rangle$ or, equivalently, $\begin{pmatrix} 1 \\ 0 \\ 0 \\ 0 \end{pmatrix}$. On the

left-hand side of Fig. 9.1, we expect an output of $U_1|0\rangle = a|0\rangle + b|1\rangle$ for the first qubit, and $U_2|0\rangle = e|0\rangle + f|1\rangle$ for the second qubit. The combined two-qubit output is $(a|0\rangle$

$+ b|1\rangle)(e|0\rangle + f|1\rangle) = ae|00\rangle + af|01\rangle + be|10\rangle + bf|11\rangle$ or, equivalently, $\begin{pmatrix} ae \\ af \\ be \\ bf \end{pmatrix}$.

On the right-hand side of Fig. 9.1, we expect an output of $U|00\rangle = (U_1 \otimes U_2)\,|00\rangle = $
$\begin{pmatrix} ae & ag & ce & cg \\ af & ah & cf & ch \\ be & bg & de & dg \\ bf & bh & df & dh \end{pmatrix} \begin{pmatrix} 1 \\ 0 \\ 0 \\ 0 \end{pmatrix} = \begin{pmatrix} ae \\ af \\ be \\ bf \end{pmatrix}$. The result is the same as the left-hand circuit.

Exercise 9.1 What is the matrix for each of the following?

$$I \otimes I$$
$$Z \otimes I$$
$$I \otimes X$$
$$X \otimes Y$$

I is the 2×2 identity matrix and X, Y and Z are the Pauli matrices.

Exercise 9.2 What is the matrix representation for the following circuit?

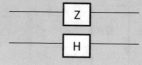

Exercise 9.3 Calculate the output of the following Bell state circuit, employing the tensor product.

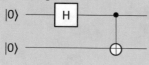

Fig. 9.2 A quantum circuit
and its equivalent operator

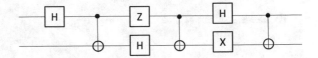

$$(CNOT)\ (H \otimes X)\ (CNOT)\ (Z \otimes H)\ (CNOT)\ (H \otimes I)$$

We can evaluate more complex circuits by using the tensor product. For example, the circuit in Fig. 9.2 is equivalent to the operator $(CNOT)\ (H \otimes X)\ (CNOT)\ (Z \otimes H)\ (CNOT)\ (H \otimes I)$. Note that time in the circuit goes from left to right, but the corresponding operators must be multiplied together from right to left, so that the earliest gates are applied to the state first.

In general, suppose we have a $n \times m$ matrix, A, and a $k \times l$ matrix, B:

$$A = \begin{pmatrix} a_{11} & a_{12} & \cdots & a_{1m} \\ a_{21} & a_{22} & \cdots & a_{2m} \\ \vdots & \vdots & \ddots & \vdots \\ a_{n1} & a_{n2} & \cdots & a_{nm} \end{pmatrix}, B = \begin{pmatrix} b_{11} & b_{12} & \cdots & b_{1l} \\ b_{21} & b_{22} & \cdots & b_{2l} \\ \vdots & \vdots & \ddots & \vdots \\ b_{k1} & b_{k2} & \cdots & b_{kl} \end{pmatrix} \tag{9.15}$$

The tensor product $A \otimes B$ is the $nk \times ml$ matrix defined as:

$$A \otimes B = \begin{pmatrix} a_{11}B & a_{12}B & \cdots & a_{1m}B \\ a_{21}B & a_{22}B & \cdots & a_{2m}B \\ \vdots & \vdots & \ddots & \vdots \\ a_{n1}B & a_{n2}B & \cdots & a_{nm}B \end{pmatrix} \tag{9.16}$$

Equation (9.16) also works for vectors by thinking of them as matrices with only one column. The tensor product obeys the properties of Table 9.1.

Table 9.1 Properties of the tensor product

Name	Property
Associative	$(A \otimes B) \otimes C = A \otimes (B \otimes C)$
No name	$(A \otimes B)(C \otimes D) = (AC) \otimes (BD)$
Distributive	$A \otimes (B + C) = A \otimes B + A \otimes C$ $(A + B) \otimes C = A \otimes C + B \otimes C$
No name	$(aA) \otimes B = A \otimes (aB) = a(A \otimes B)$
Non-commutative	$A \otimes B \neq B \otimes A$

Exercise 9.4 Show that $|26\rangle^5 = |3\rangle^2 \otimes |2\rangle^3$ where the superscript denotes the number of qubits.

Exercise 9.5 Show that $|0\rangle^n|1\rangle = |1\rangle^{n+1}$ where the superscript denotes the number of qubits.

Exercise 9.6 Show that $\text{CNOT} = |0\rangle\langle 0| \otimes I + |1\rangle\langle 1| \otimes X$.

Exercise 9.7 Show that $\text{CNOT} = (I \otimes H)\,(\text{CPHASE})\,(I \otimes H)$.

Exercise 9.8 Referring to the CHSH inequality in Chap. 5, show that quantum mechanics predicts $\langle ab \rangle = -\cos(\theta)$ where θ is the angle between any two measurement directions of electron spin. Hint: $\langle ab \rangle$ is the expectation value given by $\langle \psi | \hat{A}\hat{B} | \psi \rangle$ where $\hat{A}$ and $\hat{B}$ are the spin operators along an arbitrary direction in the measurement plane, and $|\psi\rangle = \frac{1}{\sqrt{2}}(|01\rangle - |10\rangle)$ is the entangled state as shown in Fig. 5.3.

Chapter 10
Quantum Parallelism and Computational Complexity

Quantum computers can use a superposition to run many states simultaneously through an algorithm, rather than sequentially as in a classical computer. This principle is called quantum parallelism. This chapter refines our notion of quantum parallelism and introduces computational complexity to quantify the power of quantum computers.

10.1 Achieving Superpositions

How do we achieve a superposition of states for quantum parallelism? We use the Hadamard gate. Recall from Chap. 7 that the Hadamard gate applied to a single-qubit state, $|0\rangle$, gives an output which is a superposition, $\frac{1}{\sqrt{2}}(|0\rangle + |1\rangle)$. How do we achieve a multi-qubit superposition? We use multiple Hadamard gates as shown in Fig. 10.1. For example, for the two-qubit system ($n = 2$), we get:

$$H \otimes H |00\rangle = \left[\frac{1}{\sqrt{2}}(|0\rangle + |1\rangle)\right] \otimes \left[\frac{1}{\sqrt{2}}(|0\rangle + |1\rangle)\right]$$

$$= \frac{1}{2}(|00\rangle + |01\rangle + |10\rangle + |11\rangle)$$

$$= \frac{1}{2}(|0\rangle + |1\rangle + |2\rangle + |3\rangle)$$

This is an "equally weighted superposition" of the two-qubit basis states, meaning the superposition contains all the two-qubit basis states with equal probability amplitude of 1/2 (probability of 1/4). Note that the probabilities sum to unity as required.

We can extend the superposition to more than two qubits by introducing a new notation, $H^{\otimes n}$, meaning we tensor the Hadamard transform with itself n times (Fig. 10.2). By extension of the earlier examples for $n = 1$ and $n = 2$ qubits, $H^{\otimes n}|0\rangle^{\otimes n}$ will give an output that is an equally weighted superposition of the 2^n basis states, $|x\rangle$.

© The Author(s), under exclusive license to Springer Nature Switzerland AG 2021
R. LaPierre, *Introduction to Quantum Computing*, The Materials Research Society Series,
https://doi.org/10.1007/978-3-030-69318-3_10

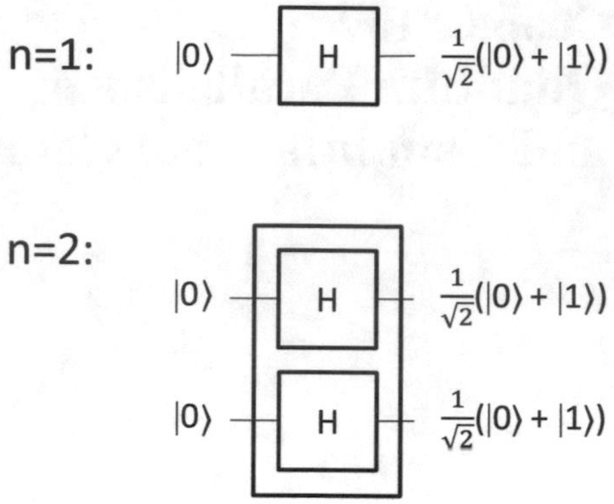

Fig. 10.1 Using Hadamard
gates to achieve
superpositions

n=1: $|0\rangle$ — H — $\frac{1}{\sqrt{2}}(|0\rangle + |1\rangle)$

n=2:

$|0\rangle$ — H — $\frac{1}{\sqrt{2}}(|0\rangle + |1\rangle)$

$|0\rangle$ — H — $\frac{1}{\sqrt{2}}(|0\rangle + |1\rangle)$

Fig. 10.2 A multi-qubit
superposition achieved by
$H^{\otimes n}$

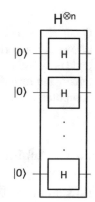

This means that we have a superposition containing all the multi-qubit basis states
with equal probability amplitudes:

$$H^{\otimes n}|0\rangle^{\otimes n} = \left[\frac{1}{\sqrt{2}}(|0\rangle + |1\rangle)\right]^{\otimes n} \tag{10.1}$$

$$= \frac{1}{\sqrt{2^n}} \sum_{x \in \{0,1\}^n} |x\rangle \tag{10.2}$$

$$= \frac{1}{\sqrt{2^n}} \sum_{x=0}^{2^n - 1} |x\rangle \tag{10.3}$$

Equation (10.2) uses the binary representation of the basis states; i.e., a string of
0's and 1's. For the two-qubit example, these would be $|00\rangle$, $|01\rangle$, $|10\rangle$ and $|11\rangle$.
Equation (10.3) uses the decimal notation of the basis states. For the two-qubit
example, these would be $|1\rangle$, $|2\rangle$, $|3\rangle$ and $|4\rangle$. Most quantum algorithms begin with all
qubits starting in the state $|0\rangle$, and then putting the qubits into an equal superposition.

Exercise 10.1 What is the matrix for $H^{\otimes 2}$?

Exercise 10.2 What is the matrix for $H^{\otimes 3}$?

Exercise 10.3 What is the output vector for $H^{\otimes 3}|0\rangle^{\otimes 3}$?

Generalizing Exercise 10.1 and 10.2, each entry of $H^{\otimes n}$ is either 1 or -1 with a constant pre-factor of $\frac{1}{\sqrt{2^n}}$. Ignoring the pre-factor, the entry in the ith row and jth column of a $H^{\otimes n}$ matrix is given by:

$$(H^{\otimes n})_{i,j} = (-1)^{i \cdot j} \tag{10.4}$$

where the indices are $i, j = 0, 1, 2 \ldots$, and $i \cdot j$ is the bitwise dot product (inner product) of the binary representation of the indices. For example, if $n = 3$, $i = 2$ and $j = 3$, then $(H^{\otimes 3})_{2,3} = (-1)^{(10) \cdot (11)} = (-1)^{1+0} = (-1)^1 = -1$. You should check this result against Exercise 10.2.

Rather than the matrix representation, it is straightforward (Exercise 7.7) to show that the Hadamard gate can be expressed in Dirac notation as:

$$H|x\rangle = \frac{1}{\sqrt{2}}|0\rangle + \frac{1}{\sqrt{2}}(-1)^x|1\rangle \tag{10.5}$$

We can also write Eq. (10.5) in the form:

$$H|x\rangle = \frac{1}{\sqrt{2}} \sum_{y \in \{0,1\}} (-1)^{xy}|y\rangle \tag{10.6}$$

For two qubits, we would obtain:

$$H^{\otimes 2}|x_1\rangle|x_2\rangle = \left[\frac{1}{\sqrt{2}} \sum_{y_1 \in \{0,1\}} (-1)^{x_1 y_1}|y_1\rangle\right]\left[\frac{1}{\sqrt{2}} \sum_{y_2 \in \{0,1\}} (-1)^{x_2 y_2}|y_2\rangle\right] \tag{10.7}$$

Equation (10.7) can be written as:

$$H^{\otimes 2}|x_1\rangle|x_2\rangle = \frac{1}{2} \sum_{y \in \{0,1\}^2} (-1)^{x_1 y_1 + x_2 y_2}|y\rangle \qquad (10.8)$$

Equation (10.8) can be generalized to:

$$H^{\otimes n}|x\rangle = \frac{1}{\sqrt{2^n}} \sum_{y \in \{0,1\}^n} (-1)^{x \cdot y}|y\rangle \qquad (10.9)$$

where $x \in \{0,1\}^n$ and $y \in \{0,1\}^n$ are n-bit strings, and $x \cdot y$ is the bitwise dot product of the two bit strings.

For example, suppose a two-qubit input state is $|x\rangle = |3\rangle$, which corresponds to $x = 11$ in binary notation. The summation of Eq. (10.9) includes the output states $|y\rangle = |0\rangle, |1\rangle, |2\rangle$ and $|3\rangle$, corresponding to $y = 00, 01, 10$ and 11 in binary, respectively. If $y = 00$, then the dot product is $x \cdot y = (11) \cdot (00) = 0 + 0 = 0$. If $y = 01$, then $x \cdot y = (11) \cdot (01) = 0 + 1 = 1$. If $y = 10$, then $x \cdot y = (11) \cdot (10) = 1 + 0 = 1$. Finally, if $y = 11$, then $x \cdot y = (11) \cdot (11) = (1 + 1) = 2$. Thus, from Eq. (10.9):

$$H^{\otimes 2}|3\rangle = \frac{1}{2}\left[(-1)^0|0\rangle + (-1)^1|1\rangle + (-1)^1|2\rangle + (-1)^2|3\rangle\right]$$

$$= \frac{1}{2}(|0\rangle - |1\rangle - |2\rangle + |3\rangle)$$

10.2   Quantum Parallelism

Let us refine the basic idea of quantum computing depicted previously in Fig. 1.16. Figure 10.3 is a refined version of Fig. 1.16. The quantum circuit, U_f, is a unitary implementation of some classical function, f. The input to U_f (after $H^{\otimes n}$) is an equally weighted superposition of the basis states, $|x\rangle$. The output to U_f is an equally weighted superposition of $U_f|x\rangle$; i.e., the function f is evaluated simultaneously for all possible states $|x\rangle$. By a single call to U_f, we have computed $f(x)$ for every possible input x! There is no classical analogue of this process, which David Deutsch (Fig. 10.4) called "quantum parallelism".

The computing power grows exponentially (2^n) with the number of qubits, n. Each additional qubit doubles the computing power. We can continue "Moore's law" by the addition of one qubit every 2 years! For example, 300 qubits are equivalent to a classical computer with $2^{300} = 10^{90}$ states. This number of states is greater than the number of particles in the universe. Where does Nature store all this information? Nobody knows.

Fig. 10.3 A circuit for quantum parallelism

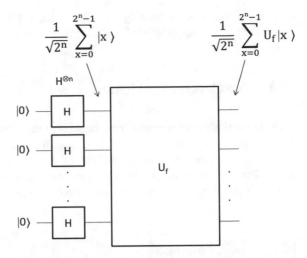

$$\frac{1}{\sqrt{2^n}} \sum_{x=0}^{2^n-1} |x\rangle \qquad \frac{1}{\sqrt{2^n}} \sum_{x=0}^{2^n-1} U_f |x\rangle$$

$H^{\otimes n}$

$|0\rangle$ — H

$|0\rangle$ — H

U_f

$|0\rangle$ — H

Fig. 10.4 David Deutsch. *Credit* Wikimedia Commons [1]

10.3 The Measurement Problem

Quantum parallelism should make quantum computers exponentially more powerful than classical computers. But are they? When we measure, we will get one of the output states $U_f|x\rangle$ with probability $\frac{1}{2^n}$. Therefore, we have limited access to the output. Can we deduce all output values by making lots of measurements? No! Remember that after the measurement, the state of the system collapses onto the result of the measurement (one of the basis states). Therefore, although we can *store* an enormous amount of information, we cannot *access* it. Quantum parallelism is not useful in cases where you want all answers to all possible inputs. Instead, we will see that quantum computing is useful for finding some "global" property of the input, such as finding the period of the input, the average, etc.

10.4   The Control Problem

Although we cannot measure all the probability amplitudes, we *can* manipulate them (this is what our quantum gates do). However, it is not feasible to go in and change each of the complex amplitudes individually (there's 2^n of them!). We can only control the n single qubits (single-qubit gates) or the interactions between them (e.g., two-qubit gates), which drives the evolution of 2^n complex probability amplitudes.

10.5   The Challenge

The challenge is to find quantum circuits that give useful outputs, given the measurement problem, and the limited control we have on n qubits. To gain the advantage of quantum parallelism, we must combine it with another quantum feature—interference. The goal is to arrange the computation such that constructive interference amplifies the desired output and destructive interference cancels the rest.

Another quantum resource is entanglement. Do we need entanglement for quantum computing? If there is no entanglement, then the state of n qubits can be written as a tensor product $(\alpha_0|0\rangle + \beta_0|1\rangle) \otimes (\alpha_1|0\rangle + \beta_1|1\rangle) \otimes \ldots (\alpha_n|0\rangle + \beta_n|1\rangle)$, which is a separable state. In this case, we only need to specify $2n$ probability amplitudes, which is polynomial in n. This could be simulated by a classical computer, and quantum computers would then have no advantage. The advantage of quantum computers is realized by more general quantum states of n qubits with 2^n probability amplitudes that includes entanglement. In general, both interference and entanglement are required for quantum computers to be more powerful than classical computers.

10.6   Computational Complexity

Even if we can build a quantum algorithm, we must demonstrate that it is more efficient than any classical algorithm. What does it mean to say a quantum algorithm is more efficient than a classical algorithm? "Computational complexity" describes the computing resources (such as time, memory, number of computational steps, etc.) that are required as the number of inputs or the length of the input (e.g., number of digits of an input) increases.

10.7 Big O Notation

To quantify computational complexity, we use the "Big O notation", which describes the limiting behaviour of a function or algorithm when the size of the input (N) tends towards infinity (Fig. 10.5). For example, $O(N^k)$ for some k is polynomial, and $O(c^N)$ for some constant c is "superpolynomial" or exponential. Big O could refer to the number of operations required, the time required to complete the algorithm, the memory resources required, the electrical power required, the cost, etc.

Computational theorists use the term "easy" to refer to a problem whose known algorithm has polynomial complexity. An "efficient" algorithm means that the computational cost increases only polynomially with the size of the input. "Hard" problems are ones where the only known algorithms have exponential complexity, which rapidly become unfeasible for large input size.

For example, find the values of the following products:

$$7 \times 5$$
$$11 \times 17$$
$$29 \times 83$$
$$337 \times 263$$

Multiplication is considered an "easy" problem. The time or number of steps that are needed to multiply two numbers increases only polynomially with the number of binary digits, N; i.e., multiplication is $O(N^k)$ where k is a constant. In fact, it can be shown that multiplying two N-bit integers requires $O(N^2)$ operations.

On the other hand, find the prime factors of the following:

$$35$$
$$187$$

Fig. 10.5 Examples of Big O notation

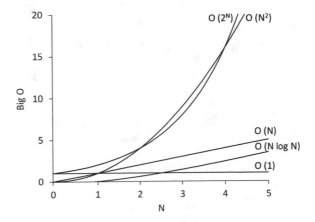

2407

88631

For some n-bit integer (N on the order of 2^n), the simplest algorithm for factoring is to check all the integers from 2 to $\sqrt{N}$ to see if any evenly divides N. This requires us to check $2^{n/2}$ numbers, which is exponential in n. The best classical algorithm, known as the general number field sieve, is still exponential in n. Factoring is a "hard" problem.

In computational complexity theory, the class of problems that can be solved in polynomial time are called "P" problems. Multiplication is a P problem. The class of problems that require exponential time to solve, but the solution is easy to verify once found, are called "NP" problems ("nondeterministic polynomial"). Factoring is an NP problem. The solution to the factoring problem is difficult; but the solution, once found, is easy to verify (multiplication is easy).

Another example is the "traveling salesperson problem" (TSP), illustrated in Fig. 10.6. Given a list of cities (A, B, C, …) and the distance between each pair of cities, is there a route with a length less than L that visits each city and returns to the origin city? The TSP is one of the most intensely studied problems in mathematics with many applications in optimization. This version of TSP is believed to be an NP problem; i.e., the worst-case running time for any algorithm for the TSP increases exponentially with the number of cities, N; but the solution, once found, is easy to verify.

One of the most important unsolved problems in mathematics is whether $P = NP$; i.e., are there solutions to NP problems that will make them P problems? If you can prove whether or not $P = NP$, the Clay Institute has a million-dollar check for you! Quantum computers offer solutions to *some* problems believed to be NP, such as factoring large numbers. It is not currently known if quantum computers can efficiently solve *all NP* problems.

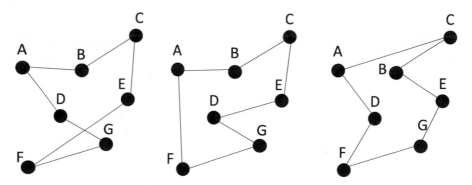

Fig. 10.6 The traveling salesperson problem with possible routes

Fig. 10.7 John Preskill.
Credit Wikimedia Commons
[5]

10.8   Church-Turing Thesis

A major foundation of classical computer science has been the Church-Turing thesis, developed by two pioneers of computer science, Alonzo Church and Alan Turing. The Church-Turing thesis provides an abstract definition of a programmable computer and forms the foundation for computer science. This thesis says that all computers are theoretically equivalent in computational power and are equivalent to a "Universal Turing Machine"—a kind of programmable computer. Note that a classical computer *can* simulate a quantum computer, but not efficiently; i.e., the required resources (time, memory, etc.) do not scale in a polynomial way with the number of input bits. Thus, the Church-Turing thesis (or its variant, the extended Church-Turing thesis) seems incorrect. It is replaced by the "quantum extended Church-Turing thesis", which states that any realistic physical computing device can be efficiently simulated by a quantum computer.

10.9   Classes of Quantum Algorithms

A quantum computer is unlikely to replace a classical computer for all tasks, but only for certain exponentially hard problems. There are currently three classes of quantum algorithms:

(1)   Quantum search algorithms: An example is Grover's algorithm for searching an unstructured database, which provides a polynomial increase in efficiency compared to classical algorithms.
(2)   Algorithms based on the quantum Fourier transform: An example is Shor's algorithm for factoring, which provides an exponential increase in efficiency.
(3)   Quantum simulation: Quantum systems used to simulate other quantum systems or exponentially hard problems; for example, the Ising model.

We will examine these in the upcoming chapters.

10.10   Quantum Advantage

"Quantum advantage" or "quantum supremacy" is a term coined by the physicist, John Preskill in 2011 (Fig. 10.7) [2]. It refers to quantum computer hardware that is demonstrably better than any classical computer. Google has already claimed quantum advantage with 53 qubits [3]. Today, we are in the era of noisy intermediate-scale quantum computing (NISQ), a term also coined by John Preskill [4], where the goal is to perform quantum algorithms and quantum simulations that achieve quantum advantage without full quantum error correction (see Chap. 25).

References

1. This file is licensed under the Creative Commons Attribution 3.0 Unported license. https://creativecommons.org/licenses/by/3.0/deed.en. File: David Deutsch.jpg. (2020, December 2). *Wikimedia Commons, the free media repository.* Retrieved 15:11, Dec 7, 2020 from https://commons.wikimedia.org/w/index.php?title=File:David_Deutsch.jpg&oldid=516285087
2. J. Preskill, Quantum computing and the entanglement frontier (2012). arXiv:1203.5813 [quant-ph]
3. F. Arute et al., Nature **574**, 505 (2019)
4. J. Preskill, Quantum **2**, 79 (2018)
5. Author: Randomaccount136631. This file is licensed under the Creative Commons Attribution-Share Alike 3.0 Unported license. https://creativecommons.org/licenses/by-sa/3.0/deed.en. File: John Preskill 2.jpg. (2020, Oct 21). *Wikimedia Commons, the free media repository.* Retrieved 15:16, Dec 7, 2020 from https://commons.wikimedia.org/w/index.php?title=File:John_Preskill_2.jpg&oldid=496587501

Chapter 11
Deutsch Algorithm

In 1992, David Deutsch (Fig. 10.4) and Richard Jozsa discovered the first problem that a quantum computer could solve in fewer steps than a classical computer [1]. The quantum algorithm, known as Deutsch's algorithm or the Deutsch-Jozsa algorithm, has no practical significance, but it demonstrates in principle how quantum computers can be better than classical computers.

11.1 Classical Unary Operator

To understand Deutsch's algorithm, we will first look at quantum versions of classical circuits. Let us begin with the simplest possible function, a unary operator. A unary operator is a function that takes 0 or 1 as input and generates 0 or 1 as output, as shown in Fig. 11.1. There are only four classical unary operators or functions, as shown in Fig. 11.2. The four unary operators are the constant 0 operator (f_0), the constant 1 operator (f_1), the NOT operator (f_2), and the identity operator (f_3). The truth tables for these operators are shown in Fig. 11.2. The output of the constant 0 operator is 0, regardless of whether the input is 0 or 1; i.e., $f_0(0) = 0$ and $f_0(1) = 0$. The output of the constant 1 operator is 1, regardless of whether the input is 0 or 1; i.e., $f_1(0) = 1$ and $f_1(1) = 1$. The output of the NOT operator is $f_2(0) = 1$ and $f_2(1) = 0$. Finally, the output of the identity operator is the same as the input; i.e., $f_2(0) = 0$ and $f_2(1) = 1$.

The matrix of each unary operator in the computational basis is shown in Fig. 11.2. The matrix for f_0 and f_1 are not unitary and therefore not reversible. You can check that the condition for unitarity, $UU^\dagger = I$, is not met by the f_0 and f_1 matrices, and are therefore not reversible. The irreversibility of f_0 and f_1 is also obvious from the truth tables. For f_0 with a constant output of 0, we cannot tell if the input was 0 or 1. Likewise, for f_1 with a constant output of 1, we cannot tell if the input was 0 or 1. The NOT operator and the identity operator are, however, unitary and reversible —given the output, we can determine the input.

© The Author(s), under exclusive license to Springer Nature Switzerland AG 2021   149
R. LaPierre, *Introduction to Quantum Computing*, The Materials Research Society Series,
https://doi.org/10.1007/978-3-030-69318-3_11

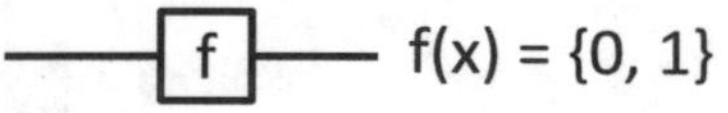

$$\underline{\quad}\boxed{f}\underline{\quad} \quad f(x) = \{0, 1\}$$

Fig. 11.1 A classical unary operator

	The constant 0 operator		The constant 1 operator		The NOT operator		The identity operator	
Truth tables:	**x**	**f_0**	**x**	**f_1**	**x**	**f_2**	**x**	**f_3**
	0	0	0	1	0	1	0	0
	1	0	1	1	1	0	1	1

Operator matrices:	$\begin{pmatrix} 1 & 1 \\ 0 & 0 \end{pmatrix}$	$\begin{pmatrix} 0 & 0 \\ 1 & 1 \end{pmatrix}$	$\begin{pmatrix} 0 & 1 \\ 1 & 0 \end{pmatrix}$	$\begin{pmatrix} 1 & 0 \\ 0 & 1 \end{pmatrix}$
	Not reversible	Not reversible	Reversible	Reversible
	Not unitary	Not unitary	Unitary	Unitary

Fig. 11.2 Truth tables and matrices for the four classical unary operators. Only the NOT and identity operators are unitary and reversible. The constant 0 and constant 1 operator are not unitary and are not reversible

In general, the classical unary operator is not reversible (not unitary); i.e., given the output, we have no way of knowing what the input was. As we saw in Chap. 7, the quantum version for this function must be unitary and therefore reversible. How do we implement a quantum version of f_0, f_1, f_2 and f_3?

Figure 11.3 shows the quantum circuit that gives a unitary implementation of the unary function f (f could be f_0, f_1, f_2 or f_3). The two-qubit input is $|x\rangle|y\rangle$ and the two-qubit output is $|x\rangle|y \oplus f(x)\rangle$ where $\oplus$ signifies binary addition, or addition modulo 2, which was introduced in Chap. 6.

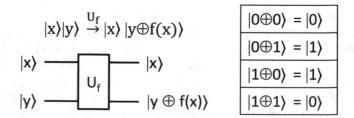

Fig. 11.3 Quantum implementation of the unary function, f, and the binary addition table

Table. 11.1 Input and output from the quantum circuit (Fig. 11.3) for unary operator, f_0

Input $\lvert x \rangle \lvert y \rangle$	Output $\lvert x \rangle \lvert y \oplus f_0(x) \rangle$
$\lvert 00 \rangle$	$\lvert 00 \rangle$
$\lvert 01 \rangle$	$\lvert 01 \rangle$
$\lvert 10 \rangle$	$\lvert 10 \rangle$
$\lvert 11 \rangle$	$\lvert 11 \rangle$

Exercise 11.1 What is the output for each possible input of $\lvert 00 \rangle$, $\lvert 01 \rangle$, $\lvert 10 \rangle$ and $\lvert 11 \rangle$ in Fig. 11.3 for each function f_0, f_1, f_2 and f_3?

For example, Table 11.1 shows the output $\lvert x \rangle \lvert y \oplus f_0(x) \rangle$ for the case of the constant 0 operator, f_0. It is easily shown that the matrix for f_0 is:

$$U_{f_0} = \begin{pmatrix} 1 & 0 & 0 & 0 \\ 0 & 1 & 0 & 0 \\ 0 & 0 & 1 & 0 \\ 0 & 0 & 0 & 1 \end{pmatrix} \tag{11.1}$$

which is the 4×4 identity matrix. The matrix for U_{f_0} in Eq. (11.1) operates on the column vector corresponding to the two-qubit basis states $\lvert 00 \rangle$, $\lvert 01 \rangle$, $\lvert 10 \rangle$ or $\lvert 11 \rangle$; i.e., $(1000)^T, (0100)^T, (0010)^T$ or $(0001)^T$. For example, the output from the quantum circuit for an input of $\lvert 00 \rangle$ will be:

$$U_{f_0} \lvert 00 \rangle = \begin{pmatrix} 1 & 0 & 0 & 0 \\ 0 & 1 & 0 & 0 \\ 0 & 0 & 1 & 0 \\ 0 & 0 & 0 & 1 \end{pmatrix} \begin{pmatrix} 1 \\ 0 \\ 0 \\ 0 \end{pmatrix} = \begin{pmatrix} 1 \\ 0 \\ 0 \\ 0 \end{pmatrix} = \lvert 00 \rangle \tag{11.2}$$

which matches Table 11.1. In a similar manner, you should verify that the other possible inputs give the outputs in Table 11.1 (Exercise 11.3). Of course, this case is simple because U_{f_0} is simply the identity matrix. You can easily show that U_{f_0} is unitary and therefore reversible (Exercise 11.2).

Exercise 11.2 Check that the U_{f_0} matrix is unitary and can therefore be implemented in a quantum circuit.

Exercise 11.3 Check that U_{f_0} gives the expected output for each possible input.

Exercise 11.4 Show that the following matrices implement f_1, f_2 and f_3, and show they are unitary. Note the short-hand notation in the right-hand matrices, where I is the 2×2 identity matrix, X is the 2×2 matrix for the X or NOT gate, and "0" is a 2×2 matrix with all elements equal to zero.

$$U_{f_1} = \begin{pmatrix} 0 & 1 & 0 & 0 \\ 1 & 0 & 0 & 0 \\ 0 & 0 & 0 & 1 \\ 0 & 0 & 1 & 0 \end{pmatrix} = \begin{pmatrix} X & 0 \\ 0 & X \end{pmatrix}$$

$$U_{f_2} = \begin{pmatrix} 0 & 1 & 0 & 0 \\ 1 & 0 & 0 & 0 \\ 0 & 0 & 1 & 0 \\ 0 & 0 & 0 & 1 \end{pmatrix} = \begin{pmatrix} X & 0 \\ 0 & I \end{pmatrix}$$

$$U_{f_3} = \begin{pmatrix} 1 & 0 & 0 & 0 \\ 0 & 1 & 0 & 0 \\ 0 & 0 & 0 & 1 \\ 0 & 0 & 1 & 0 \end{pmatrix} = \begin{pmatrix} I & 0 \\ 0 & X \end{pmatrix}$$

The exact quantum circuit will depend on the function f. As we saw in Chap. 7, we know that any classical circuit can be implemented by a quantum circuit. Since we are interested here in the quantum algorithm and not the detailed circuit implementation, we will leave the quantum circuit as a "black box", which is usually called an "Oracle". Most quantum circuits use an Oracle, which abstracts away the details of implementing some function f. You can think of the Oracle as a program subroutine. As we will see, due to quantum parallelism, the quantum algorithm reduces the number of calls needed to the Oracle as compared to the classical algorithm.

11.2 Deutsch's Problem

Based on the output, the four unary functions may be classified as "constant" (same output for $x = 0$ or 1) or "balanced" (different output for $x = 0$ or 1), as shown in Fig. 11.4. Suppose we are given some operator U_f and asked to determine if f is balanced or constant. We do not know which of the four functions, f, that U_f

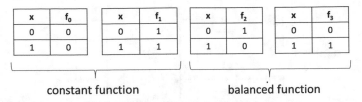

x	f_0		x	f_1		x	f_2		x	f_3
0	0		0	1		0	1		0	0
1	0		1	1		1	0		1	1

constant function   balanced function

Fig. 11.4 The four unary functions may be classified as constant or balanced

implements. We simply want to know if f is constant or balanced. This is the problem that Deutsch's algorithm solves.

To determine if f is constant or balanced, we can compute $f(0) \oplus f(1)$. If f is constant, then $f(0) \oplus f(1) = 0$; otherwise, if f is balanced, then $f(0) \oplus f(1) = 1$. Classically, we need two applications of U_f (called "queries") in Fig. 11.1 to determine $f(0) \oplus f(1)$; i.e., we need to determine $f(0)$ and $f(1)$. In the quantum circuit of Fig. 11.3, we will see that we only need one query to determine $f(0) \oplus f(1)$, by taking advantage of quantum superposition.

In the quantum circuit, we can compute $f(0)$ with an input of $|x\rangle|y\rangle = |0\rangle|0\rangle$, as shown in Fig. 11.5. The output will be $|x\rangle|y \oplus f(x)\rangle = |0\rangle|0 \oplus f(0)\rangle = |0\rangle|f(0)\rangle$, giving us $f(0)$ on the second qubit, as desired. Similarly, the quantum circuit can compute $f(1)$ with an input of $|x\rangle|y\rangle = |1\rangle|0\rangle$. The output will be $|x\rangle|y \oplus f(x)\rangle = |0\rangle|0 \oplus f(1)\rangle = |0\rangle|f(1)\rangle$, giving us $f(1)$ on the second qubit, as desired.

Now let us see what happens if we input $|x\rangle = |0\rangle$ and $|x\rangle = |1\rangle$ simultaneously as a superposition for the first qubit. As shown in Fig. 11.6, we create a superposition by using a Hadamard gate. The resulting input into the Oracle

Fig. 11.5 Computation of $f(0)$ and $f(1)$ in the quantum circuit

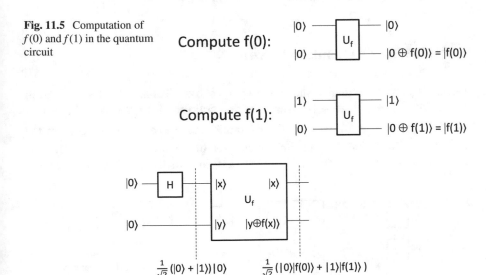

Fig. 11.6 Circuit implementing quantum parallelism on a single qubit

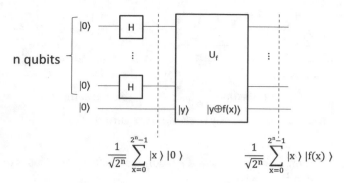

Fig. 11.7 Circuit implementing quantum parallelism on n qubits

is $\frac{1}{\sqrt{2}}(|0\rangle + |1\rangle)|0\rangle = \frac{1}{\sqrt{2}}(|0\rangle|0\rangle + |1\rangle|0\rangle)$. The output is $|x\rangle|y \oplus f(x)\rangle$ applied to each term of the superposition, giving $\frac{1}{\sqrt{2}}(|0\rangle|f(0)\rangle + |1\rangle|f(1)\rangle)$. We have evaluated $f(0)$ and $f(1)$ simultaneously. However, upon measurement we will obtain either $|0\rangle|f(0)\rangle$ or $|1\rangle|f(1)\rangle$ with equal probability. Thus, the values of both $f(0)$ and $f(1)$ are not accessible due to the measurement problem. We need another trick, which we will cover in the next section.

Note that we can extend Fig. 11.3 to n-qubits (Fig. 11.7) to evaluate a classical function $f(x)$ at many different values of x. The input is $|0\rangle^{\otimes n}|0\rangle$. After the n Hadamards, we obtain an equal superposition of the first n qubits, giving $\frac{1}{\sqrt{2^n}}\sum_{x=0}^{2^n-1}|x\rangle|0\rangle$. After the U_f transformation, we obtain $\frac{1}{\sqrt{2^n}}\sum_{x=0}^{2^n-1}|x\rangle|f(x)\rangle$. Hence, Fig. 11.7 is a refinement of Fig. 10.3. The last $|0\rangle$ input qubit is known as an ancillary qubit. It is required to make U_f unitary.

11.3 Deutsch's Quantum Circuit

Deutsch's quantum circuit is shown in Fig. 11.8. Two Hadamard gates are used to create the $|+\rangle = \frac{1}{\sqrt{2}}(|0\rangle + |1\rangle)$ superposition from the $|0\rangle$ qubit, and the $|-\rangle = \frac{1}{\sqrt{2}}(|0\rangle - |1\rangle)$ superposition from the $|1\rangle$ qubit. These two superpositions are the input to the Oracle (the quantum implementation of the unary function, f). The first qubit of the output undergoes a Hadamard transformation followed by measurement.

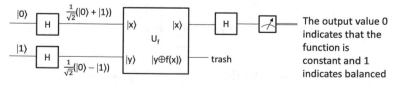

Fig. 11.8 Deutsch's quantum circuit

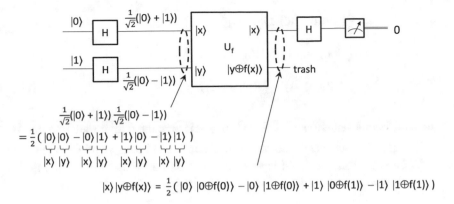

Fig. 11.9 Analysis of Deutsch's quantum circuit

The second qubit of the output is discarded (trash). An output value of 0 on the first qubit indicates that the function f is constant, and an output of 1 indicates that f is balanced. Let us prove this assertion for each of the possible unary functions f_0, f_1, f_2 and f_3.

Figure 11.9 shows a more detailed analysis of the input and output from Deutsch's quantum circuit. After the Hadamard gates, the two-qubit input state is:

$$H|0\rangle \otimes H|1\rangle = \frac{1}{\sqrt{2}}(|0\rangle + |1\rangle) \otimes \frac{1}{\sqrt{2}}(|0\rangle - |1\rangle)$$

$$= \frac{1}{2}(|00\rangle - |01\rangle + |10\rangle - |11\rangle) \tag{11.3}$$

which is the input to U_f; i.e., each of these four terms constitutes an input $|xy\rangle$ to U_f. By the principle of linear superposition, the output from U_f is determined by evaluating $|x\rangle|y \oplus f(x)\rangle$ for each of the four terms in Eq. (11.3):

$$H|0\rangle \otimes H|1\rangle \xrightarrow{U_f} \frac{1}{2}(|0\rangle|0 \oplus f(0)\rangle - |0\rangle|1 \oplus f(0)\rangle$$

$$+ |1\rangle|0 \oplus f(1)\rangle - |1\rangle|1 \oplus f(1)\rangle) \tag{11.4}$$

Now we simply need to evaluate Eq. (11.4) for each of the unary functions according to the truth tables in Fig. 11.2. For example, for the constant 0 unary function, f_0, we have $f_0(0) = 0$ and $f_0(1) = 0$. Hence, Eq. (11.4) gives:

$$H|0\rangle \otimes H|1\rangle \xrightarrow{U_f} \frac{1}{2}(|0\rangle|0 \oplus f_0(0)\rangle - |0\rangle|1 \oplus f_0(0)\rangle$$

$$+ |1\rangle|0 \oplus f_0(1)\rangle - |1\rangle|1 \oplus f_0(1)\rangle)$$

$$= \frac{1}{2}(|0\rangle|0 \oplus 0\rangle - |0\rangle|1 \oplus 0\rangle + |1\rangle|0 \oplus 0\rangle - |1\rangle|1 \oplus 0\rangle)$$

$$= \frac{1}{2}(|0\rangle|0\rangle - |0\rangle|1\rangle + |1\rangle|0\rangle - |1\rangle|1\rangle)$$

$$= \frac{1}{2}(|00\rangle - |01\rangle + |10\rangle - |11\rangle)$$

$$= \frac{1}{\sqrt{2}}(|0\rangle + |1\rangle)\frac{1}{\sqrt{2}}(|0\rangle - |1\rangle) \qquad (11.5)$$

Hence, the output state of qubit 1 is $\frac{1}{\sqrt{2}}(|0\rangle + |1\rangle)$, and the output state of qubit 2 is $\frac{1}{\sqrt{2}}(|0\rangle - |1\rangle)$. We see that the output from U_{f0} in Eq. (11.5) is identical to the input in Eq. (11.3). In fact, we could have saved ourselves a lot of work by noting that $f_0(x)$ is always 0, and thus the output from U_{f0} is:

$$|x\rangle|y \oplus f_0(x)\rangle = |x\rangle|y \oplus 0\rangle = |x\rangle|y\rangle \qquad (11.6)$$

which is identical to the input, $|x\rangle|y\rangle$. The output state of qubit 2 is completely independent of the state of qubit 1 (there is no entanglement). At this point, qubit 2 is discarded (trash) and qubit 1 is put through a Hadamard gate:

$$H|x\rangle = H\frac{1}{\sqrt{2}}(|0\rangle + |1\rangle) = |0\rangle \qquad (11.7)$$

If we measure, we will definitely obtain 0. The output of 0 indicates that the function f_0 is constant, as stated earlier. You will find the same result for the constant 1 operator, f_1 (try it).

11.4 Phase Kick-Back

Now let us examine the case for the NOT operator, f_2. Something unusual happens. Starting from Eq. (11.4), and using the truth table in Fig. 11.2 for f_2, we get:

$$H|0\rangle \otimes H|1\rangle \xrightarrow{U_f} \frac{1}{2}(|0\rangle|0 \oplus f_2(0)\rangle - |0\rangle|1 \oplus f_2(0)\rangle + |1\rangle|0 \oplus f_2(1)\rangle - |1\rangle|1 \oplus f_2(1)\rangle)$$

$$= \frac{1}{2}(|0\rangle|0 \oplus 1\rangle - |0\rangle|1 \oplus 1\rangle + |1\rangle|0 \oplus 0\rangle - |1\rangle|1 \oplus 0\rangle)$$

$$= \frac{1}{2}(|0\rangle|1\rangle - |0\rangle|0\rangle + |1\rangle|0\rangle - |1\rangle|1\rangle)$$

$$= -\frac{1}{\sqrt{2}}(|0\rangle - |1\rangle)\frac{1}{\sqrt{2}}(|0\rangle - |1\rangle) \qquad (11.8)$$

Equation (11.8) contains a negative sign in front, which is a global phase factor and can be ignored (it has no consequences for probabilities). Hence, after U_{f2}, qubit 1 is $\frac{1}{\sqrt{2}}(|0\rangle - |1\rangle)$ and qubit 2 is $\frac{1}{\sqrt{2}}(|0\rangle - |1\rangle)$. The second qubit is discarded (trash). Something interesting has happened to the first qubit. The state of qubit 1 was $\frac{1}{\sqrt{2}}(|0\rangle + |1\rangle)$ before U_{f2}, and the state is $\frac{1}{\sqrt{2}}(|0\rangle - |1\rangle)$ after U_{f2}. A phase factor, $e^{i\pi}$ $= -1$, now appears in the superposition state of qubit 1. This is known as "phase kick-back". Finally, a Hadamard gate is applied to qubit 1, resulting in:

$$H \frac{1}{\sqrt{2}}(|0\rangle - |1\rangle) = |1\rangle \qquad (11.9)$$

If we measure, we will definitely obtain 1. This output of 1 indicates that the function f_2 is balanced. You will find the same result for f_3 (try it).

Exercise 11.5 As in the previous examples, prove that the output of the Deutsch circuit for f_3 is 1, indicating it is balanced.

Although we have analyzed Deutsch's quantum circuit for each of the unary functions, we do not have to know *a priori* which of the four functions, f, that U_f implements. We simply want to know if f is constant or balanced. We need not worry about the detailed implementation of the quantum circuit for U_f since any classical circuit (function f) can be built from some universal set of quantum gates (as we learned in Chap. 7).

The classical computer must call U_f twice to distinguish a balanced function from a constant function, but a quantum computer does the job in one iteration. This is possible because the quantum computer acts on a superposition of $|0\rangle$ and $|1\rangle$. This is "quantum parallelism". Although this speed-up is not very impressive, the Deutsch algorithm was the first quantum algorithm that was theoretically faster than the classical algorithm.

Exercise 11.6 Show that $|0 \oplus f(x)\rangle - |1 \oplus f(x)\rangle = (-1)^{f(x)}(|0\rangle - |1\rangle)$. Hint: You can check that this works for each of the possible values of $f(x) = \{0, 1\}$ and using the rules for binary addition.

Exercise 11.7 Prove the following:

$$\frac{1}{\sqrt{2}}(-1)^{f(0)}|0\rangle + \frac{1}{\sqrt{2}}(-1)^{f(1)}|1\rangle$$

$$= (-1)^{f(0)}\left(\frac{1}{\sqrt{2}}|0\rangle + \frac{1}{\sqrt{2}}(-1)^{f(0)\oplus f(1)}|1\rangle\right)$$

Exercise 11.8 Show that $H\left(\frac{1}{\sqrt{2}}|0\rangle + \frac{1}{\sqrt{2}}(-1)^a|1\rangle\right) = |a\rangle$ for $a = 0$ or 1 where H is the Hadamard transformation.

11.5   General Analysis

Using the results of Exercises 11.6, 11.7 and 11.8, Deutsch's circuit can be analyzed for all possible functions $(f_0, f_1, f_2$ and $f_3)$ in a single calculation. Referring to Fig. 11.10, the input to U_f (step 1) is identical to Eq. (11.3):

$$\text{Step 1:} |\psi_1\rangle = \left[\frac{1}{\sqrt{2}}(|0\rangle + |1\rangle)\right]\left[\frac{1}{\sqrt{2}}(|0\rangle - |1\rangle)\right]$$

$$= \frac{1}{2}|0\rangle(|0\rangle - |1\rangle) + \frac{1}{2}|1\rangle(|0\rangle - |1\rangle) \tag{11.10}$$

At step 2, we get the output from U_f, which is identical to Eq. (11.4):

$$\text{Step 2:} \xrightarrow{U_f} |\psi_2\rangle = \frac{1}{2}|0\rangle(|0 \oplus f(0)\rangle - |1 \oplus f(0)\rangle)$$

$$+ \frac{1}{2}|1\rangle(|0 \oplus f(1)\rangle - |1 \oplus f(1)\rangle) \tag{11.11}$$

From the results of Exercise 11.6, Eq. (11.11) can be written as:

$$\text{Exercise 11.6} \rightarrow |\psi_2\rangle = \frac{1}{2}|0\rangle(-1)^{f(0)}(|0\rangle - |1\rangle) + \frac{1}{2}|1\rangle(-1)^{f(1)}(|0\rangle - |1\rangle)$$

$$= \left[\frac{1}{\sqrt{2}}(-1)^{f(0)}|0\rangle + \frac{1}{\sqrt{2}}(-1)^{f(1)}|1\rangle\right]\left[\frac{1}{\sqrt{2}}(|0\rangle - |1\rangle)\right] \tag{11.12}$$

The first term in square brackets of Eq. (11.12) is the state of qubit 1, while the second term in square brackets is the state of qubit 2. The U_f transformation has not changed

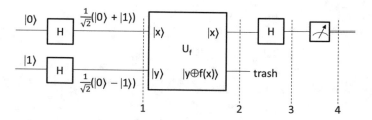

Fig. 11.10 Steps used for a general analysis of Deutsch's circuit

the state of qubit 2. However, the factors $(-1)^{f(0)}$ and $(-1)^{f(1)}$ appear in the state of qubit 1. This phenomenon is known as "phase kick-back", as described earlier. At this point, the second qubit is discarded. Its state is completely independent of the state of the first qubit (no entanglement). According to Exercise 11.7, qubit 1 of Eq. (11.12) can be written:

$$\text{Exercise 11.7} \rightarrow |\psi_2\rangle_1 = \frac{1}{\sqrt{2}}(-1)^{f(0)}|0\rangle + \frac{1}{\sqrt{2}}(-1)^{f(1)}|1\rangle$$

$$= (-1)^{f(0)}\left[\frac{1}{\sqrt{2}}|0\rangle + \frac{1}{\sqrt{2}}(-1)^{f(0)\oplus f(1)}|1\rangle\right] \quad (11.13)$$

Next, the Hadamard transformation is applied to qubit 1. According to Exercise 11.8, the state of qubit 1 at step 3 becomes:

$$|\psi_3\rangle_1 = (-1)^{f(0)}|f(0) \oplus f(1)\rangle \quad (11.14)$$

Finally, after measurement (step 4), the output is $f(0) \oplus f(1)$, which is 0 if f is constant, or 1 if f is balanced.

11.6 Deutsch-Jozsa Algorithm

The Deutsch-Jozsa algorithm is an extension of the Deutsch algorithm for multiple qubits. In this case, the function is $f(x) = \{0, 1\}$ where $x = \{0, 1\}^n$; i.e., f is a function of an n-bit input. For n bits, there are 2^n possible bit sequences; i.e., 2^n possible inputs, x, to $f(x)$ and 2^n corresponding output values (0 or 1) of f. We suppose that the function, f, is either constant (all output bits are either 0 or 1), or balanced (the number of 0's and 1's are equal; i.e., there are 2^{n-1} output bits equal to 0 and 2^{n-1} output bits equal to 1). Like the Deutsch algorithm, the Deutsch-Jozsa algorithm tells us if f is constant or balanced.

The circuit that implements the Deutsch-Jozsa algorithm is illustrated in Fig. 11.11. It is similar to the Deutsch circuit but with multiple (n) qubits in the input register plus the last ancilla qubit (like Fig. 11.7). Using the tensor notation, the circuit can be condensed as shown in Fig. 11.12.

The circuit begins with the input $|0\rangle^{\otimes n}|1\rangle$ in the quantum register. After the $n + 1$ Hadamards, we obtain:

$$\frac{1}{\sqrt{2^{n+1}}} \sum_{x=0}^{2^n-1} |x\rangle(|0\rangle - |1\rangle) \quad (11.15)$$

This is an equally weighted superposition produced from the first n input qubits of $|0\rangle$, and the state $\frac{1}{\sqrt{2}}(|0\rangle - |1\rangle)$ produced from the last $(n + 1)$ input qubit of $|1\rangle$. This is the state at "step 1" in Fig. 11.12. This can be expanded to:

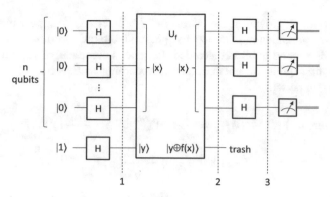

Fig. 11.11 Quantum circuit for the Deutsch-Jozsa algorithm

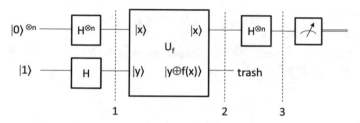

Fig. 11.12 Deutsch-Jozsa circuit using the tensor notation

$$\frac{1}{\sqrt{2^{n+1}}} \sum_{x=0}^{2^n-1} (|x\rangle|0\rangle - |x\rangle|1\rangle) \tag{11.16}$$

In the next step, U_f applies the transformation $|x\rangle|y \oplus f(x)\rangle$ to the input. Thus, after the U_f transformation, we obtain:

$$\frac{1}{\sqrt{2^{n+1}}} \sum_{x=0}^{2^n-1} (|x\rangle|0 + f(x)\rangle - |x\rangle|1 \oplus f(x)\rangle) \tag{11.17}$$

Since $0 + f(x) = f(x)$, we get:

$$\frac{1}{\sqrt{2^{n+1}}} \sum_{x=0}^{2^n-1} |x\rangle(|f(x)\rangle - |1 \oplus f(x)\rangle) \tag{11.18}$$

This is the state at step 2 in Fig. 11.12. From Exercise 11.6, Eq. (11.17) becomes:

$$\frac{1}{\sqrt{2^n}} \sum_{x=0}^{2^n-1} (-1)^{f(x)} |x\rangle \frac{1}{\sqrt{2}} (|0\rangle - |1\rangle) \tag{11.19}$$

The last qubit, $\frac{1}{\sqrt{2}}(|0\rangle - |1\rangle)$, can be ignored (trash), leaving the state of the first n qubits equal to:

$$\frac{1}{\sqrt{2^n}} \sum_{x=0}^{2^n-1} (-1)^{f(x)} |x\rangle \tag{11.20}$$

Next, a Hadamard transformation is applied to each of the qubits in Eq. (11.20), giving (step 3):

$$\frac{1}{\sqrt{2^n}} \sum_{x=0}^{2^n-1} (-1)^{f(x)} \left[\frac{1}{\sqrt{2^n}} \sum_{y=0}^{2^n-1} (-1)^{x \cdot y} |y\rangle \right] \tag{11.21}$$

where we have used Eq. (10.9). From Eq. (11.21), the probability of measuring $|y\rangle = |0\rangle^{\otimes n}$; i.e., all output qubits equal to $|0\rangle$, is:

$$P_0 = \left| \frac{1}{2^n} \sum_{x=0}^{2^n-1} (-1)^{f(x)} \right|^2 \tag{11.22}$$

Equation (11.22) gives 1 if $f(x)$ is all 0's or 1's (i.e., f is constant), or 0 is $f(x)$ is balanced. In other words, if all n measurement results give 0, the function was definitely constant. Any other output state indicates that the function was balanced. Using the Deutsch-Jozsa algorithm, we are able to determine if f is constant or balanced by a single query to the Oracle. Classically, $O(2^n)$ queries would be needed to know if f is balanced or constant. Hence, the Deutsch-Jozsa algorithm gives an exponential speed-up.

Reference

1. D. Deutsch, R. Jozsa, Proc. R. Soc. A **439**, 553 (1992)

Chapter 12
Grover Algorithm

The Grover algorithm is a quantum algorithm that provides a polynomial speedup compared to classical algorithms for searching a database. This algorithm could also be used to solve problems with *NP* complexity.

12.1 Searching a Database

Suppose we want to find the name of the person in a list with phone number 525–9140 (Fig. 12.1). Each item in the list is indexed by the integer $x = \{0, N - 1\}$, and the item we are searching for is $x = x^*$. For a classical algorithm, we may be lucky and find the item we are looking for at the first item in the list with $x^* = 0$, or we may not find the item until the last item in the list with $x^* = N - 1$. Thus, in an unstructured database, an average of $\sim N/2$ (or, using Big O notation, $O(N)$) queries is needed to find the record. More precisely, if there is a uniform probability distribution of $1/N$ that any of the N items in the list is the correct item, then the average number of queries necessary to find the correct item in the list is the weighted average of the number of queries from $i = 1$ to N:

$$\sum_{i=1}^{N} \frac{1}{N} i = \frac{1}{N} \sum_{i=1}^{N} i = \frac{1}{N} \frac{N(N + 1)}{2} = \frac{N + 1}{2} \sim \frac{N}{2} \text{ for large } N \qquad (12.1)$$

There is no better classical algorithm than $O(N)$; i.e., just searching one record at a time. Can we do better with a quantum algorithm?

© The Author(s), under exclusive license to Springer Nature Switzerland AG 2021
R. LaPierre, *Introduction to Quantum Computing*, The Materials Research Society Series,
https://doi.org/10.1007/978-3-030-69318-3_12

Index, x	Item
0	Jones, 123-4567
1	McDonald, 715-3456
2	Smith, 525-9140
3	Reni, 561-1234
...	
N-1	Gupta, 134-2890

This is the item we are searching for, labelled by x=x*=2 → 2

Fig. 12.1 Searching a list of N items. Each item is indexed by the integer $x = \{0, N - 1\}$, and the item we are searching for is labeled by $x = x^*$

12.2 Grover Algorithm

In 1996, the computer scientist, Lov Grover, found a quantum algorithm for finding an item in an unstructured database of N items [1]. Grover found a quantum search algorithm that needs $O(\sqrt{N})$ steps; i.e., a quadratic speed-up compared to the $O(N)$ steps required classically. For example, suppose $N = 10^9$. The classical algorithm would require on the order of 10^9 steps, while the quantum algorithm would require on the order of $\sqrt{N} = \sqrt{10^9} \sim 10^4$ steps. This is a significant speed-up (although only a polynomial speed-up, not an exponential speed-up). It is possible to prove that Grover is the best possible quantum search algorithm—you cannot do any better than $O(\sqrt{N})$.

Figure 12.2 illustrates the principle behind Grover's algorithm. The N elements in the list are indexed (labelled) from $x = 0$ to $N - 1$. n is the number of qubits, and $N = 2^n$ is the number of items in the list. Due to superposition, only n qubits are needed to represent $N = 2^n$ items in the list. A quantum state, $|x\rangle$, is represented by a column vector of $N = 2^n$ probability amplitudes. Each entry in the column vector is associated with the corresponding index x; i.e., α_0 is the probability amplitude associated with the first item in the list indexed by $x = 0$, α_1 is the probability

List:

Index, x	Item
0	item 0
1	item 1
2	item 2
⋮	item 3
N - 1	item N-1

Probability amplitude of finding item in list:

$$\rightarrow |x\rangle = \begin{pmatrix} \alpha_0 \\ \alpha_1 \\ \vdots \\ \alpha_{N-1} \end{pmatrix} \rightarrow \alpha_0|0\rangle + \alpha_1|1\rangle + \cdots$$

$N = 2^n$

Fig. 12.2 Principle of Grover's algorithm

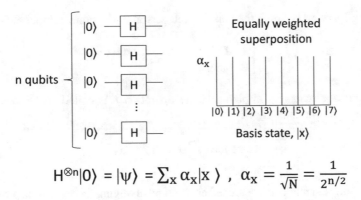

$$H^{\otimes n}|0\rangle = |\psi\rangle = \sum_x \alpha_x|x\rangle \ , \ \alpha_x = \frac{1}{\sqrt{N}} = \frac{1}{2^{n/2}}$$

Fig. 12.3 Initialization of Grover's algorithm. The probability amplitude plot shows an equally weighted superposition for $n = 3$

amplitude associated with the second item in the list indexed by $x = 1$, etc. We want the probability amplitude for the item we are searching for, with index $x = x^*$, to approach 1, and all other probability amplitudes to approach 0. This is what Grover's algorithm does.

12.3   Initialization

Like most quantum algorithms, Grover's algorithm begins by creating an equally weighted superposition of basis states. We already know how to do this from Chap. 10, using the Hadamard gate applied to each of the n qubits:

$$|\psi\rangle = H^{\otimes n}|0\rangle = \sum_{x=0}^{N-1} \alpha_x|x\rangle, \alpha_x = \frac{1}{\sqrt{N}} = \frac{1}{2^{n/2}} \qquad (12.2)$$

An example is shown in Fig. 12.3 for $n = 3$ qubits, where there are $N = 2^n = 2^3 = 8$ possible basis states with equal amplitudes $\alpha_x = \frac{1}{\sqrt{N}} = \frac{1}{\sqrt{8}}$. Grover's algorithm has been demonstrated by nuclear magnetic resonance (NMR) with $N = 8$ [2]. Of course, there would not be much benefit to using a quantum algorithm to search a list of only 8 items, but $N = 8$ serves to illustrate the basic idea behind the algorithm.

12.4   The Oracle

The item we are searching for in the list is labelled by the index $x = x^*$ (Fig. 12.1). We can construct some function $f(x)$ that tells us if we have found the item we are searching for; e.g., Is Smith the name of the person with phone number 525–9140?:

$$f(x) = \begin{cases} 1 & \text{if } x \text{ is the correct state} (x = x*) \\ 0, & \text{otherwise} \end{cases} \tag{12.3}$$

The quantum circuit that implements the operation f is called an Oracle, with a unitary operator denoted by O:

$$O|\psi\rangle = \sum_x \alpha_x (-1)^{f(x)} |x\rangle \tag{12.4}$$

The function $f(x)$ marks the searched element by changing the sign of its probability amplitude. If $f(x) = 0$, nothing happens. If $f(x) = 1$, the sign of the probability amplitude changes. We could make this operation reversible, so it could be implemented on a quantum computer.

For example, we can prepare a 3-qubit system in an equally weighted super-position state, $|\psi\rangle = \frac{1}{\sqrt{8}} (|0\rangle + |1\rangle + |2\rangle + |3\rangle + |4\rangle + |5\rangle + |6\rangle + |7\rangle)$, as shown in Fig. 12.3. Suppose the solution to the search problem is the item marked by $|x\rangle = |x*\rangle = |4\rangle$. Appying the Oracle to $|\psi\rangle$ would give $O|\psi\rangle = \frac{1}{\sqrt{8}}$ $(|0\rangle + |1\rangle + |2\rangle + |3\rangle - |4\rangle + |5\rangle + |6\rangle + |7\rangle)$. The state $|4\rangle$ is marked by a -1 phase. This is depicted in Fig. 12.4.

Although finding a solution to the search problem is hard, recognizing a solution is easy. The Oracle simply indicates if the correct answer is presented to it. Different search problems require different Oracles, so the Oracle is presented as a "black box" without specifying its circuit. Think of the Oracle as a subroutine that can be repeatedly queried. How many queries are necessary before the desired value $x*$ is found? We will see that only $O(\sqrt{N})$ is needed in the quantum algorithm, while $O(N)$ is needed classically.

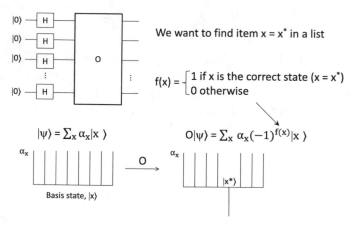

Fig. 12.4 The Oracle operation of Grover's algorithm

The key to quantum search is that we can look at all solutions simultaneously (quantum parallelism). The Oracle manipulates the probability amplitudes using a unitary operator, but the Oracle by itself is not sufficient. Changing the sign of the probability amplitude does not change the probability of the search element since probability is the square of the modulus of the probability amplitude. The next step should be to increase the amplitude of the search element while decreasing the amplitude of the others, so that a measurement will give the solution with high probability.

12.5 Amplitude Amplification

The next step of the Grover algorithm is called "inversion about the mean", which transforms the amplitude of each state so that it is as far above the average as it was below the average prior to the transformation (i.e., it flips each probability amplitude about the average). This is also known as "amplitude amplification", and is represented by the unitary transformation, D, called the "diffusion transform" (Fig. 12.5). The diffusion transform derives its name from a generalization of the mathematics of diffusion [3].

Repeated application of the Oracle (O) and the diffusion transform (D) will further amplify the desired state $|x^*\rangle$ as illustrated in Fig. 12.6. The amplitude increases by ~ $2/\sqrt{N}$ with each iteration. After $O(\sqrt{N})$ steps, the probability amplitude of the target state $|x^*\rangle$ approaches 1, while the other amplitudes shrink. A measurement will then hit $|x^*\rangle$ with high probability. It will be shown later, using a geometrical interpretation, that the number of required iterations is ~ $\frac{\pi}{4}\sqrt{2^n}$ ~ $O(\sqrt{N})$. We can't do any better than $O(\sqrt{N})$. Due to the probabilistic nature of quantum measurement, we could run Grover's algorithm twice and obtain different results. Grover's algorithm just succeeds with high probability.

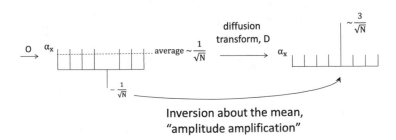

Inversion about the mean, "amplitude amplification"

Fig. 12.5 Inversion about the mean or amplitude amplification

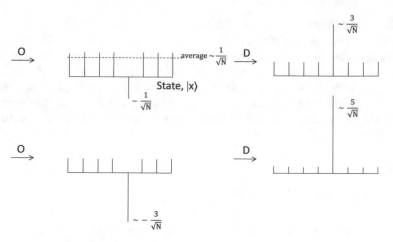

Fig. 12.6 Repeated application of the Oracle, O, and the diffusion transform, D

12.6 Diffusion Transform

We can write the diffusion transform (inversion about the mean) as follows:

$$\sum_x \alpha_x |x\rangle \overset{D}{\to} \sum_x (\bar{\alpha} + (\bar{\alpha} - \alpha_x))|x\rangle = \sum_x (2\bar{\alpha} - \alpha_x)|x\rangle \tag{12.5}$$

inversion about the mean

difference of each probability amplitude α_x from the mean $\bar{\alpha}$

$\bar{\alpha} = \frac{1}{N}\Sigma_x \alpha_x$ = average probability amplitude

The matrix operator for the diffusion transform is:

$$\underbrace{\begin{pmatrix} 2/N-1 & 2/N & \cdots & 2/N \\ 2/N & 2/N-1 & \cdots & 2/N \\ \vdots & \vdots & \ddots & \vdots \\ 2/N & 2/N & \cdots & 2/N-1 \end{pmatrix}}_{D} \begin{pmatrix} \alpha_0 \\ \vdots \\ \alpha_x \\ \vdots \\ \alpha_{N-1} \end{pmatrix}$$

$$= \begin{pmatrix} \frac{2}{N}(\alpha_0 + \alpha_1 + \cdots \alpha_{N-1}) - \alpha_0 \\ \frac{2}{N}(\alpha_0 + \alpha_1 + \cdots \alpha_{N-1}) - \alpha_1 \\ \vdots \\ \frac{2}{N}\sum_{x=0}^{N-1} \alpha_x - \alpha_x \\ \vdots \\ \frac{2}{N}(\alpha_0 + \alpha_1 + \cdots \alpha_{N-1}) - \alpha_{N-1} \end{pmatrix} \qquad (12.6)$$

Equation (12.6) shows that if we multiply any row vector in the matrix D by an input qubit column vector, then we get twice the mean minus α_x for each vector element; i.e., we achieve inversion about the mean.

Exercise 12.1 Show that D is unitary, and therefore can be implemented as a quantum circuit.

12.7   Circuit for the Diffusion Transform

We start with a quantum circuit that implements a "conditional phase shift"; i.e., it leaves $|0\rangle$ unchanged and changes all other states $|x\rangle \neq |0\rangle$ to $-|x\rangle$:

$$\text{Conditional phase shift} : |0\rangle \rightarrow |0\rangle \qquad (12.7)$$

$$|x\rangle \rightarrow -|x\rangle \text{ if } x \neq 0 \qquad (12.8)$$

The circuit that accomplishes the conditional phase shift is shown in Fig. 12.7. First, the quantum register is initialized to $|0\rangle$:

Fig. 12.7 Quantum circuit for the conditional phase shift

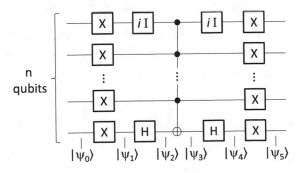

$$|\psi_0\rangle = |0\rangle|0\rangle \cdots |0\rangle|0\rangle \tag{12.9}$$

The NOT (X) gates transform the state to:

$$|\psi_1\rangle = |1\rangle|1\rangle \cdots |1\rangle|1\rangle \tag{12.10}$$

The $i\,I$ and H gate on the first and last qubit, respectively, transforms the state to:

$$|\psi_2\rangle = i|1\rangle|1\rangle \cdots |1\rangle|-\rangle \tag{12.11}$$

where $H|1\rangle = |-\rangle = \frac{1}{\sqrt{2}}(|0\rangle - |1\rangle)$. The multi-control CNOT gate transforms the state to:

$$\begin{aligned}|\psi_3\rangle &= i|1\rangle|1\rangle \cdots |1\rangle(-|-\rangle) \\ &= -i|1\rangle|1\rangle \cdots |1\rangle|-\rangle \end{aligned} \tag{12.12}$$

where we have used the fact that NOT $|-\rangle = -|-\rangle$. A second application of the $i\,I$ and H gate on the first and last qubit, respectively, transforms the state to:

$$\begin{aligned}|\psi_4\rangle &= -i \cdot i|1\rangle|1\rangle \cdots |1\rangle|1\rangle \\ &= |1\rangle|1\rangle \cdots |1\rangle|1\rangle \end{aligned} \tag{12.13}$$

Finally, the NOT gates return the state to $|0\rangle$:

$$|\psi_5\rangle = |0\rangle|0\rangle \cdots |0\rangle|0\rangle \tag{12.14}$$

Exercise 12.2 shows that any other input state, $|x\rangle$, where $x \neq 0$, results in an output of $-|x\rangle$.

Exercise 12.2 Show that the circuit of Fig. 12.7 implements an output of $-|x\rangle$ for an input of $|x\rangle$ when $x \neq 0$ (i.e., Eq. (12.8)).

For the conditional phase shift, we want a matrix which multiplies every state except $|0\rangle$ by -1. The $N \times N$ matrix ($N = 2^n$) that does this is:

$$\begin{pmatrix} 1 & 0 & \cdots & 0 \\ 0 & -1 & \cdots & 0 \\ \vdots & \vdots & \ddots & \vdots \\ 0 & 0 & \cdots & -1 \end{pmatrix} = \begin{pmatrix} 2 & & 0 \\ & 0 & \\ & & \ddots \\ 0 & & 0 \end{pmatrix} - I \tag{12.15}$$

where I is the $N \times N$ identity matrix.

Finally, let us see what happens when we apply a Hadamard transformation on each of the input and output qubits of the conditional phase shift:

$$D = H^{\otimes n} \left[\begin{pmatrix} 2 & 0 & \cdots & 0 \\ 0 & 0 & \cdots & 0 \\ \vdots & \vdots & \ddots & \vdots \\ 0 & 0 & \cdots & 0 \end{pmatrix} - I \right] H^{\otimes n} \tag{12.16}$$

$$= H^{\otimes n} \begin{pmatrix} 2 & 0 & \cdots & 0 \\ 0 & 0 & \cdots & 0 \\ \vdots & \vdots & \ddots & \vdots \\ 0 & 0 & \cdots & 0 \end{pmatrix} H^{\otimes n} - I \tag{12.17}$$

In Eq. (12.17), we have used the fact that $H^{\otimes n} I H^{\otimes n} = H^{\otimes n} H^{\otimes n} = I$ because H is its own inverse. Next, applying the Hadamard gates on the left gives:

$$D = \frac{1}{\sqrt{N}} \begin{pmatrix} 2 & 0 & \cdots & 0 \\ 2 & 0 & \cdots & 0 \\ \vdots & \vdots & \ddots & \vdots \\ 2 & 0 & \cdots & 0 \end{pmatrix} H^{\otimes n} - I \tag{12.18}$$

and applying the Hadamard gates on the right gives:

$$D = \frac{1}{N} \begin{pmatrix} 2 & 2 & \cdots & 2 \\ 2 & 2 & \cdots & 2 \\ \vdots & \vdots & \ddots & \vdots \\ 2 & 2 & \cdots & 2 \end{pmatrix} - I \tag{12.19}$$

$$= \begin{pmatrix} 2/N - 1 & 2/N & \cdots & 2/N \\ 2/N & 2/N - 1 & \cdots & 2/N \\ \vdots & \vdots & \ddots & \vdots \\ 2/N & 2/N & \cdots & 2/N - 1 \end{pmatrix} \tag{12.20}$$

You can easily prove Eqs. (12.18) and (12.19) by direct application of the Hadamard transformations (Exercise 12.3). The resulting matrix, Eq. (12.20), is the same matrix we saw before for the diffusion transform in Eq. (12.6). Thus, the diffusion transform circuit is the conditional phase shift circuit of Fig. 12.7 sandwiched by Hadamard gates.

Exercise 12.3 Prove Eqs. (12.18) and (12.19).

12.8 Solving NP Problems

Perhaps the most exciting aspect of Grover's algorithm is that it can be used for much more than just searching a database. We could solve any problem where finding a solution is hard, but recognising it is easy—that's what the Oracle does. As long as we can construct an Oracle, we can use Grover's algorithm. Hence, Grover's algorithm can be used to solve *NP* problems, such as the traveling salesperson problem (TSP) presented in Chap. 10. The TSP problem scales exponentially with N where N is the number of cities. We could construct an algorithm (the Oracle) to check whether a tour is a valid solution, and then run the Grover algorithm with $O(\sqrt{e^N})$ to explore the different paths simultaneously. This is faster but still exponential.

12.9 Geometrical Interpretation

Here, we present the so-called "geometrical interpretation" of Grover's algorithm. This analysis is a bit difficult, so consider this section as optional. It is, however, a more rigorous solution to determining the number of required iterations in Grover's algorithm.

The conditional phase shift can be represented in Dirac notation as $2|0\rangle\langle 0| - I$. This is easy to show as follows. First, if the input is $|x\rangle = 0$, then:

$$[2|0\rangle\langle 0| - I]|0\rangle = 2|0\rangle\langle 0|0\rangle - I|0\rangle = |0\rangle \tag{12.21}$$

since $\langle 0|0\rangle = 1$. For any other input, $|x\rangle \neq 0$, we have:

$$[2|0\rangle\langle 0| - I]|x\rangle = 2|0\rangle\langle 0|x\rangle - I|x\rangle = -|x\rangle \tag{12.22}$$

since $\langle 0|x\rangle = 0$ for $x \neq 0$. Equations (12.21) and (12.22) correspond to Eqs. (12.7) and (12.8), respectively. Alternatively, we could evaluate the operator $2|0\rangle\langle 0| - I$ by substituting $(1\,0\,0\ldots 0)^{\mathrm{T}}$ and $(1\,0\,0\ldots 0)$ for $|0\rangle$ and $\langle 0|$, respectively, and evaluating the tensor product in $2|0\rangle\langle 0| - I$. You can easily convince yourself that you will obtain Eq. (12.15); i.e., the matrix for the conditional phase shift (Exercise 12.4).

Exercise 12.4 Using the tensor product of the vectors $|0\rangle$ and $\langle 0|$, show the conditional phase shift is $2|0\rangle\langle 0| - I$.

Next, we apply Hadamard transforms on the left and right of the conditional phase shift as we did in Eq. (12.16):

$$D = H^{\otimes n}[2|0\rangle\langle 0| - I]H^{\otimes n} \tag{12.23}$$

$$= 2H^{\otimes n}|0\rangle\langle 0|H^{\otimes n} - I \tag{12.24}$$

$$= 2|\psi\rangle\langle\psi| - I \tag{12.25}$$

where $H^{\otimes n}|0\rangle = |\psi\rangle$ is the equally weighted superposition in Eq. (12.1), and $\langle\psi| = \langle 0|H^{\otimes n}$ since $H = H^\dagger$. Thus, we have proved that the diffusion transform can be written in the alternative form, $D = 2|\psi\rangle\langle\psi| - I$.

The initial state, $|\psi\rangle$, is a superposition of all basis states, Eq. (12.1). The target state (the item we are seeking) is $|x^*\rangle$, one particular basis state; i.e., we want to transform $|\psi\rangle$ to be as close to $|x^*\rangle$ as possible. We can divide $|\psi\rangle$ into a superposition of all non-solutions, $|x\rangle \neq |x^*\rangle$, and the solution itself, $|x^*\rangle$. The state containing a superposition of all non-solutions is a summation over all basis states, excluding $|x^*\rangle$:

$$|\alpha\rangle = \frac{1}{\sqrt{N-1}} \sum_{x \neq x^*} |x\rangle \tag{12.26}$$

There are $N - 1$ of these states (N original states minus the state $|x^*\rangle$); hence, the normalization factor in Eq. (12.26) is $\frac{1}{\sqrt{N-1}}$. The initial state, $|\psi\rangle$, can then be written in the following form:

$$|\psi\rangle = \sqrt{\frac{N-1}{N}}|\alpha\rangle + \frac{1}{\sqrt{N}}|x^*\rangle \tag{12.27}$$

If we substitute Eq. (12.26) into (12.27), we will get $|\psi\rangle$ in Eq. (12.2).

The two states, $|\alpha\rangle$ and $|x^*\rangle$, form an orthonormal basis in a two-dimensional vector space. The initial state $|\psi\rangle$ can be represented in this two-dimensional vector space as a superposition of $|\alpha\rangle$ and $|x^*\rangle$. This is depicted in Fig. 12.8. For a large number of basis states (large N), $|\psi\rangle$ is initially very close to $|\alpha\rangle$, according to Eq. (12.27).

The Oracle changes $|\psi\rangle$ to:

Fig. 12.8 $|\psi\rangle$ as a
superposition of $|\alpha\rangle$ and $|x^*\rangle$

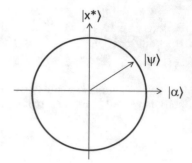

$$|\psi'\rangle = O|\psi\rangle = \sqrt{\frac{N-1}{N}}|\alpha\rangle - \frac{1}{\sqrt{N}}|x^*\rangle \qquad (12.28)$$

i.e., it marks the desired state $|x^*\rangle$ by changing the sign of its probability amplitude.
This is a reflection of $|\psi\rangle$ about the state $|\alpha\rangle$, depicted in Fig. 12.9 as a rotation by
the angle θ.

Exercise 12.5 Show that the angle in Fig. 12.9 is given by $\cos\frac{\theta}{2} = \sqrt{\frac{N-1}{N}}$
(Hint: take the scalar product between $|\psi\rangle$ and $|\alpha\rangle$).

After the Oracle, the next step of the Grover algorithm is the diffusion transform,
$D = 2|\psi\rangle\langle\psi| - I$. This reflects $|\psi'\rangle$ through the state $|\psi\rangle$, as depicted in Fig. 12.10.
This is easy to prove as follows. First, we deconstruct $|\psi'\rangle$ into components that are
parallel and perpendicular to $|\psi\rangle$:

$$|\psi'\rangle = |\psi'_{||}\rangle + |\psi'_{\perp}\rangle \qquad (12.29)$$

The reflected vector is:

Fig. 12.9 The Oracle
transform (O) reflects $|\psi\rangle$
about the state $|\alpha\rangle$

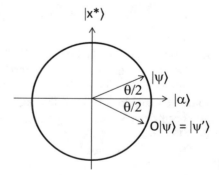

Fig. 12.10 The diffusion transform (D) reflects $|\psi'\rangle$ about the state $|\psi\rangle$

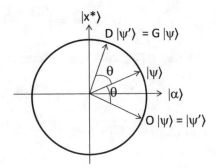

$$|\psi'\rangle - 2|\psi'_\perp\rangle \tag{12.30}$$

Substituting Eq. (12.29) gives:

$$|\psi'\rangle - 2\big(|\psi'\rangle - |\psi'_\parallel\rangle\big)$$
$$= 2|\psi'_\parallel\rangle - |\psi'\rangle \tag{12.31}$$

By definition, $|\psi'_\parallel\rangle$ is the projection of $|\psi'\rangle$ onto $|\psi\rangle$; i.e., $|\psi'_\parallel\rangle = |\psi\rangle\langle\psi|\psi'\rangle$, giving:

$$2|\psi'_\parallel\rangle - |\psi'\rangle$$
$$= 2|\psi\rangle\langle\psi|\psi'\rangle - |\psi'\rangle \tag{12.32}$$

$$= [2|\psi\rangle\langle\psi| - I]|\psi'\rangle \tag{12.33}$$

Equation (12.33) is recognized as the diffusion transform; i.e., $D = 2|\psi\rangle\langle\psi| - I$. Hence, the diffusion transform is a reflection of the state $|\psi'\rangle$ about the state $|\psi\rangle$. This reflection is equivalent to the rotation angle, θ, depicted in Figs. 12.9 and 12.10 and derived in Exercise 12.5.

$$\cos\frac{\theta}{2} = \sqrt{\frac{N-1}{N}} \tag{12.34}$$

Now, according to Fig. 12.9, we can express any state $|\psi\rangle$ as:

$$|\psi\rangle = \sin\frac{\theta}{2}|x^*\rangle + \cos\frac{\theta}{2}|\alpha\rangle \tag{12.35}$$

After doing one iteration of the Grover algorithm ($G = DO$), the state becomes:

$$G|\psi\rangle = \sin\frac{3\theta}{2}|x^*\rangle + \cos\frac{3\theta}{2}|\alpha\rangle \tag{12.36}$$

and after k iterations it is:

$$G^k|\psi\rangle = \sin\left(\frac{2k+1}{2}\theta\right)|x^*\rangle + \cos\left(\frac{2k+1}{2}\theta\right)|\alpha\rangle \qquad (12.37)$$

The final state should be as close to $|x^*\rangle$ as possible. Thus, we want to iterate until:

$$\frac{2k+1}{2}\theta \sim \pi/2 \qquad (12.38)$$

From Eq. (12.34) we have:

$$\sin\frac{\theta}{2} = \frac{1}{\sqrt{N}} \qquad (12.39)$$

For large N, we have:

$$\frac{\theta}{2} \sim \frac{1}{\sqrt{N}} \qquad (12.40)$$

Substituting Eq. (12.40) into Eq. (12.38) gives:

$$\frac{2k+1}{\sqrt{N}} \sim \pi/2 \qquad (12.41)$$

or

$$k \sim (\pi/4)\sqrt{N} \qquad (12.42)$$

Therefore, the Grover algorithm requires $O(\sqrt{N})$ steps.

References

1. L.K. Grover, *Proceedings of the 8th Annual ACM Symposium on Theory of Computing* (ACM Press, New York, USA, 1996), pp. 212–219
2. L.M.K. Vandersypen, M. Steffen, Appl. Phys. Lett. **76**, 646 (2000)
3. L.K. Grover, Am. j. Phys. **69**, 769 (2001)

Chapter 13
Shor Algorithm

In 1994, a mathematician named Peter Shor surprised the world by publishing an efficient quantum algorithm for finding the prime factors of very large numbers, which is the basis for today's data encryption. This discovery spurred the further development of quantum computing, making Shor's algorithm one of the most important algorithms.

13.1 RSA Encryption

As we saw in Chap. 10, we have efficient classical algorithms for multiplying two numbers in polynomial time, but we do not have an efficient classical algorithm for doing the reverse; i.e., for factoring. RSA (Rivest–Shamir–Adleman) is a classical data encryption method based on our inability to efficiently find the prime factors of very large numbers [1]. RSA is a public-key cryptosystem that works roughly as follows (Fig. 13.1):

1. Bob takes two large prime numbers (there are efficient classical codes to generate these randomly) and calculates their product (the key). These numbers need to be very large. It is currently recommended that RSA-2048 be used. An RSA-2048 number will have 2048 bits or 617 decimal digits.
2. Bob sends the product to Alice (public key) who uses it to encrypt her message using a mathematical function. The key is accessible to the public.
3. Bob receives the encrypted message, and only he can use his knowledge of the original prime numbers to decrypt the message.

RSA is the basis for protecting our privacy including the global financial system, internet banking, e-commerce, etc. If you could find an efficient way of factoring large numbers, you could break RSA encryption codes. In 2009, a dozen researchers and several hundred classical computers took 2 years to factor a 768-bit number (RSA-768) [2]. It is predicted that cracking an RSA-2048 number using classical

© The Author(s), under exclusive license to Springer Nature Switzerland AG 2021   177
R. LaPierre, *Introduction to Quantum Computing*, The Materials Research Society Series,
https://doi.org/10.1007/978-3-030-69318-3_13

Fig. 13.1 RSA encryption method

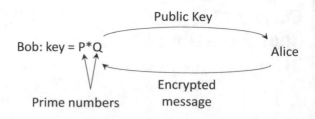

computing would require longer than the age of the universe, more energy than available on Earth, cover the planet in servers, cost a fortune, etc. Therefore, the security of today's cryptography relies on computational complexity (the exponential difficulty of factoring large numbers). For example, what are the prime factors for the following RSA-2048 number [2]?:

2519590847565789349402718324004839857142928212620403202777713783
6043662020707595556264018525880784406918290641249515082189298559
149176184502808489120072844992687392807287776735971418347270261
896375014971824691165077613379859095700097330459748808428401797
429100642458691817195118746121515172654632282216869987549182422
433637259.085141.865462043.576798423387184774447920739934236584823824
281198163815010674810451660377306056201619676256133844143603833
904414952634432190114657544454178424020924616515723350778707749
817125772467962926386356373289912154831438167899885040445364023
527381951378636564391212010397122822120720357.

13.2   Shor Algorithm

In 1994, a mathematician named Peter Shor (Fig. 13.2) discovered an efficient quantum algorithm for finding the prime factors of very large numbers, which is the basis for RSA encryption [3]. Fortunately, quantum key distribution (QKD), explored in Chap. 6, offers an alternative to RSA.

The best classical algorithm, known as the general number field sieve, still scales exponentially with the size of the input. In comparison, Shor's algorithm can factor in polynomial time. It is generally believed that there is no efficient classical algorithm for factoring large numbers. However, there is no proof that a much more efficient classical algorithm cannot be found, which is exponentially faster. Therefore, there is currently no proof that a quantum computer can factor faster than a classical computer, although this is generally believed to be true.

Fig. 13.2 Peter Shor. *Credit* Wikimedia Commons [4]

13.3 Finding Prime Factors

Shor's algorithm is based on number theory for factoring. Suppose we want to find the prime factors (Q, R) of an integer P; i.e., $P = Q * R$. Let's take the very simple example of $15 = 3 \times 5$ ($P = 15$, $Q = 3$, $R = 5$). Mathematicians have found the following algorithm for finding the prime factors of an integer:

1. Choose a random number, say $a = 7$.
2. Calculate $a^x \bmod P$ where $x = 0, 1, 2, 3, \ldots$ (i.e., the remainder of a^x divided by P). In our example, the resulting number sequence is:

$$7^0 \bmod 15, 7^1 \bmod 15, 7^2 \bmod 15, 7^3 \bmod 15,$$
$$7^4 \bmod 15, 7^5 \bmod 15, 7^6 \bmod 15, 7^7 \bmod 15, \ldots$$
$$= 1, 7, 4, 13, 1, 7, 4, 13, \ldots,$$

3. Find the periodicity of the above number sequence. In this example, $r = 4$ because the four numbers 1, 7, 4 and 13 repeats; i.e., $a^x \bmod P$ is a periodic function.
4. If r is an odd number, start over. Otherwise, if r is an even number, then Q and R are given by $Q = \gcd(a^{r/2} - 1, P)$ and $R = \gcd(a^{r/2} + 1, P)$ where gcd is the greatest common divisor. For our example:

$$Q = \gcd(a^{r/2} - 1, P) = \gcd(7^2 - 1, 15) = \gcd(48, 15)$$
$$= 3, \text{ since } 48/3 = 16 \text{ and } 15/3 = 5$$
$$R = \gcd(a^{r/2} + 1, P) = \gcd(7^2 + 1, 15) = \gcd(50, 15)$$
$$= 5, \text{ since } 50/5 = 10 \text{ and } 15/5 = 3$$

There are efficient classical algorithms (Euclid's algorithm) for efficiently finding the gcd. Therefore, the factoring problem reduces to efficiently finding the period (r) of the function $a^x \bmod P$.

13.4   Birthday Paradox

Consider the periodic function:

$$f(x) = f(x + r) = f(x + 2r) = \cdots = f(x + kr) \tag{13.1}$$

where k is an integer. $f(x)$ is said to have period r. Classically, finding r is very difficult. All we can do is test the function at many different values of x and look for a pattern. In fact, for period r, it takes $O(\sqrt{r})$ inputs to find two values of x such that $f(x)$ repeats. This is known as the "birthday problem" or "birthday paradox". For example, r might be a 617-digit number used for RSA encryption; i.e., $r \sim 10^{617}$. We would have to test on the order of $\sqrt{r} \sim 10^{309}$ inputs to $f(x)$ to find the period r. Obviously, this task is hopeless for classical computers.

Shor's algorithm is based on the fact that a quantum computer can find the period of a function more efficiently by computing $f(x)$ for many values of x in a single parallel computation (quantum parallelism). Shor's algorithm does this by using the Fourier transform. The Fourier transform (FT) is one of the most useful mathematical tools in modern science and engineering. The FT allows us to extract the underlying periodic behaviour of a function by giving us the "spectrum" of the input. It is used classically for digital signal processing (e.g., to remove noise from data), optics, spectroscopy, x-ray diffraction, etc.

13.5   Discrete Fourier Transform

The Fourier transform, named after the French mathematician and physicist, Jean-Baptiste Joseph Fourier (Fig. 13.3), describes a function as a sum of components with different amplitudes and frequencies. The discrete Fourier transform (DFT) is a version of the Fourier transform which works on discrete data sets, x_j, with $j = 0, 1, \ldots, N - 1$. The DFT is written as:

Fig. 13.3 Jean-Baptiste Joseph Fourier (1768–1830). *Credit* Wikimedia Commons [5]

$$y_k = \frac{1}{\sqrt{N}} \sum_{j=0}^{N-1} x_j e^{2\pi ijk/N} \tag{13.2}$$

where j and k are indices from 0 to $N-1$, $i = \sqrt{-1}$, x_j are complex numbers with $j = 0, 1, \ldots, N-1$, and y_k are complex numbers with $k = 0, 1, \ldots, N-1$.

For example, suppose we are given the dataset $x_j = \{1, 3\}$. Thus, $x_0 = 1$, $x_1 = 3$, and $N = 2$. Equation (13.2) gives the DFT as:

$$\text{For } k = 0, y_0 = \frac{1}{\sqrt{2}} \sum_{j=0}^{1} x_j e^{2\pi ij*0/2} = \frac{1}{\sqrt{2}}(1+3) = \frac{4}{\sqrt{2}} \tag{13.3}$$

$$\text{For } k = 1, y_1 = \frac{1}{\sqrt{2}} \sum_{j=0}^{1} x_j e^{2\pi ij*1/2} = \frac{1}{\sqrt{2}}\left(1+3e^{\pi i}\right) = -\frac{2}{\sqrt{2}} \tag{13.4}$$

Thus, the DFT is easy to calculate. The fast Fourier transform (FFT) is just another classical algorithm that allows us to compute the DFT more rapidly.

13.6 Quantum Fourier Transform

In the quantum Fourier transform (QFT), we do a DFT on the amplitudes of a quantum state. We will consider the circuit for the QFT later. For now, Fig. 13.4 shows a QFT circuit as a "black-box" with an input of n qubits on the left. The QFT transformation is denoted as QFT_N, since it operates on an N-dimensional ($N = 2^n$) input state vector $|\psi\rangle$ with elements $\alpha_0, \alpha_1, \ldots, \alpha_{N-1}$:

$$|\psi\rangle = \sum_{j=0}^{N-1} \alpha_j |j\rangle \rightarrow \begin{pmatrix} \alpha_0 \\ \cdot \\ \cdot \\ \cdot \\ \alpha_{N-1} \end{pmatrix} \tag{13.5}$$

The output of the circuit, $|\psi'\rangle$, is an N-dimensional state vector with elements $\beta_0, \beta_1, \ldots, \beta_{N-1}$, which is the QFT of the probability amplitudes:

$$|\psi'\rangle = QFT_N |\psi\rangle = \sum_{k=0}^{N-1} \beta_k |k\rangle \rightarrow \begin{pmatrix} \beta_0 \\ \cdot \\ \cdot \\ \cdot \\ \beta_{N-1} \end{pmatrix} \tag{13.6}$$

where

$$\beta_k = \frac{1}{\sqrt{N}} \sum_{j=0}^{N-1} \alpha_j e^{2\pi ijk/N} \tag{13.7}$$

For example, consider the $n = 2$ qubit state:

$$|\psi\rangle = \alpha_0|0\rangle + \alpha_1|1\rangle + \alpha_2|2\rangle + \alpha_3|3\rangle \tag{13.8}$$

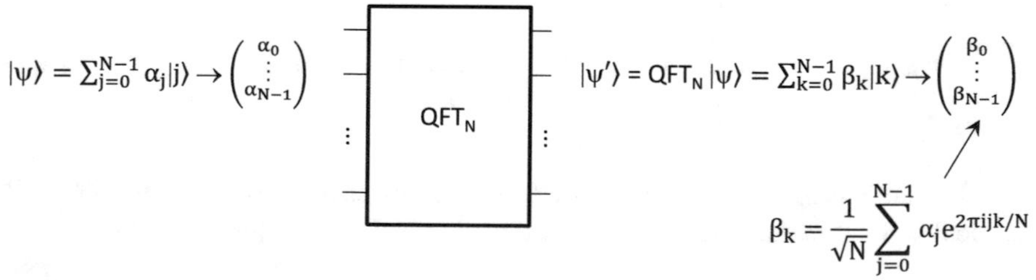

$|\psi\rangle = \sum_{j=0}^{N-1} \alpha_j|j\rangle \rightarrow \begin{pmatrix} \alpha_0 \\ \vdots \\ \alpha_{N-1} \end{pmatrix}$ QFT_N $|\psi'\rangle = QFT_N|\psi\rangle = \sum_{k=0}^{N-1} \beta_k|k\rangle \rightarrow \begin{pmatrix} \beta_0 \\ \vdots \\ \beta_{N-1} \end{pmatrix}$

$\beta_k = \frac{1}{\sqrt{N}} \sum_{j=0}^{N-1} \alpha_j e^{2\pi ijk/N}$

Fig. 13.4 Principle behind the QFT

In this example $N = 2^n = 2^2 = 4$. Thus,

$$\beta_k = \frac{1}{\sqrt{N}} \sum_{j=0}^{N-1} \alpha_j e^{\frac{2\pi ijk}{N}}$$

$$= \frac{1}{2} \sum_{j=0}^{3} \alpha_j e^{2\pi ijk/4} \qquad (13.9)$$

Equation (13.9) gives:

$$\beta_0 = \frac{1}{2} \sum_{j=0}^{3} \alpha_j = \frac{1}{2}(\alpha_0 + \alpha_1 + \alpha_2 + \alpha_3) \qquad (13.10)$$

$$\beta_1 = \frac{1}{2} \sum_{j=0}^{3} \alpha_j e^{\pi ij/2} = \frac{1}{2}\left(\alpha_0 + \alpha_1 e^{\frac{i\pi}{2}} + \alpha_2 e^{i\pi} + \alpha_3 e^{\frac{3i\pi}{2}}\right)$$

$$= \frac{1}{2}(\alpha_0 + \alpha_1 i - \alpha_2 - \alpha_3 i) \qquad (13.11)$$

$$\beta_2 = \frac{1}{2} \sum_{j=0}^{3} \alpha_j e^{\pi ij} = \frac{1}{2}\left(\alpha_0 + \alpha_1 e^{i\pi} + \alpha_2 e^{2i\pi} + \alpha_3 e^{3i\pi}\right)$$

$$= \frac{1}{2}(\alpha_0 - \alpha_1 + \alpha_2 - \alpha_3) \qquad (13.12)$$

$$\beta_3 = \frac{1}{2} \sum_{j=0}^{3} \alpha_j e^{3\pi ij/2} = \frac{1}{2}\left(\alpha_0 + \alpha_1 e^{\frac{3i\pi}{2}} + \alpha_2 e^{3i\pi} + \alpha_3 e^{\frac{9i\pi}{2}}\right)$$

$$= \frac{1}{2}(\alpha_0 - \alpha_1 i - \alpha_2 + \alpha_3 i) \qquad (13.13)$$

Exercise 13.1 Derive the QFT$_4$ output vector for each of the two-qubit basis states. Show that the resulting states are orthonormal.

13.7 QFT Matrix

In the previous example, we can express the operator QFT$_4$ in terms of a transformation matrix, using Eqs. (13.10)–(13.13):

$$QFT_4 = \frac{1}{2} \begin{pmatrix} 1 & 1 & 1 & 1 \\ 1 & i & -1 & -i \\ 1 & -1 & 1 & -1 \\ 1 & -i & -1 & i \end{pmatrix} \qquad (13.14)$$

Exercise 13.2 Show QFT_4 is a unitary operator, and therefore can be implemented.

Exercise 13.3 Using the matrix for QFT_4, what is the output vector for an input state of $|0\rangle$, $|1\rangle$, $|2\rangle$ and $|3\rangle$?

If we generalize the previous example to n qubits, then QFT_N can be shown to be an $N \times N$ matrix given by:

$$QFT_N = \frac{1}{\sqrt{N}} \begin{pmatrix} 1 & 1 & 1 & 1 & \cdots & 1 \\ 1 & \omega & \omega^2 & \omega^3 & \cdots & \omega^{N-1} \\ 1 & \omega^2 & \omega^4 & \omega^6 & \cdots & \omega^{2(N-1)} \\ \vdots & \vdots & \vdots & \vdots & \ddots & \vdots \\ 1 & \omega^{N-1} & \omega^{2(N-1)} & \omega^{3(N-1)} & \cdots & \omega^{(N-1)(N-1)} \end{pmatrix} \qquad (13.15)$$

where we define $\omega = e^{2\pi i/N}$. Note that for two qubits, $N = 4$ and $\omega = e^{i\pi/2} = i$, $\omega^2 = -1$, $\omega^3 = -i$, $\omega^4 = 1$, etc., and Eq. (13.15) simplifies to Eq. (13.14). Also, note that QFT_N has rows (j) from 0 to $N - 1$, and columns (k) from 0 to $N - 1$. Ignoring the pre-factor of $\frac{1}{\sqrt{N}}$, the (j, k) entry of QFT_N is ω^{jk}.

Exercise 13.4 Show that QFT_N is unitary.

Consider the single-qubit state ($n = 1$, $N = 2^n = 2$). Equation (13.15) becomes:

$$QFT_2 = \frac{1}{\sqrt{2}} \begin{pmatrix} 1 & 1 \\ 1 & \omega \end{pmatrix} \qquad (13.16)$$

where $\omega = e^{2\pi i/N} = e^{2\pi i/2} = e^{i\pi} = -1$. Thus,

$$QFT_2 = \frac{1}{\sqrt{2}} \begin{pmatrix} 1 & 1 \\ 1 & -1 \end{pmatrix} \qquad (13.17)$$

We recognize QFT$_2$ as the Hadamard transform. The Hadamard transform is a single-qubit QFT.

13.8 Period Finding

The QFT is useful when there is an underlying periodicity to the wavefunction. Figure 13.5 illustrates a wavefunction whose probabilities, $|\alpha_j|^2$ with $j = 0, 1, \ldots, N - 1$, show a periodicity, r. The wavefunction can be written as:

$$|\psi\rangle = \frac{1}{\sqrt{m}} \sum_{k=0}^{m-1} |x_o + kr\rangle \qquad (13.18)$$

where $k = 0, 1, 2, \ldots, m - 1$ and $\frac{1}{\sqrt{m}}$ is a normalization factor. The QFT gives a wavefunction:

$$|\psi'\rangle = \frac{1}{\sqrt{m}} \sum_{k=0}^{m-1} \frac{1}{\sqrt{N}} \sum_{y=0}^{N-1} e^{2\pi i(x_0 + kr)y/N} |y\rangle \qquad (13.19)$$

or

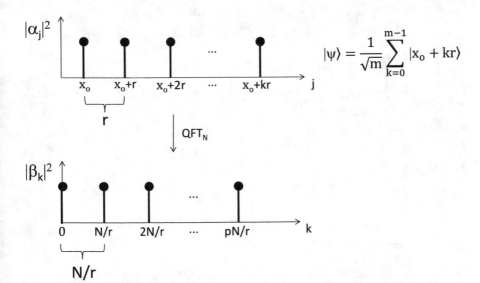

Fig. 13.5 QFT of a periodic function

$$|\psi'\rangle = \frac{1}{\sqrt{Nm}} \sum_{y=0}^{N-1} e^{2\pi i x_0 y/N} \sum_{k=0}^{m-1} e^{2\pi i k r y/N} |y\rangle \qquad (13.20)$$

The term $e^{2\pi i x_0 y/N}$ is a global phase factor for each state $|y\rangle$ and can be ignored (it does not affect probabilities); i.e., we have effectively eliminated the constant, x_o. The probability for each state $|y\rangle$ is then:

$$P_y = \left| \frac{1}{\sqrt{Nm}} \sum_{k=0}^{m-1} e^{2\pi i k r y/N} \right|^2 \qquad (13.21)$$

Equation (13.21) gives maxima in P_y when y is an integer multiple of N/r; i.e., when the exponent becomes 1. In that case, a measurement is likely to give a state $|pN/r\rangle$, $p = 0, 1, 2, \ldots$, as illustrated in Fig. 13.5. Hence, a wavefunction with period r gives a frequency N/r in the QFT.

Exercise 13.5 Consider the 5-qubit state:

$$|\psi\rangle = \frac{1}{\sqrt{3}}(|01000\rangle + |10000\rangle + |11000\rangle) = \frac{1}{\sqrt{3}}(|8\rangle + |16\rangle + |24\rangle)$$

Show that $\text{QFT}_{32} |\psi\rangle$ has frequency N/r as illustrated below.

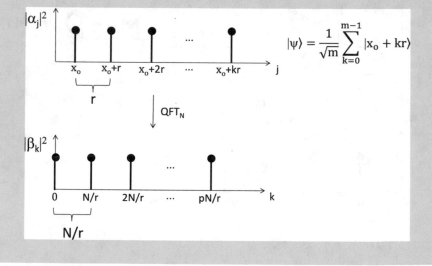

13.9 QFT Circuit

For the QFT circuit, we need to introduce a new gate called the controlled-phase gate, R_k, which implements a phase shift, $e^{2\pi i/2^k}$, on $|1\rangle$ only if the control qubit is $|1\rangle$. Based on Fig. 7.8, the matrix for the controlled-phase gate can be written as:

$$R_k = \begin{pmatrix} 1 & 0 & 0 & 0 \\ 0 & 1 & 0 & 0 \\ 0 & 0 & 1 & 0 \\ 0 & 0 & 0 & e^{2\pi i/2^k} \end{pmatrix} \tag{13.22}$$

The circuit representation of the controlled-phase gate is shown in Fig. 13.6.

The circuit which implements QFT_N in Fig. 13.7 uses the controlled-phase gate, the Hadamard gate, and the SWAP gate. Although we won't do the proof here, it can be shown that the circuit of Fig. 13.7 implements QFT_N.

For example, for the single-qubit (QFT_2), $n = 1$ and Fig. 13.7 simplifies to a single gate operation, the Hadamard gate (the last qubit of Fig. 13.7). For the two-qubit QFT (QFT_4), Fig. 13.7 simplifies to the circuit shown in Fig. 13.8 (the last two qubits of Fig. 13.7). Based on Fig. 13.8, the transformation matrix is:

$$QFT_4 = (SWAP)(I \otimes H)(R_2)(H \otimes I) \tag{13.23}$$

Fig. 13.6 Circuit representation of the controlled-phase gate

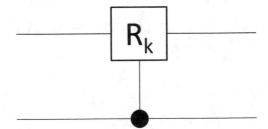

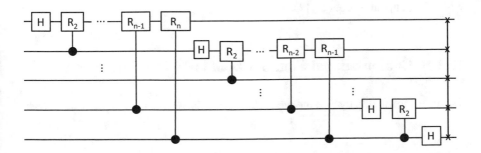

Fig. 13.7 Quantum circuit for QFT_N

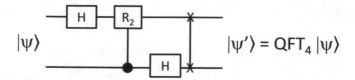

$$|\psi\rangle \qquad\qquad\qquad |\psi'\rangle = QFT_4 |\psi\rangle$$

Fig. 13.8 Quantum circuit for the two-qubit QFT

where it is easily shown that:

$$H \otimes I = \frac{1}{\sqrt{2}} \begin{pmatrix} 1 & 0 & 1 & 0 \\ 0 & 1 & 0 & 1 \\ 1 & 0 & -1 & 0 \\ 0 & 1 & 0 & -1 \end{pmatrix} \qquad (13.24)$$

$$I \otimes H = \frac{1}{\sqrt{2}} \begin{pmatrix} 1 & 1 & 0 & 0 \\ 1 & -1 & 0 & 0 \\ 0 & 0 & 1 & 1 \\ 0 & 0 & 1 & -1 \end{pmatrix} \qquad (13.25)$$

$$\text{SWAP} = \begin{pmatrix} 1 & 0 & 0 & 0 \\ 0 & 0 & 1 & 0 \\ 0 & 1 & 0 & 0 \\ 0 & 0 & 0 & 1 \end{pmatrix} \qquad (13.26)$$

and R_2 is given by Eq. (13.22). Substituting Eqs. (13.22), (13.24), (13.25) and (13.26) into Eq. (13.23) gives:

$$QFT_4 = \frac{1}{2} \begin{pmatrix} 1 & 1 & 1 & 1 \\ 1 & i & -1 & -i \\ 1 & -1 & 1 & -1 \\ 1 & -i & -1 & i \end{pmatrix} \qquad (13.27)$$

which is identical to Eq. (13.14).

13.10   Computational Complexity of QFT

We have shown already that the matrix for QFT_N is:

$$QFT_N = \frac{1}{\sqrt{N}} \begin{pmatrix} 1 & 1 & 1 & 1 & \cdots & 1 \\ 1 & \omega & \omega^2 & \omega^3 & \cdots & \omega^{N-1} \\ 1 & \omega^2 & \omega^4 & \omega^6 & \cdots & \omega^{2(N-1)} \\ \vdots & \vdots & \vdots & \vdots & \cdots & \vdots \\ 1 & \omega^{N-1} & \omega^{2(N-1)} & \omega^{3(N-1)} & \cdots & \omega^{(N-1)(N-1)} \end{pmatrix}$$

Note that this matrix also implements the classical DFT, obtained by multiplication of the QFT_N matrix by the $N \times 1$ column vector that contains the classical dataset. This multiplication would require N^2 operations. Therefore, we would expect the classical DFT to require $O(N^2) = O(2^{2n})$ operations, which is exponential in n. The classical Fast Fourier Transform (FFT) algorithm can compute the DFT in $O(N \log N)$ or $O(n2^n)$ operations, which is faster but still exponential in n.

How many operations (gates) does the QFT use? We simply count the number of gate operations in Fig. 13.7:

$$1\text{st row}: 1\,H \text{ gate} + (n-1)R \text{ gates} = n \text{ gates}$$
$$2\text{nd row}: 1\,H \text{ gate} + (n-2)R \text{ gates} = n-1 \text{ gates}$$
$$\vdots$$
$$(n-1)\text{th row}: 1\,H \text{ gate} + 1\,R \text{ gate} = 2 \text{ gates}$$
$$n\text{th row}: 1\,H \text{ gate} = 1 \text{ gate}$$

Adding the gate count from each row gives $n + (n-1) + (n-2) + \cdots + 1$, or $O(n^2)$ gates, which is polynomial in n. Therefore, the QFT is exponentially faster than the DFT or FFT.

Unfortunately, we cannot extract the Fourier transform directly from the QFT circuit due to the measurement problem (the output is a superposition of states, and we can't know each of the amplitudes of the states because the output collapses to only one of the states). The QFT is useful only as a piece of another algorithm.

13.11 Shor Circuit

Now let us return to Shor's algorithm. Recall that we want to find the prime factors of $P = Q * R$. This requires us to find the periodicity, r, of $a^x \mod P$. The prime factors are then $Q = \gcd(a^{r/2} - 1, P)$ and $R = \gcd(a^{r/2} + 1, P)$. The gcd can be found by efficient classical algorithms (Euclid's algorithm). The difficult part is finding r, but we can do it by applying the QFT.

The Shor circuit is shown in Fig. 13.9. The circuit contains two registers. The circuit is initialized by forming an equally weighted superposition in the first register and $|0\rangle$'s in the second register:

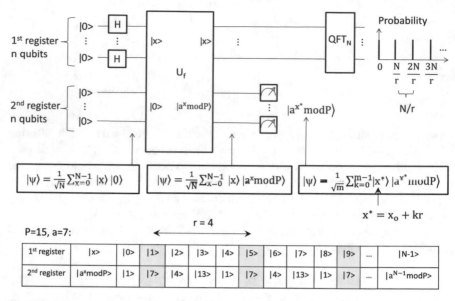

Fig. 13.9 The Shor circuit and table showing the circuit output from U_f. The shaded columns show a possible collapse of the wavefunction after measurement on the 2nd register. The QFT then gives the period ($r = 4$) of the first register

$$|\psi\rangle = \frac{1}{\sqrt{N}} \sum_{x=0}^{N-1} |x\rangle |0\rangle \tag{13.28}$$

$$= \frac{1}{\sqrt{N}}(|0\rangle + |1\rangle + |2\rangle + |3\rangle + \cdots |N-1\rangle)|0\rangle, N = 2^n \tag{13.29}$$

Next, we use an Oracle that takes a value x from the first register and writes $f(x) = a^x \bmod P$ into the second register, forming an entangled state:

$$|\psi\rangle = \frac{1}{\sqrt{N}} \sum_{x=0}^{N-1} |x\rangle |a^x \bmod P\rangle \tag{13.30}$$

For example, if $a = 7$ and $P = 15$:

$$|\psi\rangle = \frac{1}{\sqrt{N}}\left(|0\rangle|1\rangle + |1\rangle|7\rangle + |2\rangle|4\rangle + |3\rangle|13\rangle + \ldots + |N-1\rangle|a^{N-1}\bmod P\rangle\right) \tag{13.31}$$

Next, we make a measurement on the second register. The second register must collapse to some state $|a^{x^*} \bmod P\rangle$ (e.g., $|7\rangle$) for some specific values x^*. The first register then collapses into the superposition of all states $|x^*\rangle$ that give $|a^{x^*} \bmod P\rangle$

in the second register (e.g., $|x^*\rangle = |1\rangle, |5\rangle, |9\rangle, |13\rangle$):

$$|\psi\rangle = \frac{1}{\sqrt{N}}(|1\rangle + |5\rangle + |9\rangle + |13\rangle + \cdots)|7\rangle \qquad (13.32)$$

More generally,

$$|\psi\rangle = \frac{1}{\sqrt{N}} \sum |x^*\rangle |a^{x^*} \bmod P\rangle \qquad (13.33)$$

The first register, containing $|x^*\rangle$, now has period r (e.g., $r = 4$). Thus, $x^* = x_o + kr$ where k is an integer ($k = 0, 1, \ldots$). Thus, Eq. (13.33) can be written:

$$|\psi\rangle = \frac{1}{\sqrt{m}} \sum_{k=0}^{m-1} |x_o + kr\rangle |a^{x^*} \bmod P\rangle \qquad (13.34)$$

where $\frac{1}{\sqrt{m}}$ is a normalization factor. Continuing our example,

$$|\psi\rangle = \frac{1}{\sqrt{m}}(|1\rangle + |5\rangle + |9\rangle + |13\rangle + \cdots)|7\rangle \qquad (13.35)$$

$$= \frac{1}{\sqrt{m}} \sum_{k=0}^{m-1} |1 + 4k\rangle |7\rangle \qquad (13.36)$$

To determine r, we apply the QFT to the first register, giving us a spectrum (Eq. (13.21) with frequency N/r; i.e., a measurement is likely to give some multiple of N/r. For example, for $r = 4$, $n = 7$, and $N = 2^n = 2^7 = 128$, we get a QFT frequency of $N/r = 128/4 = 32$. This means we would likely measure $|0\rangle$, $|32\rangle$, $|68\rangle$ or $|104\rangle$ with high probability from the QFT. After a few measurements, we can quickly determine the period, r. Finally, the prime factors in our example are determined by a classical computer as $Q = \gcd(a^{r/2} - 1, P) = 3$ and $R = \gcd(a^{r/2} + 1, P) = 5$.

13.12 Outlook

In 2001, Shor's algorithm was famously demonstrated by a group at IBM, who factored 15 into 3×5, using an NMR implementation (Chap. 19) of a quantum computer with 7 qubits [6]. Shor's algorithm has also been implemented using superconducting qubits to factor the number $15 = 3 \times 5$ [7]. The number $21 = 3 \times 7$ has been factored using optics [8].

How many qubits are needed in Shor's algorithm? To find the period r, we want many instances of it within the domain $[0, N-1]$. If we are trying to factor a number with m bits, it can be shown that we need $n > 2\,m$ qubits. For RSA-2048, this means that $m = 2048$ and $n > 4096$. If we add error correction (Chap. 25), then about 100–1000 physical qubits may be needed per logical qubit; i.e., $>10^6$ physical qubits may be needed. As of the year 2020, the largest quantum computer built contains about 50 qubits. Thus, RSA encryption, which can be broken by Shor's algorithm, is likely safe for a while longer!

References

1. R.L. Rivest, A. Sharmir, L. Adleman, Commun. ACM **21**, 120 (1978)
2. https://en.wikipedia.org/wiki/RSA_numbers#RSA-2048
3. P.W. Shor, SIAM J. Comput. **26**, 1484 (1994)
4. Author: International Centre for Theoretical Physics. This file is licensed under the Creative Commons Attribution 3.0 Unported license (https://creativecommons.org/licenses/by/3.0/deed.en). File: Peter Shor 2017 Dirac Medal Award Ceremony.png. (2020, October 7). *Wikimedia Commons, the free media repository*. Retrieved 15:32, December 7, 2020 from https://commons.wikimedia.org/w/index.php?title=File:Peter_Shor_2017_Dirac_Medal_Award_Ceremony.png&oldid=483495773
5. File: Fourier2.jpg. (2020, September 14). *Wikimedia Commons, the free media repository*. Retrieved 15:34, December 7, 2020 from https://commons.wikimedia.org/w/index.php?title=File:Fourier2.jpg&oldid=459132355
6. M.K.L. Vandersypen et al., Nature **414**, 883 (2001)
7. E. Lecero, Nat. Phys. **8**, 719 (2012)
8. E. Martín-López et al., Nat. Photon. **6**, 773 (2012)

Chapter 14
Precession

In the next few chapters, we are going to examine how to build quantum gates. First, let us consider single-qubit gates. Single-qubit gates can be implemented by spin precession in a magnetic field. We begin with a mathematical description of spin rotation, followed by implementation using magnetic fields.

14.1 Rotation Matrices

Recall that we can express any single-qubit state using the Bloch vector form or its vector representation:

$$|\psi\rangle = \cos\frac{\theta}{2}|0\rangle + e^{i\phi}\sin\frac{\theta}{2}|1\rangle \rightarrow \begin{pmatrix} \cos\frac{\theta}{2} \\ e^{i\phi}\sin\frac{\theta}{2} \end{pmatrix} \tag{14.1}$$

A single-qubit gate corresponds to a rotation of the qubit in the Bloch sphere. It can be shown (Exercise 14.1) that $\hat{R}_x(\theta)$, $\hat{R}_y(\theta)$ and $\hat{R}_z(\theta)$ are unitary transformation matrices that implement a counter-clockwise rotation of a single qubit by the angle θ about the x, y, and z-axis of the Bloch sphere, respectively:

$$\hat{R}_x(\theta) = \begin{pmatrix} \cos\frac{\theta}{2} & -i\sin\frac{\theta}{2} \\ -i\sin\frac{\theta}{2} & \cos\frac{\theta}{2} \end{pmatrix} \tag{14.2}$$

$$\hat{R}_y(\theta) = \begin{pmatrix} \cos\frac{\theta}{2} & -\sin\frac{\theta}{2} \\ \sin\frac{\theta}{2} & \cos\frac{\theta}{2} \end{pmatrix} \tag{14.3}$$

$$\hat{R}_z(\theta) = \begin{pmatrix} e^{-i\theta/2} & 0 \\ 0 & e^{i\theta/2} \end{pmatrix} \tag{14.4}$$

© The Author(s), under exclusive license to Springer Nature Switzerland AG 2021
R. LaPierre, *Introduction to Quantum Computing*, The Materials Research Society Series,
https://doi.org/10.1007/978-3-030-69318-3_14

Exercise 14.1 Show that the rotation matrices $\widehat{R}_x(\theta)$, $\widehat{R}_y(\theta)$ and $\widehat{R}_z(\theta)$ result in a rotation angle of θ about the x, y and z axes, respectively, in the Bloch sphere.

Exercise 14.2 Show that the rotation matrices are unitary.

Exercise 14.3 Show the transformation matrix for the following circuit is
$$\begin{pmatrix} 1 & 0 & 0 & 0 \\ 0 & 1 & 0 & 0 \\ 0 & 0 & 0 & -1 \\ 0 & 0 & -1 & 0 \end{pmatrix}$$; i.e., a CNOT gate but with a phase shift of -1.

Recall from Chap. 3 that the Pauli operators ($\hat{\sigma}_x$, $\hat{\sigma}_y$, $\hat{\sigma}_z$) and the identity operator ($\hat{I}$) form a basis for any 2×2 matrix. Hence, the rotation matrices are a linear combination of $\hat{\sigma}_x$, $\hat{\sigma}_y$, $\hat{\sigma}_z$ and $\hat{I}$. This can be shown as follows:

$$\widehat{R}_x(\theta) = \begin{pmatrix} \cos\frac{\theta}{2} & -i\sin\frac{\theta}{2} \\ -i\sin\frac{\theta}{2} & \cos\frac{\theta}{2} \end{pmatrix} = \cos\frac{\theta}{2}\begin{pmatrix} 1 & 0 \\ 0 & 1 \end{pmatrix} - i\sin\frac{\theta}{2}\begin{pmatrix} 0 & 1 \\ 1 & 0 \end{pmatrix}$$
$$= \cos\frac{\theta}{2}\hat{I} - i\sin\frac{\theta}{2}\hat{\sigma}_x \tag{14.5}$$

$$\widehat{R}_y(\theta) = \begin{pmatrix} \cos\frac{\theta}{2} & -\sin\frac{\theta}{2} \\ \sin\frac{\theta}{2} & \cos\frac{\theta}{2} \end{pmatrix} = \cos\frac{\theta}{2}\begin{pmatrix} 1 & 0 \\ 0 & 1 \end{pmatrix} - i\sin\frac{\theta}{2}\begin{pmatrix} 0 & -i \\ i & 0 \end{pmatrix}$$
$$= \cos\frac{\theta}{2}\hat{I} - i\sin\frac{\theta}{2}\hat{\sigma}_y \tag{14.6}$$

$$\hat{R}_z(\theta) = \begin{pmatrix} e^{-\frac{i\theta}{2}} & 0 \\ 0 & e^{\frac{i\theta}{2}} \end{pmatrix} = \begin{pmatrix} \cos\frac{\theta}{2} - i\sin\frac{\theta}{2} & 0 \\ 0 & \cos\frac{\theta}{2} + i\sin\frac{\theta}{2} \end{pmatrix}$$

$$= \cos\frac{\theta}{2}\begin{pmatrix} 1 & 0 \\ 0 & 1 \end{pmatrix} - i\sin\frac{\theta}{2}\begin{pmatrix} 1 & 0 \\ 0 & -1 \end{pmatrix} = \cos\frac{\theta}{2}\hat{I} - i\sin\frac{\theta}{2}\hat{\sigma}_z \quad (14.7)$$

The rotation operators are also expressible in exponential form:

$$\hat{R}_x(\theta) = \cos\frac{\theta}{2}\hat{I} - i\sin\frac{\theta}{2}\hat{\sigma}_x = e^{-i\frac{\theta}{2}\hat{\sigma}_x} \quad (14.8)$$

$$\hat{R}_y(\theta) = \cos\frac{\theta}{2}\hat{I} - i\sin\frac{\theta}{2}\hat{\sigma}_y = e^{-i\frac{\theta}{2}\hat{\sigma}_y} \quad (14.9)$$

$$\hat{R}_z(\theta) = \cos\frac{\theta}{2}\hat{I} - i\sin\frac{\theta}{2}\hat{\sigma}_z = e^{-i\frac{\theta}{2}\hat{\sigma}_z} \quad (14.10)$$

Equations (14.8) to (14.10) should appear familiar—they look like the Euler relation, $e^{i\theta} = \cos\theta + i\sin\theta$. Before we prove Eqs. (14.8) to (14.10), we should ask what it means to have an operator in an exponent. The meaning is made clear by the familiar Taylor expansion:

$$e^x = \sum_{n=0}^{\infty} \frac{x^n}{n!} = 1 + x + \frac{x^2}{2!} + \frac{x^3}{3!} + \dots \quad (14.11)$$

Likewise, for an operator $\hat{O}$, we can write:

$$e^{\hat{O}} = 1 + \hat{O} + \frac{1}{2!}\left(\hat{O}\right)^2 + \frac{1}{3!}\left(\hat{O}\right)^3 + \dots \quad (14.12)$$

where 1 is the identity operator. Two other useful Taylor expansions are:

$$\cos(x) = \sum_{n=0}^{\infty} \frac{(-1)^n x^{2n}}{(2n)!} \quad (14.13)$$

and

$$\sin(x) = \sum_{n=0}^{\infty} \frac{(-1)^n x^{2n+1}}{(2n+1)!} \quad (14.14)$$

Let us use these Taylor expansions to prove Eq. (14.8). From Eq. (14.11), we can write:

$$e^{-i\frac{\theta}{2}\hat{\sigma}_x} = \sum_{n=0}^{\infty} \frac{\left(-\frac{i\theta\hat{\sigma}_x}{2}\right)^n}{n!} \tag{14.15}$$

We can separate this summation into even and odd integers:

$$e^{-i\frac{\theta}{2}\hat{\sigma}_x} = \sum_{n=0}^{\infty} \frac{\left(-\frac{i\theta\hat{\sigma}_x}{2}\right)^{2n}}{(2n)!} + \sum_{n=0}^{\infty} \frac{\left(-\frac{i\theta\hat{\sigma}_x}{2}\right)^{2n+1}}{(2n+1)!} \tag{14.16}$$

It can easily be proven that $\hat{\sigma}_x^2 = \hat{I}$, $\hat{\sigma}_x^{2n} = \hat{I}$ and $\hat{\sigma}_x^{2n+1} = \hat{\sigma}_x$. Using these relations, Eq. (14.16) becomes:

$$e^{-i\frac{\theta}{2}\hat{\sigma}_x} = \left[\sum_{n=0}^{\infty} \frac{(-1)^n \left(\frac{\theta}{2}\right)^{2n}}{(2n)!}\right]\hat{I} - i\left[\sum_{n=0}^{\infty} \frac{(-1)^n \left(\frac{\theta}{2}\right)^{2n+1}}{(2n+1)!}\right]\hat{\sigma}_x \tag{14.17}$$

Finally, we recognize the terms in square brackets in Eq. (14.17) as the Taylor expansions in Eqs. (14.13) and (14.14). Thus,

$$e^{-i\frac{\theta}{2}\hat{\sigma}_x} = \cos\frac{\theta}{2}\hat{I} - i\sin\frac{\theta}{2}\hat{\sigma}_x \tag{14.18}$$

In fact, the proof outlined above is similar to that used to prove the Euler relation. In an identical manner, you can prove Eqs. (14.9) and (14.10).

In summary, we have the following important relations for single-qubit rotations:

$$\hat{R}_x(\theta) = \begin{pmatrix} \cos\frac{\theta}{2} & -i\sin\frac{\theta}{2} \\ -i\sin\frac{\theta}{2} & \cos\frac{\theta}{2} \end{pmatrix} \quad \hat{R}_y(\theta) = \begin{pmatrix} \cos\frac{\theta}{2} & -\sin\frac{\theta}{2} \\ \sin\frac{\theta}{2} & \cos\frac{\theta}{2} \end{pmatrix}$$

$$\hat{R}_z(\theta) = \begin{pmatrix} e^{-i\theta/2} & 0 \\ 0 & e^{i\theta/2} \end{pmatrix}$$

$$\hat{R}_x(\theta) = \cos\frac{\theta}{2}\hat{I} - i\sin\frac{\theta}{2}\hat{\sigma}_x \quad \hat{R}_y(\theta) = \cos\frac{\theta}{2}\hat{I} - i\sin\frac{\theta}{2}\hat{\sigma}_y$$

$$\hat{R}_z(\theta) = \cos\frac{\theta}{2}\hat{I} - i\sin\frac{\theta}{2}\hat{\sigma}_z$$

$$\hat{R}_x(\theta) = e^{-i\frac{\theta}{2}\hat{\sigma}_x} \quad \hat{R}_y(\theta) = e^{-i\frac{\theta}{2}\hat{\sigma}_y}$$

$$\hat{R}_z(\theta) = e^{-i\frac{\theta}{2}\hat{\sigma}_z}$$

$$\hat{\sigma}_x = \hat{X} = \begin{pmatrix} 0 & 1 \\ 1 & 0 \end{pmatrix} \quad \hat{\sigma}_y = \hat{Y} = \begin{pmatrix} 0 & -i \\ i & 0 \end{pmatrix}$$

$$\hat{\sigma}_z = \hat{Z} = \begin{pmatrix} 1 & 0 \\ 0 & -1 \end{pmatrix} \quad \hat{I} = \begin{pmatrix} 1 & 0 \\ 0 & 1 \end{pmatrix}$$

14.2 Time Evolution of a Quantum System

Let us consider the time evolution of a quantum system as described by the time-independent Schrodinger equation:

$$\widehat{H}|\psi\rangle = i\hbar\frac{\partial|\psi\rangle}{\partial t} \tag{14.19}$$

where $\widehat{H}$ is the Hamiltonian operator. If $\widehat{H}$ is time-independent (e.g., a constant electric or magnetic field), then the solution to Eq. (14.19) is:

$$|\psi(t)\rangle = \underbrace{e^{-\frac{i}{\hbar}\widehat{H}(t-t_0)}}_{\widehat{U}=\text{evolution operator}} |\psi(t_o)\rangle \tag{14.20}$$

The exponential in Eq. (14.20) corresponds to a unitary transformation and is called the evolution operator, $\widehat{U}$. $\widehat{U}$ transforms an initial state $|\psi(t_o)\rangle$ to a final state $|\psi(t)\rangle$. Hence, quantum gates are implemented by applying the correct $\widehat{H}$ for the correct amount of time.

Exercise 14.4 Show that $(\widehat{U}(t))^{-1} = \left(\widehat{U}(t)\right)^{\dagger} = \widehat{U}(-t)$ for a time-independent $\widehat{H}$; i.e., the evolution operator is unitary and reversible.

What is the correct $\widehat{H}$ corresponding to the single-qubit rotation? The evolution operator in Eq. (14.20) is almost in the same format as that for the exponential form of the rotation operators in Eqs. (14.8) to (14.10):

$$\widehat{R}_i(\theta) = e^{-i\frac{\theta}{2}\hat{\sigma}_i} \tag{14.21}$$

where the index i corresponds to x, y or z. Recall from Chap. 3 that the spin angular momentum operator is $\hat{S}_i = \frac{\hbar}{2}\hat{\sigma}_i$. Hence,

$$\widehat{R}_i(\theta) = e^{-\frac{i}{\hbar}\theta\hat{S}_i} \tag{14.22}$$

Let us assume that the qubit rotation occurs at a uniform angular frequency, ω_o. Thus, after some time t, the rotation angle θ in the Bloch sphere is $\theta = \omega_o t$, and Eq. (14.22) can be written as:

$$\widehat{R}_i(\theta) = e^{-\frac{i}{\hbar}\omega_o\hat{S}_i t} \tag{14.23}$$

Finally, we see that $\widehat{R}_i(\theta)$ is in the same form as the evolution operator $\widehat{U}$ in Eq. (14.20) with the Hamiltonian given by $\widehat{H} = \omega_o \widehat{S}_i = \frac{\hbar \omega_o}{2} \hat{\sigma}_i$:

$$\widehat{R}_i(\theta) = e^{-\frac{i}{\hbar} \widehat{H} t} \tag{14.24}$$

$$\widehat{H} = \omega_o \widehat{S}_i = \frac{\hbar \omega_o}{2} \hat{\sigma}_i \tag{14.25}$$

Intuitively, it makes senses that the Hamiltonian, $\widehat{H} = \omega_o \widehat{S}_i$, corresponding to a qubit rotation, is related to the spin angular momentum operator, $\widehat{S}_i$.

Exercise 14.5 Show that $e^{i\frac{\pi}{2}\hat{\sigma}_z} e^{i\frac{\pi}{4}\hat{\sigma}_y} = \frac{i}{\sqrt{2}} \begin{pmatrix} 1 & 1 \\ 1 & -1 \end{pmatrix}$; i.e., it is equal to the Hadamard gate within a global phase factor of i.

14.3   Classical Description of Precession

How do we physically implement a qubit rotation, corresponding to $\widehat{H} = \omega_o \widehat{S}_i$? Let us consider a magnet in a magnetic field, $\mathbf{B}_o$. What happens to the magnet? The magnet will align to the B field. Now what happens if the magnet is spinning? From classical physics, the magnet will precess like a gyroscope at some angular frequency, ω_o. We know that the electron, with its intrinsic spin angular momentum, $\mathbf{S}$, also possesses a magnetic moment, μ, and is thus analogous to a spinning magnet. The electron spin will undergo precession about a B-field.

As we saw in Chap. 3, the magnetic moment and spin angular momentum are related:

$$\mu = \gamma \, \mathbf{S} \tag{14.26}$$

where γ is the gyromagnetic ratio:

$$\gamma = g \frac{q}{2m_e} = \frac{q}{m_e} \text{for the electron } (g = 2) \tag{14.27}$$

Recall that g is the g-factor with a value that is very close to 2 for the electron. Since the charge q on the electron is negative, Eqs. (14.26) and (14.27) indicate that the vectors μ and $\mathbf{S}$ point in opposite directions for the electron, as shown in Fig. 14.1. Classically, the rate of change of angular momentum $\frac{d\mathbf{S}}{dt}$ is related to the torque $\mathbf{N}$:

Fig. 14.1 Spin precession for an electron

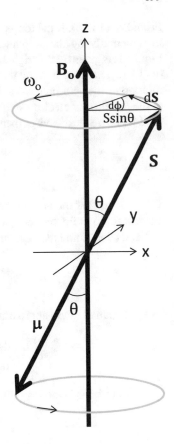

$$\frac{dS}{dt} = N \tag{14.28}$$

Also, from classical physics, the torque on a magnetic moment μ in a magnetic field B_o is:

$$N = \mu \times B_o \tag{14.29}$$

Combining Eqs. (14.26), (14.28) and (14.29) gives:

$$\frac{dS}{dt} = \gamma\, S \times B_o \tag{14.30}$$

Rearranging, we obtain:

$$\frac{dS}{dt} = -\gamma\, B_o \times S \tag{14.31}$$

From Eq. (14.28), if the torque is constant, then the angular momentum will change incrementally by the amount dS in a time dt. The direction of dS, indicated in Fig. 14.1, is given by the "right hand rule" according to the cross product in Eq. (14.30) or (14.31), keeping in mind that γ is negative for the electron. Thus, the direction of rotation of the spin precession is counter-clockwise around the z-axis (the direction of the B-field) for electron spin. Due to the nature of the cross product in Eq. (14.30) or (14.31), dS is always perpendicular to both S and $\mathbf{B_o}$. Thus,

$$\frac{dS}{dt} \cdot \mathbf{B_o} = \frac{d(S \cdot \mathbf{B_o})}{dt} = 0 \qquad (14.32)$$

This tells us that the angle θ between S and $\mathbf{B_o}$ is constant. This is consistent with the description of precession, as depicted in Fig. 14.1.

Let us determine the angular frequency of rotation, ω_o. By definition,

$$\omega_o = \frac{d\phi}{dt} \qquad (14.33)$$

Using the small angle approximation in Fig. 14.1, $d\phi = \frac{dS}{S \sin \theta}$. Hence,

$$\omega_o = \frac{d\phi}{dt} = \frac{1}{dt} \frac{dS}{S \sin \theta} = \frac{1}{S \sin \theta} \frac{dS}{dt} \qquad (14.34)$$

$\frac{dS}{dt}$ in Eq. (14.34) is the magnitude of the torque from Eq. (14.28). Thus,

$$\omega_o = \frac{N}{S \sin \theta} \qquad (14.35)$$

From the definition of the cross product, the magnitude of the torque is $N = \mu B_o \sin \theta$. Thus, Eq. (14.35) reduces to:

$$\omega_o = \frac{\mu B_o}{S} \qquad (14.36)$$

Recognizing μ/S as the gyromagnetic ratio, we get:

$$\omega_o = -\gamma B_o \qquad (14.37)$$

where the negative sign is introduced since μ and S point in opposite directions for the electron. Since γ is negative for the electron, we get $\omega_o > 0$, which indicates counter-clockwise precession of the electron spin. Conversely, spin precession is clockwise for protons.

The angular frequency, ω_o, is called the Larmor frequency, named after Joseph Larmor (Fig. 14.2), and the precession is called Larmor precession. From Eqs. (14.27)

Fig. 14.2 Joseph Larmor (1857–1942). *Credit* Wikimedia Commons [1]

and (14.37), the electron will precess at an angular frequency $\omega_o = eB_o/m_e$ where e is the fundamental electron charge.

14.4 Hamiltonian of an Electron in a Magnetic Field

The classical expression for the potential energy of a magnetic moment in a magnetic field is $U = -\boldsymbol{\mu} \cdot \mathbf{B}$. Correspondingly, the Hamiltonian is:

$$\widehat{H} = -\boldsymbol{\mu} \cdot \mathbf{B} \tag{14.38}$$

For the electron, the magnitude of $\boldsymbol{\mu}$ is the Bohr magneton, μ_B, introduced in Eq. (3.23). From Eqs. (14.26), (14.38) may be written as:

$$\widehat{H} = -\gamma \, \mathbf{B} \cdot \hat{\mathbf{S}} = -\gamma \left(B_x \hat{S}_x + B_y \hat{S}_y + B_z \hat{S}_z \right) \tag{14.39}$$

Suppose we have a magnetic field, B_o, along the z-direction. Equation (14.39) then simplifies to:

$$\widehat{H} = -\gamma \, B_o \, \hat{S}_z \tag{14.40}$$

or, using Eq. (14.37),

$$\widehat{H} = \omega_o \, \hat{S}_z = \frac{\hbar}{2} \omega_o \hat{\sigma}_z \tag{14.41}$$

Equation (14.41) is identical to Eq. (14.25), which we derived from the rotation matrices and the time-dependent Schrodinger equation. Of course, there is nothing special about the z-direction. A magnetic field along the x, y or z direction results in Larmor precession of the spin around that direction, allowing us to achieve any single-qubit rotation in the Bloch sphere.

14.5 Zeeman Effect

From Eq. (14.41), we may write the Hamiltonian for an electron in a constant magnetic field as:

$$\widehat{H} = \frac{\hbar\omega_o}{2}\hat{\sigma}_z = \frac{\hbar\omega_o}{2}\begin{pmatrix} 1 & 0 \\ 0 & -1 \end{pmatrix} = \mu_B B_o \begin{pmatrix} 1 & 0 \\ 0 & -1 \end{pmatrix} \tag{14.42}$$

where μ_B is the Bohr magneton from Eq. (3.23).

Exercise 14.6 Show that the dual vector representation for $\widehat{H}$ in the computational basis is $\widehat{H} = \mu_B B_o \left(|0\rangle\langle 0| - |1\rangle\langle 1|\right)$.

Knowing the Hamiltonian, we can now determine the energy levels for an electron in a constant magnetic field, using the time-independent Schrodinger equation:

$$\widehat{H}|\psi\rangle = E|\psi\rangle \tag{14.43}$$

Thus, the energies of a spin particle in a magnetic field are the eigenvalues of the $\widehat{H}$ operator. Let us substitute Eq. (14.42) into (14.43) and use the computational basis. Hence, for the $|0\rangle$ state (spin up):

$$\widehat{H}|0\rangle = \frac{\hbar\omega_o}{2}\begin{pmatrix} 1 & 0 \\ 0 & -1 \end{pmatrix}\begin{pmatrix} 1 \\ 0 \end{pmatrix} = \frac{\hbar\omega_o}{2}\begin{pmatrix} 1 \\ 0 \end{pmatrix} = \frac{\hbar\omega_o}{2}|0\rangle = \mu_B B_o|0\rangle \tag{14.44}$$

Similarly, for the $|1\rangle$ state (spin down):

$$\widehat{H}|1\rangle = \frac{\hbar\omega_o}{2}\begin{pmatrix} 1 & 0 \\ 0 & -1 \end{pmatrix}\begin{pmatrix} 0 \\ 1 \end{pmatrix} = -\frac{\hbar\omega_o}{2}\begin{pmatrix} 0 \\ 1 \end{pmatrix} = -\frac{\hbar\omega_o}{2}|1\rangle = -\mu_B B_o|1\rangle \tag{14.45}$$

Thus, the time-independent Schrodinger equation is satisfied with energy:

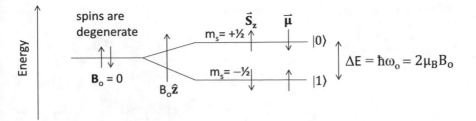

Fig. 14.3 The Zeeman effect for a free electron

$$E_\uparrow = +\mu_B B_o = \frac{1}{2}\hbar\omega_o \qquad (14.46)$$

for spin up, and

$$E_\downarrow = -\mu_B B_o = -\frac{1}{2}\hbar\omega_o \qquad (14.47)$$

for spin down. These two energy levels correspond to the stationary states, $|0\rangle$ and $|1\rangle$, for the electron spin in a magnetic field. The resulting energy diagram for an electron in a B-field is depicted in Fig. 14.3. Note that the energy levels are proportional to the magnetic field. In the absence of a magnetic field ($B_o = 0$), the energy for spin up and spin down are degenerate. The spin-degenerate energy levels are split due to the B field. Remember that the μ and S vectors point in opposite directions for the electron. Thus, the energy is lowest $\left(E_\downarrow = -\mu_B B_o = -\frac{1}{2}\hbar\omega_o\right)$ for spin down when the magnetic moment is parallel to B_o; and the energy is highest $\left(E_\uparrow = +\mu_B B_o = \frac{1}{2}\hbar\omega_o\right)$ for spin up when the magnetic moment is anti-parallel to B_o. This result could have been predicted directly from the classical equation, $U = -\mu \cdot B$. The energy splitting between spin up and spin down is $2\mu_B B_o = \hbar\omega_o$. This energy splitting due to a magnetic field is known as the Zeeman effect, named after the Dutch physicist, Pieter Zeeman (Fig. 14.4). The electron magnetic moment will be aligned or anti-aligned with the magnetic field according to the probability of occupation of the two energy levels given by Boltzmann statistics, with a higher probability in the lower than the upper level. Therefore, the magnetic field induces a high probability of alignment of the magnetic moment in the field direction, like we expect classically.

Note that the literature usually ascribes $|0\rangle$ to the ground state and $|1\rangle$ to the excited state, in contradiction to Fig. 14.3. However, Fig. 14.3 is consistent with the usual definition of $|0\rangle = \begin{pmatrix} 1 \\ 0 \end{pmatrix}$, representing a spin up electron, and $|1\rangle = \begin{pmatrix} 0 \\ 1 \end{pmatrix}$, representing a spin down electron. Of course, the labels inside the Dirac notation are arbitrary. We could just as well have labeled spin down as $|0\rangle$ and spin up as $|1\rangle$. Also, the g-factor (Eq. (3.19)) is negative (γ is positive) for electrons in many important semiconductors (e.g., InAs, InSb) due to spin–orbit interaction [2], which would flip the energy levels; i.e., $|0\rangle = \begin{pmatrix} 1 \\ 0 \end{pmatrix}$ would correspond to the lowest energy level and

$|1\rangle = \begin{pmatrix} 0 \\ 1 \end{pmatrix}$ would correspond to the highest energy level. Also, γ is positive for the proton, so again the energy levels would flip for proton spins.

Exercise 14.7 Calculate the Zeeman splitting for an electron in a 10 T magnetic field.

Exercise 14.8 Calculate the Zeeman splitting for a proton in a 10 T magnetic field.

Exercise 14.9 Estimate the magnetic field that would be required for the electron Zeeman splitting to exceed the thermal energy at room temperature? At 4 K?

14.6 Quantum Description of Precession

Remember from Chap. 2 that stationary states have a time-dependence of the form:

$$|\psi(t)\rangle = |\psi(0)\rangle e^{-iEt/\hbar} \tag{14.48}$$

Let us look at an example using electron spin. Any spin state, say at $t = 0$, can be represented by the following Bloch vector (equivalent to Eq. (14.1) within a global phase factor):

$$|\psi(0)\rangle = e^{-i\frac{\phi}{2}} \cos\frac{\theta}{2}|\uparrow\rangle + e^{+i\frac{\phi}{2}} \sin\frac{\theta}{2}|\downarrow\rangle \tag{14.49}$$

where $|\uparrow\rangle$ represents the spin up state and $|\downarrow\rangle$ represents the spin down state. Suppose we apply a B-field along the z-direction. Then each stationary state ($|\uparrow\rangle$, $|\downarrow\rangle$) in the superposition of Eq. (14.49) will evolve according to Eq. (14.48). Adding the known time dependence from Eq. (14.48) into (14.49), the spin state at some later time t will be:

$$|\psi(t)\rangle = e^{-i\frac{\phi}{2}} e^{-iE_\uparrow \frac{t}{\hbar}} \cos\frac{\theta}{2}|\uparrow\rangle + e^{+i\frac{\phi}{2}} e^{-iE_\downarrow \frac{t}{\hbar}} \sin\frac{\theta}{2}|\downarrow\rangle \tag{14.50}$$

where $E_\uparrow$ and $E_\downarrow$ are the electron energy for spin up and spin down, respectively. Substituting Eqs. (14.46) and (14.47) gives:

$$|\psi(t)\rangle = e^{-i(\phi+\omega_o t)/2} \cos\frac{\theta}{2}|\uparrow\rangle + e^{+i(\phi+\omega_o t)/2} \sin\frac{\theta}{2}|\downarrow\rangle \tag{14.51}$$

Comparing Eq. (14.51) with Eq. (14.49), we see that the phase ϕ evolves at the constant rate $\omega_o t$, while the angle θ is unaffected. The spin rotates (precesses) around the z-axis in the Bloch sphere at the Larmor frequency, ω_o, similar to the classical description. The state is not stationary; this is expected—remember from Chap. 2 that a superposition is not a stationary state.

Using the vector representation, we can write Eq. (14.51) as:

$$|\psi(t)\rangle = \hat{R}_z(\omega_o t)|\psi(0)\rangle \tag{14.52}$$

where

$$\hat{R}_z(\omega_o t) = \begin{pmatrix} e^{-\frac{i\omega_o t}{2}} & 0 \\ 0 & e^{\frac{i\omega_o t}{2}} \end{pmatrix} \tag{14.53}$$

is the rotation matrix, and

$$|\psi(0)\rangle = \begin{pmatrix} e^{-i\frac{\phi}{2}} \cos\frac{\theta}{2} \\ e^{+i\frac{\phi}{2}} \sin\frac{\theta}{2} \end{pmatrix} \tag{14.54}$$

is the initial qubit state.

For example, suppose we have a spin oriented along the x-direction. From Eq. (3.26), we know this can be written as a superposition of spin states along z:

$$|\psi(0)\rangle = \frac{1}{\sqrt{2}}(|\uparrow_z\rangle + |\downarrow_z\rangle) = |\uparrow_x\rangle \tag{14.55}$$

The time evolution of this wavefunction in a B-field is:

$$|\psi(t)\rangle = \frac{1}{\sqrt{2}}\left(e^{-\frac{i\omega_0 t}{2}}|\uparrow_z\rangle + e^{+\frac{i\omega_0 t}{2}}|\downarrow_z\rangle\right) \tag{14.56}$$

Suppose we implement the B-field for some duration, t, corresponding to $\omega_o t = \pi$. This is referred to as a π pulse. Substituting $\omega_o t = \pi$ into Eq. (14.56) gives:

$$|\psi(t)\rangle = \frac{1}{\sqrt{2}}(-i|\uparrow_z\rangle + i|\downarrow_z\rangle) \tag{14.57}$$

From Eq. (3.26), this can be written as a spin state along x:

$$|\psi(t)\rangle = -i|\downarrow_x\rangle \tag{14.58}$$

Equation (14.58) is equal to $|\downarrow_x\rangle$ within the arbitrary phase factor of $-i$. We ignore the arbitrary phase factor of $-i$ since it doesn't affect probabilities. Thus, the B-field pulse (of duration $\omega_o t = \pi$) along the z direction flips the spin from $|\uparrow_x\rangle$ to $|\downarrow_x\rangle$. The spin precesses in the x-y plane of the Bloch sphere around the B-field pointing along the z-axis.

Using magnetic fields (B_x, B_y, B_z) applied for a suitable time, you can rotate a qubit around the x, y or z axis in the Bloch sphere $(\widehat{R}_x, \widehat{R}_y, \widehat{R}_z)$; i.e., we can evolve any spin state into any other spin state by precession. This forms any single-qubit gate operation, $\widehat{U}(\theta, \phi)$.

References

1. File: Joseph Larmor.jpeg. (2020, June 1). *Wikimedia Commons, the free media repository.* Retrieved 15:38, December 7, 2020 from https://commons.wikimedia.org/w/index.php?title= File:Joseph_Larmor.jpeg&oldid=423068824
2. H. Kosaka et al., Electronics Lett. **37**, 464 (2001)

3. This is an image from the Nationaal Archief, the Dutch National Archives, and Spaarnestad Photo. This file is licensed under the Creative Commons Attribution-Share Alike 3.0 Netherlands license (https://creativecommons.org/licenses/by-sa/3.0/nl/deed.en). File: Pieter Zeeman.jpg. (2020, September 19). *Wikimedia Commons, the free media repository.* Retrieved 15:42, December 7, 2020 from https://commons.wikimedia.org/w/index.php?title=File:Pieter_Zeeman.jpg&oldid=464278095

Chapter 15
Electron Spin Resonance

The analysis of the preceding Chapter was for a constant magnetic field. From Exercise 14.9, we see that the magnetic field is very large (>1 T) for the Zeeman splitting to exceed the thermal energy. This is desired to prevent unwanted thermal excitation between the Zeeman split energy levels from affecting our qubit states. However, it is difficult to change such large static B-fields rapidly enough for quantum gates. Also, the resulting Larmor frequency ω_o is very fast ($\gg$GHz). We would like to have a slower rotation to get better control. To use smaller fields and slow down the precession, we use a technique called "electron spin resonance" (ESR).

15.1 ESR

ESR, developed by Isidor Rabi (Fig. 15.1) [1], uses both a static magnetic field, $B_o\hat{z}$ (called the longitudinal field), and an oscillating magnetic field, $B_1 \cos(\omega_o t)\hat{x}$ (called the transverse field):

$$B(t) = B_o\hat{z} + B_1 \cos(\omega_o t)\hat{x} \tag{15.1}$$

where ω_o is the Larmor frequency. The magnitude of the B_1 field is much less than that for the B_o field. Typically, $B_1 \sim 10^{-3} B_o$. Thus, as depicted in Fig. 15.2, a large B_o field vector along z oscillates back and forth along x due to the small B_1 field. From Eqs. (14.38) and (14.39), the Hamiltonian for the ESR field is:

$$\widehat{H}(t) = -\mathbf{\mu} \cdot \mathbf{B} = -\gamma \mathbf{B} \cdot \hat{\mathbf{S}} = -\gamma \left[B_o\hat{S}_z + B_1 \cos(\omega_o t)\hat{S}_x \right] \tag{15.2}$$

where $\gamma = \frac{q}{m_e}$ for the electron. We could plug Eq. (15.2) into the time-dependent Schrödinger equation and solve for $|\psi(t)\rangle$ to obtain the dynamics of this system.

© The Author(s), under exclusive license to Springer Nature Switzerland AG 2021
R. LaPierre, *Introduction to Quantum Computing*, The Materials Research Society Series,
https://doi.org/10.1007/978-3-030-69318-3_15

Fig. 15.1 Isidor Isaac Rabi
(1898–1988); Nobel Prize in
Physics in 1944. *Credit*
Wikimedia Commons [2]

Fig. 15.2 The ESR field

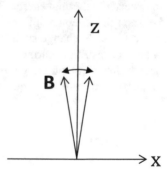

This rigorous derivation is done in the last section of this chapter. However, there is an easier approach as follows.

The large $B_o\hat{z}$ field results in Larmor precession around $\hat{z}$ with the Larmor frequency:

$$\omega_o = \frac{eB_o}{m_e} = \frac{2\mu_B B_o}{\hbar} \tag{15.3}$$

We can divide the transverse $B_1 \cos(\omega_o t)\hat{x}$ field into two counter-rotating components, rotating at the frequency ω_o, as depicted in Fig. 15.3. One component, B_{ccw}, is rotating counter-clockwise in the x-y plane, while the other component, B_{cw}, is rotating clockwise. Referring to Fig. 15.3, if we superimpose the B_{ccw} and B_{cw} fields at any given time, the y-components will cancel, while the x components add, giving us the desired transverse field, $B_1 \cos(\omega_o t)\hat{x}$.

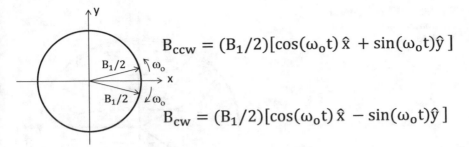

Fig. 15.3 The transverse field is divided into two counter-rotating components

Now let us change our frame of reference, so we rotate counter-clockwise at the same frequency as ω_o. The Larmor precession around $\hat{z}$ due to the longitudinal B_o field will disappear in the rotating frame of reference, and the B_{ccw} component will appear to be static, giving a field of $\frac{B_1}{2}\hat{x}$. The spin will now precess slowly around $\hat{x}$ due to the small $\frac{B_1}{2}\hat{x}$ field. The B_{cw} component appears to rotate very rapidly and time averages to zero (this is called the "rotating wave approximation" (RWA)).

The rate of precession around $\hat{x}$ due to $\frac{B_1}{2}\hat{x}$ will be:

$$\Omega = \frac{eB_1}{2m_e} = \frac{\mu_B B_1}{\hbar} \tag{15.4}$$

Equation (15.4) is obtained by replacing B_o in Eq. (15.3) with $B_1/2$. Ω is called the Rabi frequency. By comparing Eqs. (15.3) and (15.4), we see that $\Omega \ll \omega_o$, since $B_1 \ll B_o$. Hence, while ω_o is in the GHz range, Ω is typically in the MHz (radio frequency (RF)) range. The slower rotation gives better control of the qubit.

The Hamiltonian in the rotating frame due to the ESR field is:

$$\widehat{H}' = \frac{\hbar}{2}\Omega\hat{\sigma}_x = \begin{pmatrix} 0 & \frac{\hbar}{2}\Omega \\ \frac{\hbar}{2}\Omega & 0 \end{pmatrix} = \frac{\hbar}{2}\Omega(|0\rangle\langle 1| + |1\rangle\langle 0|) \tag{15.5}$$

which is similar to Eq. (14.41) but with ω_o replaced by Ω, and $\hat{\sigma}_z$ replaced by $\hat{\sigma}_x$. We see that $\widehat{H}'$ couples the states $|0\rangle$ and $|1\rangle$.

We can visualize the effect of the ESR field on a single qubit in the Bloch sphere. If we move back into the static frame of reference, the qubit will precess rapidly around $\hat{z}$ at the Larmor frequency ω_o due to the longitudinal $B_o\hat{z}$ field, and also precess slowly around $\hat{x}$ at the Rabi frequency Ω due to the transverse $B_1\cos(\omega_o t)\hat{x}$ field. The combined rotation appears as a helical or corkscrew trajectory on the surface of the Bloch sphere, as depicted in Fig. 15.4.

At time $t = 0$, the single qubit can be written in the usual Bloch vector form:

$$|\psi(0)\rangle = \cos\left(\frac{\theta_o}{2}\right)|0\rangle + e^{i\phi_o}\sin\left(\frac{\theta_o}{2}\right)|1\rangle \tag{15.6}$$

Fig. 15.4 ESR trajectory of electron spin depicted in the Bloch sphere

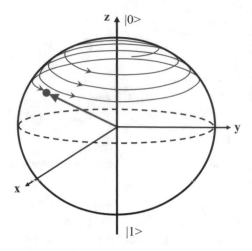

If we turn on the ESR field, the single qubit at some later time, t, becomes:

$$|\psi(t)\rangle = \cos\left(\frac{\theta_o + \Omega t}{2}\right)|0\rangle + e^{i(\phi_o + \omega_o t)}\sin\left(\frac{\theta_o + \Omega t}{2}\right)|1\rangle \tag{15.7}$$

The qubit state is rotating (precessing) rapidly around the z-axis in the Bloch sphere at the Larmor frequency with azimuthal angle $\phi = \phi_o + \omega_o t$. The qubit is simultaneously traveling down the Bloch sphere at polar angle $\theta = \theta_o + \Omega t$, much slower compared to ω_o. Although ω_o is very large, Ω can be very small, allowing a greater degree of control. The total rotation angle depends on the B_1 pulse duration. To make a NOT gate, we apply the oscillating B_1 field for a total time of $t = \pi/\Omega$. If we continue applying the oscillating field, the qubit will transition back to $|0\rangle$. To achieve a superposition state, just rotate $|0\rangle$ to the x-y plane of the Bloch sphere with a $\pi/2$ pulse, then turn off B_1. The electron thus oscillates back and forth between $|0\rangle$ and $|1\rangle$ at a frequency equal to Ω. This oscillatory behavior in response to the oscillating field is called Rabi oscillation or Rabi flopping.

Thus, an oscillating magnetic field along the x-axis allows us to realize rotation about the x-axis in the Bloch sphere. Similarly, an oscillating magnetic field along the y or z-axis allows us to realize rotation about the y or z-axis in the Bloch sphere. A theorem from geometry tells us that any rotation about a vector $\hat{n}$ can be performed by doing three rotations about two fixed axes. If we take our two axes to be x and y, for example, then any rotation operator around an arbitrary direction $\hat{n}$ can be written as:

$$\hat{R}_n(\theta) = \hat{R}_y(\alpha)\hat{R}_x(\beta)\hat{R}_y(\delta) \tag{15.8}$$

for some angles α, β, and δ. Therefore, we only need two of the 3 rotation matrices. Furthermore, as will be proven below, we can implement a rotation around the y-axis simply by adjusting the phase of the RF pulse along the x-axis. Thus, we only need an RF field along one direction. In this manner, any arbitrary unitary transformation can be made on a single qubit; i.e., any arbitrary rotation on the Bloch sphere. In principle, the fields should be switched off when the gate is achieved. The weak alternating field B_1 is easy to switch on and off, hence the term 'pulse'. The large static field B_o is not so easy to turn off and one must then keep track of the accumulated phase on the qubit resulting from Larmor precession.

Exercise 15.1 What B_o field is required for electron spin resonance at 10 T? Suppose $B_1 = 10^{-3} B_o$. Starting in the $|0\rangle$ state, describe how you would achieve the $\frac{1}{\sqrt{2}}(|0\rangle + |1\rangle)$ state. What Larmor frequency and ESR pulse duration would be required?

15.2 Rigorous Derivation of ESR

This section may be considered optional. We will prove that an oscillating magnetic field, $B_1 \cos(\omega t)\hat{x}$, causes precession around the x-axis. We need to solve (where we have dropped the ket notation):

$$i\hbar \frac{\partial \psi}{\partial t} = \widehat{H}\psi \tag{15.9}$$

with the Hamiltonian:

$$\widehat{H} = -\gamma \mathbf{B}(t) \cdot \widehat{\mathbf{S}} = -\gamma \left[B_o \widehat{S}_z + B_1 \widehat{S}_x \cos(\omega t) \right] \tag{15.10}$$

This is tricky to solve due to the time-dependence of $\widehat{H}$. Let us apply a rotating frame of reference. A state in a static (lab) frame will appear to be rotating clockwise in a counter-clockwise rotating frame. To describe this, let us apply the transformation:

$$\psi' = e^{-\frac{i}{\hbar}(-\omega_o)\widehat{S}_z t} \psi \tag{15.11}$$

state in rotating frame   cw rotation   state in static frame

The exponential in Eq. (15.11) is the unitary transformation for clockwise rotation of the state at the frequency ω_o. We rearrange Eq. (15.11) for ψ, and insert ψ into Eq. (15.9). Using the rotating wave approximation for the oscillatory B_1 field, we get:

$$i\hbar\frac{\partial}{\partial t}\left(e^{-\frac{i}{\hbar}\omega_o\hat{S}_z t}\psi'\right)$$

$$= -\gamma\left[B_o\hat{S}_z + \frac{B_1}{2}\left(\hat{S}_x\cos(\omega_o t) + \hat{S}_y\sin(\omega_o t)\right)\right]\left(e^{-\frac{i}{\hbar}\omega_o\hat{S}_z t}\psi'\right) \quad (15.12)$$

Note that Eq. (15.12) is also exact for a circularly polarized wave, $\frac{B_1}{2}\left(\cos(\omega_o t)\hat{x} + \sin(\omega_o t)\hat{y}\right)$. Substituting $\omega_o = -\gamma B_o$ and $\Omega = -\gamma B_1/2$ gives:

$$i\hbar\frac{\partial}{\partial t}\left(e^{-\frac{i}{\hbar}\omega_o\hat{S}_z t}\psi'\right) = \left[\omega_o\hat{S}_z + \Omega\left(\hat{S}_x\cos(\omega_o t) + \hat{S}_y\sin(\omega_o t)\right)\right]\left(e^{-\frac{i}{\hbar}\omega_o\hat{S}_z t}\psi'\right)$$

$$(15.13)$$

Evaluating the partial derivative, introducing the Pauli operators, and rearranging gives:

$$\frac{\partial\psi'}{\partial t} = -\frac{i}{2}e^{+\frac{i\omega_o\hat{\sigma}_z t}{2}}\left[\Omega\left(\hat{\sigma}_x\cos(\omega_o t) + \hat{\sigma}_y\sin(\omega_o t)\right)\right]e^{-\frac{i\omega_o\hat{\sigma}_z t}{2}}\psi' \quad (15.14)$$

Equation (15.14) can be simplified by the following relations:

$$e^{+\frac{i\omega_o\hat{\sigma}_z t}{2}}\hat{\sigma}_x e^{-\frac{i\omega_o\hat{\sigma}_z t}{2}} = \begin{pmatrix} e^{+\frac{i\omega_o t}{2}} & 0 \\ 0 & e^{-\frac{i\omega_o t}{2}} \end{pmatrix}\begin{pmatrix} 0 & 1 \\ 1 & 0 \end{pmatrix}\begin{pmatrix} e^{-\frac{i\omega_o t}{2}} & 0 \\ 0 & e^{+\frac{i\omega_o t}{2}} \end{pmatrix}$$

$$= \begin{pmatrix} 0 & e^{+i\omega_o t} \\ e^{-i\omega_o t} & 0 \end{pmatrix} \quad (15.15)$$

$$e^{+\frac{i\omega_o\hat{\sigma}_z t}{2}}\hat{\sigma}_y e^{-\frac{i\omega_o\hat{\sigma}_z t}{2}} = \begin{pmatrix} e^{+\frac{i\omega_o t}{2}} & 0 \\ 0 & e^{-\frac{i\omega_o t}{2}} \end{pmatrix}\begin{pmatrix} 0 & -i \\ i & 0 \end{pmatrix}\begin{pmatrix} e^{-\frac{i\omega_o t}{2}} & 0 \\ 0 & e^{+\frac{i\omega_o t}{2}} \end{pmatrix}$$

$$= \begin{pmatrix} 0 & -ie^{+i\omega_o t} \\ ie^{-i\omega_o t} & 0 \end{pmatrix} \quad (15.16)$$

Substituting these into Eq. (15.14) gives:

$$\frac{\partial\psi'}{\partial t} = -\frac{i}{2}\left[\Omega\cos(\omega_o t)\begin{pmatrix} 0 & e^{+i\omega_o t} \\ e^{-i\omega_o t} & 0 \end{pmatrix} + \Omega\sin(\omega_o t)\begin{pmatrix} 0 & -ie^{+i\omega_o t} \\ ie^{-i\omega_o t} & 0 \end{pmatrix}\right]\psi'$$

$$(15.17)$$

or

$$\frac{\partial \psi'}{\partial t} = -\frac{i}{2}\left[\Omega\left(\begin{array}{cc} 0 & [\cos(\omega_o t) - i\sin(\omega_o t)]e^{+i\omega_o t} \\ [\cos(\omega_o t) + i\sin(\omega_o t)]e^{-i\omega_o t} & 0 \end{array}\right)\right]\psi'$$

$$(15.18)$$

Next, we substitute the Euler relation, resulting in:

$$\frac{\partial \psi'}{\partial t} = -\frac{i}{2}\left[\Omega\left(\begin{array}{cc} 0 & 1 \\ 1 & 0 \end{array}\right)\right]\psi' \tag{15.19}$$

or

$$\frac{\partial \psi'}{\partial t} = -\frac{i}{2}\Omega\hat{\sigma}_x\psi' \tag{15.20}$$

Finally, after rearranging:

$$i\hbar\frac{\partial \psi'}{\partial t} = \Omega\hat{S}_x\psi' \tag{15.21}$$

Equation (15.21) is the time-independent Schrodinger equation with a Hamiltonian given by:

$$\widehat{H}' = \Omega\hat{S}_x \tag{15.22}$$

Hence, our transformation, Eq. (15.11), transformed the time-dependent Hamiltonian (Eq. (15.10)) to a time-independent one. We know how to solve the time-dependent Schrodinger equation with a time-independent Hamiltonian:

$$\psi'(t) = e^{-\frac{i}{\hbar}\widehat{H}'t}\psi'(0) = e^{-\frac{i}{\hbar}\Omega\hat{S}_x t}\psi'(0) \tag{15.23}$$

Finally, we apply Eq. (15.11) again to transform back to the static frame:

$$\psi(t) = e^{-\frac{i}{\hbar}\omega_o\hat{S}_z t}e^{-\frac{i}{\hbar}\Omega\hat{S}_x t}\psi(0) = e^{-\frac{i}{\hbar}\left(\omega_o\hat{S}_z + \Omega\hat{S}_x\right)t}\psi(0) \tag{15.24}$$

The exponential term corresponds to counter-clockwise rotation around the x-axis at frequency Ω, and counter-clockwise rotation around the z-axis at frequency ω_o.

Now what happens if we add a phase φ to the incident B-field. Equation (15.18) becomes:

$$\frac{\partial \psi'}{\partial t}$$

$$= -\frac{i}{2}\left[\Omega\begin{pmatrix} 0 & [\cos(\omega_o t + \varphi) - i\sin(\omega_o t + \varphi)]e^{+i\omega_o t} \\ [\cos(\omega_o t + \varphi) + i\sin(\omega_o t + \varphi)]e^{-i\omega_o t} & 0 \end{pmatrix}\right]\psi'$$

(15.25)

Substituting the Euler relation gives:

$$\frac{\partial \psi'}{\partial t} = -\frac{i}{2}\left[\Omega\begin{pmatrix} 0 & e^{-i\varphi} \\ e^{+i\varphi} & 0 \end{pmatrix}\right]\psi'$$

(15.26)

Note that:

$$\begin{pmatrix} 0 & e^{-i\varphi} \\ e^{+i\varphi} & 0 \end{pmatrix} = \begin{pmatrix} 0 & \cos\varphi \\ \cos\varphi & 0 \end{pmatrix} + \begin{pmatrix} 0 & -i\sin\varphi \\ i\sin\varphi & 0 \end{pmatrix} = \cos\varphi\hat{\sigma}_x + \sin\varphi\hat{\sigma}_y$$

(15.27)

Hence, we obtain:

$$i\hbar\frac{\partial \psi'}{\partial t} = \left(\Omega\cos\varphi\hat{S}_x + \Omega\sin\varphi\hat{S}_y\right)\psi'$$

(15.28)

which is simply the time-dependent Schrodinger equation with a Hamiltonian:

$$\widehat{H}' = \Omega\cos\varphi\hat{S}_x + \Omega\sin\varphi\hat{S}_y$$

(15.29)

By a suitable phase, one can have rotation around the x or y-axis.

References

1. I.I. Rabi, Phys. Rev. **51**, 652 (1937)
2. File: II Rabi.jpg. (2020, October 21). *Wikimedia Commons, the free media repository*. Retrieved 15:45, December 7, 2020 from https://commons.wikimedia.org/w/index.php?title=File:II_Rabi.jpg&oldid=496554939.d

Chapter 16
Two-State Dynamics

In the previous chapter, we examined electron spin resonance (ESR) used to implement single-qubit gates. In this chapter, the dynamics of a two-level quantum system and ESR will be examined in more detail. We consider how the electron cycles back and forth between two energy levels.

16.1 Hamiltonian of a Two-Level System

As we saw in Chap. 15, if we apply a B_1 field continuously at frequency ω_o to a two-level system with energy separation $\hbar\omega_o$ (this is known as resonance), then the system cycles back and forth between $|0\rangle$ and $|1\rangle$ deterministically at a well-defined rate with Rabi frequency Ω. Starting in the ground state, an electron can transition to the excited state by photon absorption. If we continue to apply the B_1 field, then the electron can transition from the excited to the ground state by "stimulated emission". Stopping the field at the appropriate time can produce quantum gates.

The Hamiltonian for the two-level system is:

$$\widehat{H} = \begin{pmatrix} E_0 & 0 \\ 0 & E_1 \end{pmatrix} \tag{16.1}$$

Thus, the time-independent Schrodinger equation, $\widehat{H}|\psi\rangle = E|\psi\rangle$, for the two states becomes:

$$\widehat{H}|0\rangle = \begin{pmatrix} E_0 & 0 \\ 0 & E_1 \end{pmatrix}\begin{pmatrix} 1 \\ 0 \end{pmatrix} = E_0\begin{pmatrix} 1 \\ 0 \end{pmatrix} = E_0|0\rangle \tag{16.2}$$

and

© The Author(s), under exclusive license to Springer Nature Switzerland AG 2021
R. LaPierre, *Introduction to Quantum Computing*, The Materials Research Society Series,
https://doi.org/10.1007/978-3-030-69318-3_16

$$\widehat{H}|1\rangle = \begin{pmatrix} E_0 & 0 \\ 0 & E_1 \end{pmatrix} \begin{pmatrix} 0 \\ 1 \end{pmatrix} = E_1 \begin{pmatrix} 0 \\ 1 \end{pmatrix} = E_1|1\rangle \tag{16.3}$$

Thus, the states $|0\rangle$ and $|1\rangle$ are stationary states of this system. Note that the matrix, Eq. (16.1), is a diagonal matrix, meaning it contains elements only along its diagonal. The eigenvalues are equal to the diagonal matrix elements.

A general quantum state is the superposition:

$$|\psi(t)\rangle = \alpha|0\rangle + \beta|1\rangle \rightarrow \begin{pmatrix} \alpha \\ \beta \end{pmatrix} \tag{16.4}$$

As we saw in Chap. 2, a superposition of stationary states is not a stationary state. The time-dependent Schrodinger equation, $i\hbar\frac{\partial}{\partial t}|\psi(t)\rangle = \widehat{H}|\psi(t)\rangle$, becomes:

$$i\hbar\frac{\partial}{\partial t}\begin{pmatrix} \alpha \\ \beta \end{pmatrix} = \begin{pmatrix} E_0 & 0 \\ 0 & E_1 \end{pmatrix}\begin{pmatrix} \alpha \\ \beta \end{pmatrix} \tag{16.5}$$

We know from Chap. 2 that the solution to Eq. (16.5) is:

$$|\psi(t)\rangle = \begin{pmatrix} \alpha e^{-i\frac{E_0}{\hbar}t} \\ \beta e^{-i\frac{E_1}{\hbar}t} \end{pmatrix} \tag{16.6}$$

Now suppose we add a time-dependent perturbation $\widehat{H}'$ (also called the interaction Hamiltonian). For example, $\widehat{H}'$ could be an oscillating magnetic field as used, for example, in electron spin resonance (ESR) . The time-dependent Schrodinger equation becomes:

$$(\widehat{H} + \widehat{H}')|\psi(t)\rangle = i\hbar\frac{\partial}{\partial t}|\psi(t)\rangle \tag{16.7}$$

or

$$\left[\begin{pmatrix} E_0 & 0 \\ 0 & E_1 \end{pmatrix} + \begin{pmatrix} W_{11} & W_{12} \\ W_{21} & W_{22} \end{pmatrix}\right]\begin{pmatrix} \alpha \\ \beta \end{pmatrix} = i\hbar\frac{\partial}{\partial t}\begin{pmatrix} \alpha \\ \beta \end{pmatrix} \tag{16.8}$$

where we have introduced the matrix for $\widehat{H}'$ with matrix elements W_{11}, W_{21}, W_{12} and W_{22}:

$$\widehat{H}' = \begin{pmatrix} W_{11} & W_{12} \\ W_{21} & W_{22} \end{pmatrix} \tag{16.9}$$

Exercise 16.1 Show that the matrix elements of $\widehat{H}'$ are $W_{11} = \langle 0|\widehat{H}'|0\rangle$, $W_{12} = \langle 0|\widehat{H}'|1\rangle$, $W_{21} = \langle 1|\widehat{H}'|0\rangle$, and $W_{22} = \langle 1|\widehat{H}'|1\rangle$.

16.2 Spin Qubits

The explicit form of $\widehat{H}'$ depends on the perturbation. Let us examine the case of electron spin resonance (ESR) where we know from Chap. 15 that:

$$\widehat{H} + \widehat{H}' = -\gamma\left[B_o\widehat{S}_z + B_1\cos(\omega_o t)\widehat{S}_x\right] = \mu_B\begin{pmatrix} B_o & B_1\cos(\omega_o t) \\ B_1\cos(\omega_o t) & -B_o \end{pmatrix} \quad (16.10)$$

The time-dependent Schrodinger equation becomes:

$$i\hbar\frac{\partial}{\partial t}|\psi(t)\rangle = \mu_B\begin{pmatrix} B_o & B_1\cos(\omega_o t) \\ B_1\cos(\omega_o t) & -B_o \end{pmatrix}|\psi(t)\rangle \quad (16.11)$$

Using Eq. (16.6) for inspiration, let us try the following solution to Eq. (16.11):

$$|\psi(t)\rangle = \begin{pmatrix} \alpha(t)e^{-i\frac{E_0}{\hbar}t} \\ \beta(t)e^{-i\frac{E_1}{\hbar}t} \end{pmatrix} \quad (16.12)$$

where the coefficients, $\alpha(t)$ and $\beta(t)$, now have a time dependence. Equation (16.11) contains two coupled differential equations:

$$i\hbar\frac{\partial}{\partial t}\left[\alpha(t)e^{-i\frac{E_0}{\hbar}t}\right] = \mu_B B_o\left[\alpha(t)e^{-i\frac{E_0}{\hbar}t}\right] + \mu_B B_1\cos(\omega_o t)\left[\beta(t)e^{-i\frac{E_1}{\hbar}t}\right] \quad (16.13)$$

and

$$i\hbar\frac{\partial}{\partial t}\left[\beta(t)e^{-i\frac{E_1}{\hbar}t}\right] = \mu_B B_1\cos(\omega_o t)\left[\alpha(t)e^{-i\frac{E_0}{\hbar}t}\right] - \mu_B B_o\left[\beta(t)e^{-i\frac{E_1}{\hbar}t}\right] \quad (16.14)$$

Thus, we see from Eqs. (16.13) and (16.14) that the amplitude $\beta(t)$ to be found in the $|1\rangle$ state will change based on the amplitude $\alpha(t)$ to be in the $|0\rangle$ state, and vice versa. The two states are coupled.

We need to solve Eqs. (16.13) and (16.14) to find the time-dependence of $|\psi\rangle$. Let us start with Eq. (16.13). Evaluating the derivative on the left-hand side, we obtain:

$$i\hbar\frac{\partial\alpha(t)}{\partial t}e^{-i\frac{E_0}{\hbar}t} + E_0\alpha(t)e^{-i\frac{E_0}{\hbar}t} = \mu_B B_o\alpha(t)e^{-i\frac{E_0}{\hbar}t} + \mu_B B_1\cos(\omega_o t)\beta(t)e^{-i\frac{E_1}{\hbar}t}$$

$$(16.15)$$

Recognizing that $E_0 = \mu_B B_o$, we get:

$$i\hbar\frac{\partial\alpha(t)}{\partial t} = \mu_B B_1\beta(t)\cos(\omega_o t)e^{-i\frac{(E_1-E_0)}{\hbar}t}$$

$$(16.16)$$

Substituting $E_0 - E_1 = 2\mu_B B_o = \hbar\omega_o$, we get:

$$i\hbar\frac{\partial\alpha(t)}{\partial t} = \mu_B B_1\beta(t)\cos(\omega_o t)e^{+i\omega_o t}$$

$$(16.17)$$

We can expand $\cos(\omega_o t)$, obtaining:

$$i\hbar\frac{\partial\alpha(t)}{\partial t} = \mu_B B_1\beta(t)\frac{e^{i\omega t} + e^{-i\omega t}}{2}e^{+i\omega_o t}$$

$$(16.18)$$

or

$$i\hbar\frac{\partial\alpha(t)}{\partial t} = \frac{\mu_B B_1\beta(t)}{2}\left(e^{i(\omega+\omega_o)t} + e^{-i(\omega-\omega_o)t}\right)$$

$$(16.19)$$

At resonance ($\omega = \omega_o$), the second term in brackets is 1, and the first term is changing very rapidly compared to the second term and time averages to zero. This is the rotating wave approximation (RWA) mentioned in Chap. 15. Finally, Eq. (16.19) becomes:

$$\frac{\partial\alpha(t)}{\partial t} = \frac{\mu_B B_1}{2i\hbar}\beta(t)$$

$$(16.20)$$

A similar derivation applied to Eq. (16.14) gives:

$$\frac{\partial\beta(t)}{\partial t} = \frac{\mu_B B_1}{2i\hbar}\alpha(t)$$

$$(16.21)$$

If we assume an initial condition of $\alpha(0) = 1$ and $\beta(0) = 0$ (state starts in $|0\rangle$), then the solution is:

$$\alpha(t) = \cos\left(\frac{\Omega t}{2}\right)$$

$$(16.22)$$

$$\beta(t) = -i\sin\left(\frac{\Omega t}{2}\right)$$

$$(16.23)$$

where Ω is the Rabi frequency:

$$\Omega = \frac{\mu_B B_1}{\hbar} \tag{16.24}$$

Thus, from Eq. (16.12):

$$|\psi(t)\rangle = \begin{pmatrix} \cos\left(\frac{\Omega t}{2}\right) e^{-i\frac{E_0}{\hbar}t} \\ -i\sin\left(\frac{\Omega t}{2}\right) e^{-i\frac{E_1}{\hbar}t} \end{pmatrix} \tag{16.25}$$

Equation (16.25) gives the probability amplitudes for the states $|0\rangle$ and $|1\rangle$. The corresponding probabilities are:

$$P_0 = |\alpha(t)|^2 = \left| \cos\left(\frac{\Omega t}{2}\right) e^{-i\frac{E_0}{\hbar}t} \right|^2 = \cos^2\left(\frac{\Omega t}{2}\right) \tag{16.26}$$

and

$$P_1 = |\beta(t)|^2 = \left| -i\sin\left(\frac{\Omega t}{2}\right) e^{-i\frac{E_1}{\hbar}t} \right|^2 = \sin^2\left(\frac{\Omega t}{2}\right) \tag{16.27}$$

P_0 and P_1 are shown in Fig. 16.1 as a function of time. The probabilities oscillate between $|0\rangle$ and $|1\rangle$ at the frequency Ω. At $t = (2n + 1)\,\pi/\Omega$ where n is an integer ($n = 0, 1, 2, \ldots$), the system exists in the state $|1\rangle$. At $t = 2\pi n/\Omega$, the system exists in the state $|0\rangle$. At other times, we have a superposition. These are the Rabi oscillations discussed in the previous chapter.

Exercise 16.2 Write out the explicit form of the state $|\psi\rangle$ at a time $t = \pi/2\Omega$.

Note that for the single-qubit system described above, we had to solve two coupled differential equations, Eqs. (16.13) and (16.14). For two qubits, we would need to solve four differential equations for the time-dependence of the four basis states, $|00\rangle$, $|01\rangle$, $|10\rangle$ and $|11\rangle$. In general, for n qubits, we would need to solve 2^n differential equations. This is the difficult aspect of simulating quantum systems using classical computers. Instead, we could use a quantum computer to efficiently simulate another quantum system, which was Feynman's original idea [1]. Using Eq. (16.25), we can write $|\psi\rangle$ in the form:

Fig. 16.1 Rabi oscillations

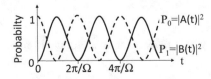

$$|\psi(t)\rangle = \cos\left(\frac{\Omega t}{2}\right)e^{-i\frac{E_0}{\hbar}t}|0\rangle - i\sin\left(\frac{\Omega t}{2}\right)e^{-i\frac{E_1}{\hbar}t}|1\rangle \qquad (16.28)$$

If we take out a global phase factor of $e^{-i\frac{E_0}{\hbar}t}$ from Eq. (16.28), we obtain:

$$|\psi(t)\rangle = e^{-i\frac{E_0}{\hbar}t}\left[\cos\left(\frac{\Omega t}{2}\right)|0\rangle - i\sin\left(\frac{\Omega t}{2}\right)e^{-i\frac{E_1-E_0}{\hbar}t}|1\rangle\right] \qquad (16.29)$$

Ignoring the global phase factor (because it does not affect probabilities), substituting $\omega_o = \frac{E_0-E_1}{\hbar}$, and realizing $-i = e^{-i\pi/2}$, we obtain:

$$|\psi(t)\rangle = \cos\left(\frac{\Omega t}{2}\right)|0\rangle + e^{+i\left(\omega_o t-\frac{\pi}{2}\right)}\sin\left(\frac{\Omega t}{2}\right)|1\rangle \qquad (16.30)$$

Equation (16.30) is in the familiar Bloch vector form.

Let us derive the unitary matrix, $\widehat{U}(t)$, for the rotations due to the ESR field. For the state starting in $|0\rangle = \begin{pmatrix} 1 \\ 0 \end{pmatrix}$, we previously derived (Eq. (16.25):

$$|\psi(t)\rangle = \begin{pmatrix} e^{-i\frac{\omega_o}{2}t}\cos\left(\frac{\Omega t}{2}\right) \\ -ie^{+i\frac{\omega_o}{2}t}\sin\left(\frac{\Omega t}{2}\right) \end{pmatrix} \qquad (16.31)$$

For the state starting in $|1\rangle$, we can similarly derive (Exercise 16.3):

$$|\psi(t)\rangle = \begin{pmatrix} -ie^{-i\frac{\omega_o}{2}t}\sin\left(\frac{\Omega t}{2}\right) \\ e^{+i\frac{\omega_o}{2}t}\cos\left(\frac{\Omega t}{2}\right) \end{pmatrix} \qquad (16.32)$$

Combining these results, we obtain:

$$\widehat{U}(t) = \begin{pmatrix} e^{-i\frac{\omega_o}{2}t}\cos\left(\frac{\Omega t}{2}\right) & -ie^{-i\frac{\omega_o}{2}t}\sin\left(\frac{\Omega t}{2}\right) \\ -ie^{+i\frac{\omega_o}{2}t}\sin\left(\frac{\Omega t}{2}\right) & e^{+i\frac{\omega_o}{2}t}\cos\left(\frac{\Omega t}{2}\right) \end{pmatrix} \qquad (16.33)$$

We can decompose $\widehat{U}(t)$ as follows:

$$\widehat{U}(t) = \begin{pmatrix} e^{-i\frac{\omega_o}{2}t}\cos\left(\frac{\Omega t}{2}\right) & -ie^{-i\frac{\omega_o}{2}t}\sin\left(\frac{\Omega t}{2}\right) \\ -ie^{+i\frac{\omega_o}{2}t}\sin\left(\frac{\Omega t}{2}\right) & e^{+i\frac{\omega_o}{2}t}\cos\left(\frac{\Omega t}{2}\right) \end{pmatrix}$$

$$= \begin{pmatrix} e^{-i\frac{\omega_o}{2}t} & 0 \\ 0 & e^{+i\frac{\omega_o}{2}t} \end{pmatrix}\begin{pmatrix} \cos\left(\frac{\Omega t}{2}\right) & -i\sin\left(\frac{\Omega t}{2}\right) \\ -i\sin\left(\frac{\Omega t}{2}\right) & \cos\left(\frac{\Omega t}{2}\right) \end{pmatrix} \qquad (16.34)$$

which is identical to a product of rotation matrices:

$$\widehat{U}(t) = R_z(\phi)R_x(\theta) \qquad (16.35)$$

with $\phi = \omega_o t$ and $\theta = \Omega t$, as expected.

Exercise 16.3 Derive Eq. (16.32).

16.3 Charge Qubits

Let us examine the case of "charge qubits" where, for example, we represent $|0\rangle$ and $|1\rangle$ by an electron in different energy levels where the spin is not necessarily changing; e.g., an electron in a H atom or a quantum well. In this case, we can achieve Rabi oscillations using oscillating electric fields coupled to the dipole moment of the charge distribution. In the classical model, the electric field causes oscillations of the electric dipole moment, **p**. How do we treat this quantum mechanically?

Remember that we defined a perturbation Hamiltonian, $\widehat{H}'$, in Eq. (16.9) with matrix elements defined in Exercise 16.1:

$$H'_{ij} = \langle \psi_i | \widehat{H}' | \psi_j \rangle = \int \psi_i^* \widehat{H}' \psi_j dxdydz \qquad (16.36)$$

Let us suppose $\widehat{H}'$ refers to an oscillating electric field, ε (e.g., a light wave). The electric field causes electric dipole oscillations with a classical potential energy $U = -\mathbf{p} \cdot \varepsilon$. If we arbitrarily choose the x-axis as the polarization direction of incident light, then $\varepsilon_x = \varepsilon_o \cos(\omega_o t)$. Further, if we consider a single electron, then the energy of the dipole moment is $U = -p_x \varepsilon_x = -qx\varepsilon_o \cos(\omega_o t)$ where q is the electron charge and x is the charge displacement from the positively charged nucleus. Analogously, the Hamiltonian can be written as $\widehat{H}' = -qx\varepsilon_o \cos(\omega_o t)$, resulting in the matrix elements:

$$H'_{ij} = \int \psi_i^* [-qx\varepsilon_o \cos(\omega_o t)] \psi_j dxdydz \qquad (16.37)$$

We can use the symmetry of the wavefunction to simplify Eq. (16.37). ψ is usually an even or odd function of x; e.g., think about the infinite quantum well, or the H atom. If ψ is an even or odd function of x, $|\psi|^2$ is always an even function of x, and $x|\psi|^2$ is always an odd function of x. Thus, the integral in Eq. (16.37) becomes zero when $i = j$, and Eq. (16.9) becomes:

$$\widehat{H}' = \begin{pmatrix} 0 & W_{11} \\ W_{21} & 0 \end{pmatrix} \qquad (16.38)$$

For the case of charge qubits, we have:

$$\widehat{H}' = \begin{pmatrix} 0 & U\cos(\omega_o t) \\ U\cos(\omega_o t) & 0 \end{pmatrix} \tag{16.39}$$

where

$$U = -\varepsilon_o(q \int \psi_0^* x \psi_1 \, dxdydz) \tag{16.40}$$

The term in brackets in Eq. (16.40) is simply the charge multiplied by the expectation value of x; i.e., it is a dipole matrix element. Finally, we can write:

$$\widehat{H} = \widehat{H} + \widehat{H}' = \begin{pmatrix} E_0 & U\cos(\omega_o t) \\ U\cos(\omega_o t) & E_1 \end{pmatrix} \tag{16.41}$$

Thus, we have the same form for the Hamiltonian as we had before for spin qubits (Eq. (16.10)) but with a Rabi frequency now given by:

$$\Omega = U/\hbar \tag{16.42}$$

Actually, Eq. (16.41) is often taken as the general form of the Hamiltonian in the presence of an oscillating perturbation. You only need to substitute the correct form of U associated with the oscillating field. In the case of spin qubits, U is the energy associated with a magnetic moment in a magnetic field ($U = \mu_B B_1$). In the case of charge qubits, U is the energy associated with an electric dipole in an electric field (Eq. (16.40)).

16.4 Avoided Crossing

Suppose we have a two-level system with:

$$\widehat{H}|\psi_0\rangle = E_0|\psi_0\rangle \tag{16.43}$$

and

$$\widehat{H}|\psi_1\rangle = E_1|\psi_1\rangle \tag{16.44}$$

The energies, E_0 and E_1, are eigenvalues of the system, and $|\psi_0\rangle$ and $|\psi_1\rangle$ are stationary eigenstates. If the system is prepared in the state $|\psi_0\rangle$ with energy E_0, or the state $|\psi_1\rangle$ with energy E_1, then it will stay there forever. The Hamiltonian can be written as:

$$\widehat{H} = \begin{pmatrix} E_0 & 0 \\ 0 & E_1 \end{pmatrix} \tag{16.45}$$

which is the same as Eq. (16.1).

Now suppose we add a perturbation, $\widehat{H}'$, to the system, so the total Hamiltonian becomes $\widehat{H} + \widehat{H}'$. Let us suppose that $\widehat{H}'$ has the general form:

$$\widehat{H}' = \begin{pmatrix} W_{11} & W_{12} \\ W_{21} & W_{22} \end{pmatrix} \tag{16.46}$$

which is identical to Eq. (16.9). As discussed in Chap. 7, $\widehat{H}'$ must be Hermitian, which tells us that W_{11} and W_{22} are real, and $W_{12} = W_{21}^*$. In general, due to the perturbation, E_0 and E_1 are no longer eigenvalues of the system, and $|\psi_0\rangle$ and $|\psi_1\rangle$ are no longer stationary states. If the system starts in the state $|\psi_0\rangle$ at time $t = 0$, there is a probability $P_1(t)$ that it will transition to the state $|\psi_1\rangle$ at some later time, t. The perturbation, $\widehat{H}'$, couples the two states. The total Hamiltonian is:

$$\widehat{H}' + \widehat{H}' = \begin{pmatrix} E_0 + W_{11} & W_{12} \\ W_{21} & E_1 + W_{22} \end{pmatrix} \tag{16.47}$$

The Schrodinger equation with the perturbation becomes:

$$(\widehat{H} + \widehat{H}')|\psi_+\rangle = E_+|\psi_+\rangle \tag{16.48}$$

and

$$(\widehat{H} + \widehat{H}')|\psi_-\rangle = E_-|\psi_-\rangle \tag{16.49}$$

where E_+ and E_- are the new eigenenergies of the system, and $|\psi_+\rangle$ and $|\psi_-\rangle$ are the new eigenstates.

A general state in the $|\psi_0\rangle$ and $|\psi_1\rangle$ basis is $\alpha|\psi_0\rangle + \beta|\psi_1\rangle$. Hence, the equation we need to solve is:

$$\begin{pmatrix} E_0 + W_{11} & W_{12} \\ W_{21} & E_1 + W_{22} \end{pmatrix} \begin{pmatrix} \alpha \\ \beta \end{pmatrix} = E \begin{pmatrix} \alpha \\ \beta \end{pmatrix} \tag{16.50}$$

Equation (16.50) gives us the two eigenvalues:

$$E_+ = \frac{1}{2}(E_0 + W_{11} + E_1 + W_{22}) + \frac{1}{2}\sqrt{(E_0 + W_{11} - E_1 - W_{22})^2 + |2W_{21}|^2} \tag{16.51}$$

$$E_- = \frac{1}{2}(E_0 + W_{11} + E_1 + W_{22}) - \frac{1}{2}\sqrt{(E_0 + W_{11} - E_1 - W_{22})^2 + |2W_{21}|^2} \tag{16.52}$$

and the corresponding eigenstates:

$$|\psi_+\rangle = e^{-i\phi/2} \cos\frac{\theta}{2}|\psi_0\rangle + e^{+i\phi/2}\sin\frac{\theta}{2}|\psi_1\rangle \tag{16.53}$$

$$|\psi_-\rangle = -e^{-i\phi/2}\sin\frac{\theta}{2}|\psi_0\rangle + e^{+i\phi/2}\cos\frac{\theta}{2}|\psi_1\rangle \tag{16.54}$$

where

$$\theta = \tan^{-1}\left(\frac{2|W_{21}|}{E_0 + W_{11} - E_1 - W_{22}}\right) \tag{16.55}$$

and

$$W_{21} = |W_{21}|e^{-i\phi} \tag{16.56}$$

The wavefunctions, $|\psi_+\rangle$ and $|\psi_-\rangle$ in Eqs. (16.53) and (16.54), respectively, are in the familiar Bloch vector form. $|\psi_+\rangle$ and $|\psi_-\rangle$ are orthonormal and form a new basis for any qubit.

Exercise 16.4 Derive Eqs. (16.51) to (16.54).

Let us examine the consequences of the above equations. First, if there is no coupling ($\widehat{H}' = 0$), then $W_{11} = W_{21} = W_{12} = W_{22} = 0$ and we recover the energies E_0 and E_1 from Eqs. (16.51) and (16.52), as expected. Also, the wavefunctions, $|\psi_+\rangle$ and $|\psi_-\rangle$, become $|\psi_0\rangle$ and $|\psi_1\rangle$. If there are no off-diagonal terms ($W_{21} = W_{12} = 0$, $W_{11} \neq 0$, $W_{22} \neq 0$), then the stationary states of the system become the same as those of the unperturbed system ($|\psi_0\rangle$ and $|\psi_1\rangle$) and the eigenenergies become $E_0 + W_{11}$ and $E_1 + W_{22}$. Thus, the energies just shift. Thus, we will assume for simplicity that the diagonal terms of $\widehat{H}'$ equal zero ($W_{11} = W_{22} = 0$). In fact, this was the case for the oscillatory magnetic and electric fields considered earlier for spin and charge qubits. With this simplification, the Hamiltonian becomes:

$$\widehat{H} + \widehat{H}' = \begin{pmatrix} E_0 & W_{12} \\ W_{21} & E_1 \end{pmatrix} \tag{16.57}$$

and the energy levels become:

$$E_+ = E_a + \sqrt{\Delta^2 + |W_{21}|^2} \tag{16.58}$$

$$E_- = E_a - \sqrt{\Delta^2 + |W_{21}|^2} \tag{16.59}$$

where E_a is the average energy:

Fig. 16.2 Eigenenergies
(E_+: upper curve; E_-: lower
curve) versus Δ. The dashed
lines are the energies without
the perturbation

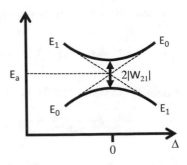

$$E_a = \frac{1}{2}(E_0 + E_1) \tag{16.60}$$

and Δ is defined as:

$$\Delta = \frac{1}{2}(E_0 - E_1) \tag{16.61}$$

Figure 16.2 shows the eigenenergies, E_+ (solid upper curve) and E_- (solid lower curve) as a function of Δ. The dashed lines show the eigenenergies (E_0 and E_1) in the absence of the perturbation. $\Delta = 0$ corresponds to a resonance where $E_0 = E_1$ (degeneracy). The coupling, W_{21}, lifts the degeneracy, as seen in Fig. 16.2. Far away from the resonance ($\Delta \gg 0$ or $\Delta \ll 0$), the eigenenergies become E_0 and E_1. When Δ approaches 0, the eigenenergies become E_+ and E_-, which are split by $2|W_{21}|$. This effect is known as level repulsion, anti-crossing, or avoided crossing. This effect is familiar from other problems in physics. For example, when two mechanical oscillators (e.g., pendulums) or two electronic oscillators (e.g., LC circuits) are coupled, the identical resonance frequencies of the individual systems become split into two different frequencies of the coupled system [2]. Level repulsion will be an important concept when we consider certain qubit systems, such as superconducting circuits.

16.5 Rabi Formula

Suppose the system is initially in the state, $|\psi(0)\rangle = |\psi_0\rangle$. We can rearrange Eqs. (16.53) and (16.54) to solve for $|\psi_0\rangle$:

$$|\psi(0)\rangle = |\psi_0\rangle = e^{+i\phi/2}\left(\cos\frac{\theta}{2}|\psi_+\rangle - \sin\frac{\theta}{2}|\psi_-\rangle\right) \tag{16.62}$$

To find the state at some later time, t, we apply the usual time-dependence:

$$|\psi(t)\rangle = e^{+i\phi/2}\left(e^{-iE_+t/\hbar}\cos\frac{\theta}{2}|\psi_+\rangle - e^{-iE_-t/\hbar}\sin\frac{\theta}{2}|\psi_-\rangle\right) \tag{16.63}$$

The probability amplitude of finding the system in the state $|\psi_1\rangle$ at some time t is:

$$\langle\psi_1|\psi(t)\rangle = e^{+i\phi/2}\left(e^{-iE_+t/\hbar}\cos\frac{\theta}{2}\langle\psi_1|\psi_+\rangle - e^{-iE_-t/\hbar}\sin\frac{\theta}{2}\langle\psi_1|\psi_-\rangle\right) \quad (16.64)$$

Again, using Eqs. (16.53) and (16.54), we obtain:

$$\langle\psi_1|\psi(t)\rangle = e^{+i\phi}\sin\frac{\theta}{2}\cos\frac{\theta}{2}\left(e^{-iE_+t/\hbar} - e^{-iE_-t/\hbar}\right) \quad (16.65)$$

The probability of finding the system in the state $|\psi_1\rangle$ at time t is:

$$P_1(t) = |\langle\psi_1|\psi(t)\rangle|^2 = \sin^2\theta\,\sin^2\left(\frac{E_+ - E_-}{2\hbar}t\right) \quad (16.66)$$

We can rewrite this equation as:

$$P_1(t) = \frac{|2W_{21}|^2}{(E_0 - E_1)^2 + |2W_{21}|^2}\sin^2\left(\frac{\sqrt{(E_0 - E_1)^2 + |2W_{21}|^2}}{2\hbar}t\right) \quad (16.67)$$

or

$$P_1(t) = \frac{|W_{21}|^2}{\Delta^2 + |W_{21}|^2}\sin^2\left(\frac{\sqrt{\Delta^2 + |W_{21}|^2}}{\hbar}t\right) \quad (16.68)$$

or

$$P_1(t) = \frac{|W_{21}|^2}{\Delta^2 + |W_{21}|^2}\sin^2\left(\frac{E_+ - E_-}{2\hbar}t\right) \quad (16.69)$$

Exercise 16.5 Derive Eq. (16.67).

According to Eq. (16.69), the coupled system oscillates at the Rabi frequency:

$$\Omega = (E_+ - E_-)/\hbar \quad (16.70)$$

That is, the system oscillates between the two energy levels, E_+ and E_-, in Fig. 16.2. With strong coupling ($|W_{12}| \gg E_0 - E_1$) or resonance ($\Delta = 0$), Eq. (16.70) gives:

$$\Omega = 2|W_{21}|/\hbar \quad (16.71)$$

According to Eq. (16.69), the Rabi frequency and probability amplitude are proportional to the strength of the coupling. Also, in the case of strong coupling, Eq. (16.55) gives $\theta = \pi/2$ and the wavefunctions become:

$$|\psi_{\pm}\rangle = \frac{1}{\sqrt{2}}\left(e^{+i\phi/2}|\psi_1\rangle \pm e^{-i\phi/2}|\psi_0\rangle\right) \tag{16.72}$$

which are called the symmetric (+) and anti-symmetric (−) wavefunctions. The system will oscillate between these two states at the frequency Ω.

16.6 Driven Rabi Oscillations

The Hamiltonian of the two-level quantum system considered above is:

$$\widehat{H} + \widehat{H}' = \begin{pmatrix} E_0 & W_{12} \\ W_{21} & E_1 \end{pmatrix} \tag{16.73}$$

Suppose $W_{12} = |W_{12}|e^{-i\omega t}$; i.e., we have driven oscillations like in ESR. Then $W_{21} = W_{12}^* = |W_{12}|e^{+i\omega t}$ and the Hamiltonian becomes:

$$\widehat{H} + \widehat{H}' = \begin{pmatrix} E_0 & |W_{12}|e^{-i\omega t} \\ |W_{12}|e^{+i\omega t} & E_1 \end{pmatrix} \tag{16.74}$$

It is convenient to rewrite Eq. (16.74) in terms of frequencies [3]:

$$\widehat{H} + \widehat{H}' = \frac{\hbar}{2} \begin{pmatrix} \omega_o & \omega_d e^{-i\omega t} \\ \omega_d e^{+i\omega t} & -\omega_o \end{pmatrix} \tag{16.75}$$

where we define the amplitude of the driving field as $\omega_d = \frac{2|W_{21}|}{\hbar}$. Also, note that:

$$H + H' = \frac{\hbar}{2} \begin{pmatrix} \omega_o & \omega_d e^{-i\omega t} \\ \omega_d e^{+i\omega t} & -\omega_o \end{pmatrix} = \frac{\hbar\omega_o}{2}\hat{\sigma}_z$$
$$+ \frac{\hbar\omega_d}{2}\cos(\omega t)\hat{\sigma}_x + \frac{\hbar\omega_d}{2}\sin(\omega t)\hat{\sigma}_y \tag{16.76}$$

Hence, with an appropriate phase of the driving field, we can achieve rotation in the Bloch sphere about the x or y-axis, in addition to the free precession (Larmor precession) around the z-axis, as we saw in Chap. 15. Also, as we saw in Chap. 15, the driving field corresponds to a rotating field in the x-y plane. Let us apply a rotating frame of reference at the same frequency as the driving field, ω, such that the driving field appears stationary. Then the Hamiltonian becomes:

$$H + H' = \frac{\hbar}{2}\begin{pmatrix} -\Delta\omega & \omega_d \\ \omega_d & \Delta\omega \end{pmatrix} \tag{16.77}$$

where $\Delta\omega = \omega - \omega_o$. The eigenenergies become:

$$E_{\pm} = \pm\frac{\hbar}{2}\sqrt{\Delta\omega^2 + \omega_d^2} \tag{16.78}$$

and the eigenfunctions become:

$$|\psi_+\rangle = \cos\frac{\theta}{2}|\psi_0\rangle + \sin\frac{\theta}{2}|\psi_1\rangle \tag{16.79}$$

$$|\psi_-\rangle = -\sin\frac{\theta}{2}|\psi_0\rangle + \cos\frac{\theta}{2}|\psi_1\rangle \tag{16.80}$$

Following the same derivation as Eq. (16.66), we obtain:

$$P_1(t) = \frac{\omega_d^2}{\Delta\omega^2 + \omega_d^2}\sin^2\left(\frac{\sqrt{\Delta\omega^2 + \omega_d^2}}{2}t\right) \tag{16.81}$$

which is the Rabi formula for driven oscillations. Hence, the Rabi frequency becomes:

$$\Omega = \sqrt{\Delta\omega^2 + \omega_d^2} = \sqrt{(\omega - \omega_o)^2 + |2W_{21}/\hbar|^2} \tag{16.82}$$

At resonance ($\omega = \omega_o$), we again get:

$$\Omega = 2|W_{21}|/\hbar \tag{16.83}$$

Also, at resonance, Eq. (16.81) reproduces Eq. (16.27) which we previously derived for the special case of resonance. Equation (16.81) is more general and includes the off-resonance ($\omega \neq \omega_o$) case. Equation (16.81) indicates that when we are off resonance, the qubit oscillates faster between $|0\rangle$ and $|1\rangle$ but the amplitude is reduced, as shown in Fig. 16.3.

Fig. 16.3 Rabi oscillations at resonance and off-resonance condition

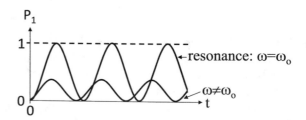

Exercise 16.6   Derive Eqs. (16.78), (16.79) and (16.80).

Exercise 16.7   Derive Eq. (16.81).

References

1. R.P. Feynman, Int. J. Theor. Phys. **21**, 467 (1982)
2. W. Frank, P. von Brentano, Am. J. Phys. **62**, 706 (1994)
3. R. Gross, A. Marx, F. Deppe, Applied superconductivity: Josephson effect and superconducting electronics (Walter De Gruyter Inc., 2016)

Chapter 17
Implementing Two-Qubit Gates

The previous chapters were about single-qubit operations. Single-qubit rotations and a two-qubit gate (such as a CNOT gate) are required to form a set of universal gates. Therefore, let us consider how to implement a two-qubit CNOT gate.

17.1   CNOT Gate

The CNOT gate is a two-qubit operation that flips the second qubit state if the first qubit state is $|1\rangle$ (Fig. 17.1):

$$\alpha_{00}|00\rangle + \alpha_{01}|01\rangle + \alpha_{10}|10\rangle + \alpha_{11}|11\rangle \overset{\text{CNOT}}{\rightarrow} \alpha_{00}|00\rangle$$
$$+ \alpha_{01}|01\rangle + \alpha_{10}|11\rangle + \alpha_{11}|10\rangle \qquad (17.1)$$

Notice that we just want to exchange the amplitude between $|10\rangle$ and $|11\rangle$.

To implement a CNOT gate, we need to arrange the Hamiltonian so that $|10\rangle$ and $|11\rangle$ energy levels have separation $\hbar\omega_o$ that is different than the other energy separations (Fig. 17.1). We then drive a Rabi oscillation (π-pulse) at the $|10\rangle$ to $|11\rangle$ resonant frequency, ω_o. The $|10\rangle$ and $|11\rangle$ amplitudes will flip, while $|00\rangle$ and $|01\rangle$ states are unchanged. Note that Fig. 17.1 shows $|00\rangle$ (electron spins parallel to the B-field) with the highest energy level, as described previously in Fig. 14.3.

17.2   Spin Qubit Implementation

Consider two neighboring spins, as depicted in Fig. 17.2. Spin 1 forms the control qubit and spin 2 forms the target qubit of a CNOT gate. The magnetic field lines associated with the magnetic moment of each spin is shown in Fig. 17.2 (remember that the magnetic moment is opposite the spin angular momentum for an electron).

© The Author(s), under exclusive license to Springer Nature Switzerland AG 2021
R. LaPierre, *Introduction to Quantum Computing*, The Materials Research Society Series,
https://doi.org/10.1007/978-3-030-69318-3_17

Input state	Output state
$|00\rangle$	$|00\rangle$
$|01\rangle$	$|01\rangle$
$|10\rangle$	$|11\rangle$
$|11\rangle$	$|10\rangle$

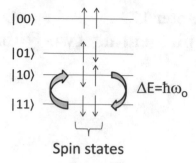

Spin states

Fig. 17.1 Truth table for CNOT gate, and energy levels for CNOT implementation

Fig. 17.2 Spin qubit implementation of a CNOT gate and corresponding circuit representation

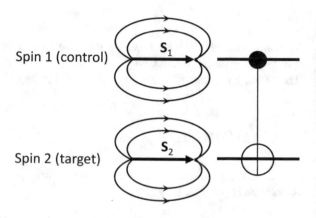

The magnetic field of spin 1 (control qubit) can influence the energy of spin 2 (target qubit), forming the basis for a CNOT gate. The energy associated with the magnetic moment ($\mathbf{\mu}_2$) of spin 2 in the presence of the magnetic field ($\mathbf{B}_1$) from spin 1 is $U_2 = -\mathbf{\mu}_2 \cdot \mathbf{B}_1$. The magnetic moment of spin 2 is $\mathbf{\mu}_2 = \gamma \mathbf{S}_2$ where γ is the gyromagnetic ratio and $\mathbf{S}_2$ is the spin angular momentum. $\mathbf{B}_1$ is proportional to the magnetic moment of spin 1 and therefore $\mathbf{S}_1$. Hence, the energy of spin 2 is proportional to both $\mathbf{S}_1$ and $\mathbf{S}_2$:

$$U_2 = -\mathbf{\mu}_2 \bullet \mathbf{B}_1 \tag{17.2}$$

$$\mathbf{\mu}_2 = \gamma \mathbf{S}_2 \qquad \mathbf{B}_1 \propto \mathbf{S}_1$$

From Eq. (17.2), the Hamiltonian for the spin-spin interaction is proportional to $\mathbf{S}_1$ and $\mathbf{S}_2$, or $\hat{\sigma}_{z1}$ and $\hat{\sigma}_{z2}$:

$$\widehat{H}_{\text{int}} = J\hat{\sigma}_{z1} \otimes \hat{\sigma}_{z2} \tag{17.3}$$

where J is called the "J-coupling".

The total Hamiltonian is:

$$\widehat{H} = \widehat{H}_1 + \widehat{H}_2 + \widehat{H}_{int} \tag{17.4}$$

where $\widehat{H}_1$ is the Hamiltonian for spin 1 alone, $\widehat{H}_2$ is the Hamiltonian for spin 2 alone, and $\widehat{H}_{int}$ is the interaction Hamiltonian. Equation (17.4) is known as the Ising model for two interacting spins. Hence, for the two-qubit system:

$$\widehat{H} = \left(\frac{\hbar\omega_1}{2}\hat{\sigma}_{z1}\right) \otimes \hat{I} + \hat{I} \otimes \left(\frac{\hbar\omega_2}{2}\hat{\sigma}_{z2}\right) + J\hat{\sigma}_{z1} \otimes \hat{\sigma}_{z2} \tag{17.5}$$

Now we can evaluate Eq. (17.5) using the following:

$$\hat{\sigma}_{z1} \otimes \hat{I} = \begin{pmatrix} 1 & 0 \\ 0 & -1 \end{pmatrix} \otimes \begin{pmatrix} 1 & 0 \\ 0 & 1 \end{pmatrix} = \begin{pmatrix} 1 & 0 & 0 & 0 \\ 0 & 1 & 0 & 0 \\ 0 & 0 & -1 & 0 \\ 0 & 0 & 0 & -1 \end{pmatrix} \tag{17.6}$$

$$\hat{I} \otimes \hat{\sigma}_{z2} = \begin{pmatrix} 1 & 0 \\ 0 & 1 \end{pmatrix} \otimes \begin{pmatrix} 1 & 0 \\ 0 & -1 \end{pmatrix} = \begin{pmatrix} 1 & 0 & 0 & 0 \\ 0 & -1 & 0 & 0 \\ 0 & 0 & 1 & 0 \\ 0 & 0 & 0 & -1 \end{pmatrix} \tag{17.7}$$

$$\hat{\sigma}_{z1} \otimes \hat{\sigma}_{z2} = \begin{pmatrix} 1 & 0 \\ 0 & -1 \end{pmatrix} \otimes \begin{pmatrix} 1 & 0 \\ 0 & -1 \end{pmatrix} = \begin{pmatrix} 1 & 0 & 0 & 0 \\ 0 & -1 & 0 & 0 \\ 0 & 0 & -1 & 0 \\ 0 & 0 & 0 & 1 \end{pmatrix} \tag{17.8}$$

Substituting Eqs. (17.6)–(17.8) into (17.5) gives:

$$\widehat{H} = \frac{\hbar}{2} \begin{pmatrix} \omega_1 + \omega_2 + 2J/\hbar & 0 & 0 & 0 \\ 0 & \omega_1 - \omega_2 - 2J/\hbar & 0 & 0 \\ 0 & 0 & -\omega_1 + \omega_2 - 2J/\hbar & 0 \\ 0 & 0 & 0 & -\omega_1 - \omega_2 + 2J/\hbar \end{pmatrix} \tag{17.9}$$

Using the time-independent Schrodinger equation, $\widehat{H}|\psi\rangle = E|\psi\rangle$, we can now evaluate the energy (E_{00}, E_{01}, E_{10} and E_{11}) of each basis state $|\psi\rangle = \begin{pmatrix} 1 \\ 0 \\ 0 \\ 0 \end{pmatrix}, \begin{pmatrix} 0 \\ 1 \\ 0 \\ 0 \end{pmatrix}, \begin{pmatrix} 0 \\ 0 \\ 1 \\ 0 \end{pmatrix}, \begin{pmatrix} 0 \\ 0 \\ 0 \\ 1 \end{pmatrix}$, corresponding to $|00\rangle$, $|01\rangle$, $|10\rangle$ and $|11\rangle$, respectively. In fact, Eq. (17.9) is a diagonal matrix, so the diagonal entries of $\widehat{H}$ give the eigenenergies directly. Finally, the energy separations between energy states can be determined. For example, the energy separation between E_{00} and E_{01} is:

$$
\begin{aligned}
E_{00-01} &= E_{00} - E_{01} \\
&= \frac{\hbar}{2}[(\omega_1 + \omega_2 + 2J/\hbar) - (\omega_1 - \omega_2 - 2J/\hbar)] \\
&= \frac{\hbar}{2}(2\omega_2 + 4J/\hbar) \\
&= \hbar(\omega_2 + 2J/\hbar)
\end{aligned}
\tag{17.10}
$$

Exercise 17.1 Derive the following energy separations:

$$
\begin{aligned}
E_{00-10} &= \hbar(\omega_1 + 2J/\hbar) \\
E_{01-11} &= \hbar(\omega_1 - 2J/\hbar) \\
E_{10-11} &= \hbar(\omega_2 - 2J/\hbar)
\end{aligned}
$$

Figures 17.3 and 17.4 shows the energy separations between the $|00\rangle$, $|01\rangle$, $|10\rangle$ and $|11\rangle$ states as derived above. The J-coupling causes a shift of the energy levels and, hence, the resonance frequencies between states. The energy levels move according to the J-coupling; i.e., parallel spins move up in energy, and anti-parallel spins move down in energy. Without J-coupling, the E_{00-01} energy separation is identical to the E_{10-11} separation. Therefore, we cannot implement a CNOT gate that requires a unique energy separation for the $|10\rangle$ to $|11\rangle$ transition. With the J-coupling, we

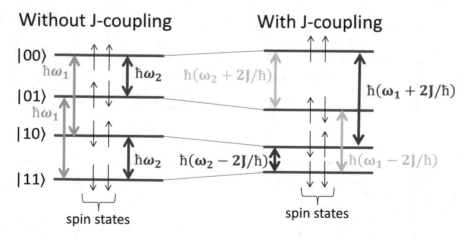

Fig. 17.3 Energy separations between spin states with and without J-coupling. The electron spins in each energy level are indicated

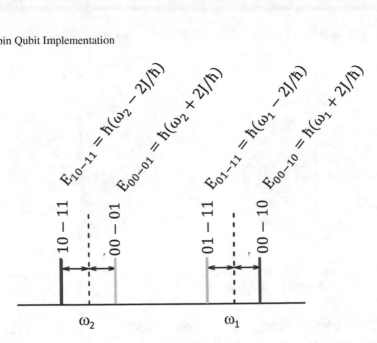

Fig. 17.4 Frequency spectrum due to the J-coupling. The horizontal arrows indicate the frequency shifts due to the J-coupling. The color of each frequency line matches the colors in Fig. 17.3

achieve a unique energy difference for each possible transition. We can selectively excite the $|10\rangle$ to $|11\rangle$ transition with the energy $E_{10-11} = \hbar(\omega_2 - 2J/\hbar)$. A π-pulse at the frequency $\hbar(\omega_2 - 2J/\hbar)$ creates a CNOT gate.

17.3 Another Way to Implement Two-Qubit Gates

Another way to implement two-qubit gates is by a series of single-qubit pulses and "free precession" due to the J-coupling. In Fig. 17.5, blue is the control qubit (spin 1), and red is the target qubit (spin 2). The unitary transformation of spin 2 is:

$$\hat{U}(t) = e^{-\frac{i}{\hbar}\hat{H}t} \qquad (17.11)$$

where $\hat{H} = J\hat{\sigma}_{z1} \otimes \hat{\sigma}_{z2}$. Due to the sign of the J-coupling, if the control qubit is spin up, then precession of the target qubit occurs counter-clockwise; if the control qubit is spin down, then precession of the target qubit occurs clockwise, as shown in the Bloch sphere of Fig. 17.5.

For a CNOT gate, we implement a sequence of single-qubit rotations, as shown in Fig. 17.6. The blue and red arrows represent spin 1 (control qubit) and spin 2 (target qubit), respectively. The first column has both spins initially in the $|0\rangle$ state, represented by two arrows pointing up in the Bloch sphere. The second column has the first spin (control qubit) initially in the $|1\rangle$ state and the second spin (target qubit)

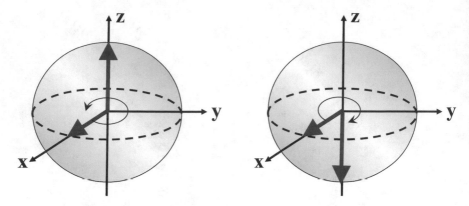

Fig. 17.5 The direction of free precession of spin 2 (target qubit, red) due to the J-coupling depends on the orientation of spin 1 (control qubit, blue)

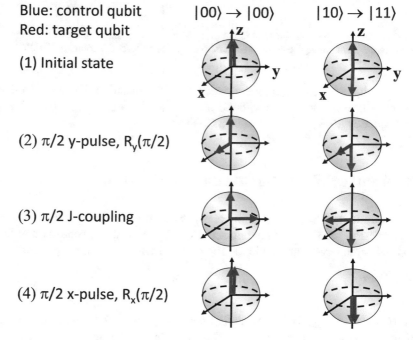

Fig. 17.6 CNOT gate implemented by a series of single-qubit gates and J-coupling on target qubit

in the $|0\rangle$ state. The configurations are shown as follows: (1) at the beginning of the sequence, (2) after a 90° y-rotation ($\pi/2$-pulse) of spin 2, (3) after a 90° z-rotation of spin 2 due to the J-coupling, and (4) after a 90° x-rotation ($\pi/2$-pulse) of spin 2. The J-coupling causes a rotation, called "free evolution", of spin 2 in different directions, depending on the orientation of spin 1. Note that the J coupling is on all the time but is a weak effect. Typically, J is on the order of 100 Hz, while the single-qubit gates are on the order of MHz (RF pulses). Therefore, the free evolution due to the J-coupling can be ignored during the other more rapid pulses.

To implement a $\pi/2$-pulse around z due to free evolution of the J-coupling, we set the time t such that:

$$\frac{Jt}{\hbar} = \frac{\pi}{4} \tag{17.12}$$

Thus,

$$\hat{U}(t) = e^{-\frac{i}{\hbar}\hat{H}t} = e^{-i\frac{\pi}{4}\hat{\sigma}_{z1}\otimes\hat{\sigma}_{z2}} \tag{17.13}$$

Exercise 17.2 Show the following:

$$\hat{U}\left(t = \frac{\pi}{4}\frac{\hbar}{J}\right) = \sqrt{i}\begin{pmatrix} -i & 0 & 0 & 0 \\ 0 & 1 & 0 & 0 \\ 0 & 0 & 1 & 0 \\ 0 & 0 & 0 & -i \end{pmatrix} \tag{17.14}$$

We can check that the sequence of operations described above, and depicted in Figs. 17.6 and 17.7, yields a CNOT gate:

$$\hat{I} \otimes \hat{R}_y(\pi/2) = \begin{pmatrix} 1 & 0 \\ 0 & 1 \end{pmatrix} \otimes \frac{1}{\sqrt{2}}\begin{pmatrix} 1 & -1 \\ 1 & 1 \end{pmatrix} = \frac{1}{\sqrt{2}}\begin{pmatrix} 1 & -1 & 0 & 0 \\ 1 & 1 & 0 & 0 \\ 0 & 0 & 1 & -1 \\ 0 & 0 & 1 & 1 \end{pmatrix} \tag{17.15}$$

$$\hat{I} \otimes \hat{R}_x(\pi/2) = \begin{pmatrix} 1 & 0 \\ 0 & 1 \end{pmatrix} \otimes \frac{1}{\sqrt{2}}\begin{pmatrix} 1 & -i \\ -i & 1 \end{pmatrix} = \frac{1}{\sqrt{2}}\begin{pmatrix} 1 & -i & 0 & 0 \\ -i & 1 & 0 & 0 \\ 0 & 0 & 1 & -i \\ 0 & 0 & -i & 1 \end{pmatrix} \tag{17.16}$$

Using Eqs. (17.14)–(17.16), we get:

$$\hat{U}_{\text{tot}} = [\hat{I} \otimes \hat{R}_x(\pi/2)][\hat{U}(\pi/2)][\hat{I} \otimes \hat{R}_y(\pi/2)] = \sqrt{i}\begin{pmatrix} -i & 0 & 0 & 0 \\ 0 & 1 & 0 & 0 \\ 0 & 0 & 0 & -1 \\ 0 & 0 & -i & 0 \end{pmatrix} \tag{17.17}$$

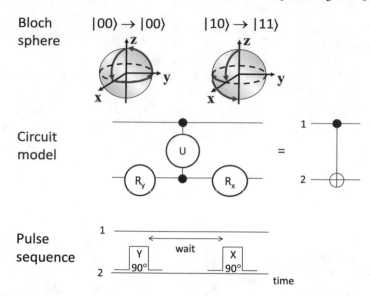

Fig. 17.7 Bloch sphere rotations, circuit model, and pulse sequence for a CNOT gate

Let us apply $\widehat{U}_{tot}$ to each of the four basis input states ($|00\rangle$, $|01\rangle$, $|10\rangle$ and $|11\rangle$). Ignoring pre-factors that don't affect probabilities, we obtain $U_{tot}|00\rangle = |00\rangle$, $U_{tot}|01\rangle = |01\rangle$, $U_{tot}|10\rangle = |11\rangle$, and $U_{tot}|11\rangle = |10\rangle$; i.e., we obtain a CNOT gate within some arbitrary phase factors.

Chapter 18
DiVincenzo Criteria

The DiVincenzo criteria, named after the physicist, David DiVincenzo, is a list of criteria that every quantum computer should satisfy [1, 2]. The DiVincenzo criteria are:

1. A scalable physical system with well-defined qubits.
2. Initialization of qubit register.
3. Read-out of qubits.
4. Ability to implement universal quantum gates.
5. Long qubit lifetimes.

Let us examine each of these criteria in more detail.

18.1 A Scalable Physical System with Well-Defined Qubits

A method of implementing well-defined qubits is required for quantum computing (Fig. 18.1). For example, a qubit could be a superposition of:

- Electron with spin up or spin down
- Electron in ground or excited state of an atom
- Photon in "path 1" or "path 2"
- Photon with horizontal or vertical polarization
- Photon with $+45°$ or $-45°$ polarization
- Photon with left or right circular polarization
- Superconducting current flowing clockwise or counter-clockwise.

A large number (millions) of qubits is needed for most quantum algorithms. The biggest challenge for quantum computing is going from one qubit to a large-scale system of many qubits. The challenge lies not just in fabricating more qubits, but also in connecting each qubit to the external world for gate control without introducing dissipation or decoherence.

R. LaPierre, *Introduction to Quantum Computing*, The Materials Research Society Series,
https://doi.org/10.1007/978-3-030-69318-3_18

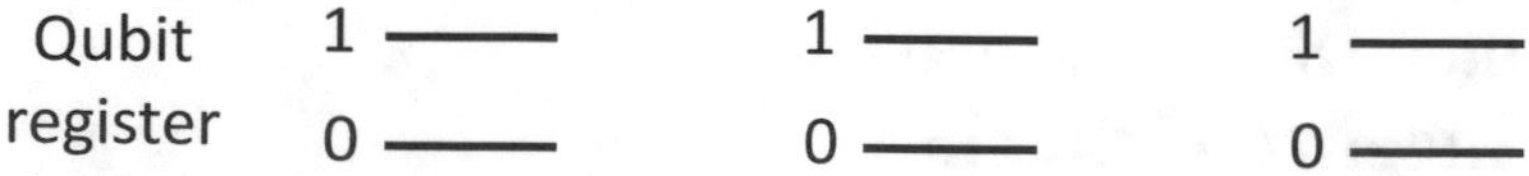

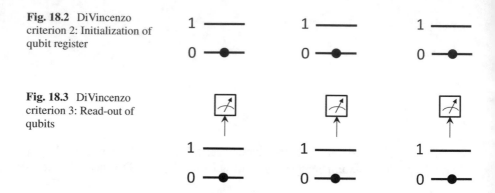

Qubit register

Fig. 18.1 DiVincenzo criterion 1: A scalable physical system with well-defined qubits

Fig. 18.2 DiVincenzo criterion 2: Initialization of qubit register

Fig. 18.3 DiVincenzo criterion 3: Read-out of qubits

18.2 Initialization of Qubit Register

The first step in operating a quantum computer is initialization (Fig. 18.2) to prepare all qubits in a well-defined state, typically $|0\rangle|0\rangle \ldots |0\rangle$. For example, initialization of spin qubits may be done by cooling the system so that the qubits relax to their ground state energy within a time T_1 (called the relaxation time). Alternatively, active methods such as optical pumping (discussed in Chap. 21) may be used to initialize qubits.

18.3 Read-Out of Qubits

A method is required for reading $|0\rangle$ or $|1\rangle$ from the qubit state (Fig. 18.3). For example, we could use a Stern-Gerlach apparatus to read-out an atomic spin state, although this method is not very practical. In practice, we will see that qubits can be read-out by more practical methods such as spin-dependent fluorescence or spin-to-charge conversion.

18.4 Ability to Implement Universal Quantum Gates

We need to be able to set an arbitrary superposition of $|0\rangle$ and $|1\rangle$ for each qubit (Fig. 18.4). Recall that we can do this by using a set of universal gates. As discussed

Fig. 18.4 DiVincenzo
criterion 4: Ability to
implement universal
quantum gates

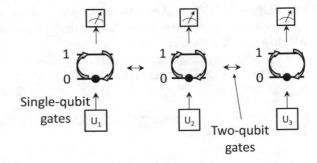

in Chap. 7, a universal gate set could be a single qubit rotation together with a
two-gate operation such as a CNOT gate.

18.5 Long Qubit Lifetimes

The qubit lifetimes should be long compared to the duration of the calculation. This
allows many gate operations to occur before decoherence. More on decoherence and
quantum error correction will be presented in Chap. 25.

18.6 Extended DiVincenzo Criterion

We may need to establish links between qubit chips, like a "quantum internet"
(Fig. 18.5). This criterion may not be essential if the goal is to build a stationary
quantum computer but would be necessary for some other applications including

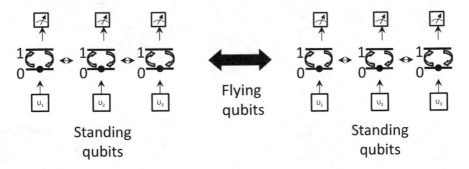

Fig. 18.5 Extended DiVincenzo criterion: ability to create "flying" qubits

quantum networks. The qubits of the stationary quantum computer are called "standing" qubits. The qubits that connect the quantum computers are called "flying qubits", which are usually photons.

As we explore quantum computing technologies in the next chapters, think about them in terms of the DiVincenzo criteria.

References

1. D.P. DiVincenzo, D. Loss, Superlattices Microstruct. **23**, 419 (1998)
2. D.P. DiVincenzo, Fortshritte der Physik **48**, 771 (2000)

Chapter 19
Nuclear Magnetic Resonance

Nuclear magnetic resonance (NMR) uses a large ensemble of molecules in a test tube (a liquid). The spin states of nuclei within the atoms of the molecules act as the qubits. Each molecule is a quantum computer.

19.1 Nuclear Spin

A selection of molecules that have been used to implement NMR quantum computing experiments are shown in Fig. 19.1. The number of qubits (n) is shown in blue, while the atomic nuclei used as qubits are indicated in pink.

Protons and neutrons are fermions and have spin 1/2. How can a neutron, which has no charge, exhibit a magnetic moment? Neutrons have spin and a magnetic moment because they are a composite particle comprised of quarks, which do have charge.

Each nucleus has a spin quantum number, usually denoted I, which is determined by its nucleus (number of protons and neutrons):

$$S^2 = I(I + 1)\hbar^2 \tag{19.1}$$

Equation (19.1) is analogous to Eq. (3.14). The spin quantum number, I, of selected nuclei is given in Table 19.1.

Protons and neutrons form pairs of opposite spin angular momentum in atomic nuclei. Thus, the magnetic moment of a nucleus with an even number of both protons and neutrons is zero. All isotopes that contain an odd number of protons and/or neutrons have an intrinsic nuclear magnetic moment and spin angular momentum.

The spin projection along z is given by:

$$S_z = m\hbar \tag{19.2}$$

© The Author(s), under exclusive license to Springer Nature Switzerland AG 2021
R. LaPierre, *Introduction to Quantum Computing*, The Materials Research Society Series,
https://doi.org/10.1007/978-3-030-69318-3_19

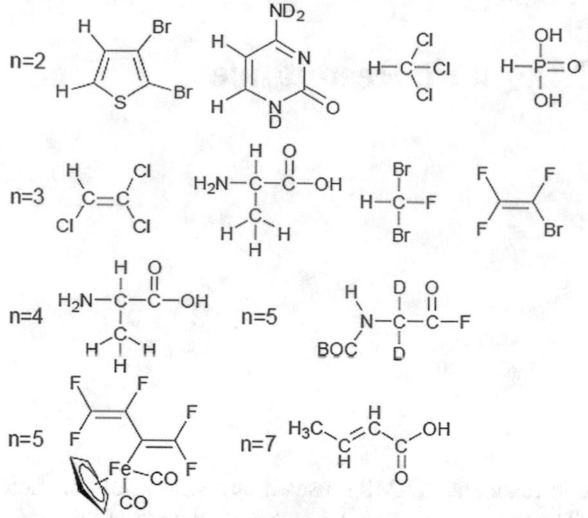

Fig. 19.1 Some molecules used in quantum computing. Republished with permission of Royal Society of Chemistry, from J.A. Jones, Phys. Chem. Comm. 4 (2001) 49, https://doi.org/10.1039/B103231N; permission conveyed through Copyright Clearance Center, Inc. [1]

Table 19.1 Number of protons, p, number of neutrons, n, and spin quantum number, I, of selected nuclei

Nucleus	p	n	I
^{1}H	1	0	$\frac{1}{2}$
^{2}H	1	1	1
^{3}H	1	2	$\frac{1}{2}$
^{3}He	2	1	$\frac{1}{2}$
^{7}Li	3	4	$\frac{3}{2}$
^{11}B	5	6	$\frac{3}{2}$
^{12}C	6	6	0
^{13}C	6	7	$\frac{1}{2}$
^{15}N	7	8	$\frac{1}{2}$
^{16}O	8	8	0
^{17}O	8	9	$\frac{5}{2}$
^{19}F	9	10	$\frac{1}{2}$
^{23}Na	11	12	$\frac{3}{2}$
^{29}Si	14	15	$\frac{1}{2}$
^{31}P	15	16	$\frac{1}{2}$
^{35}Cl	17	18	$\frac{3}{2}$

I	m	Examples
0	0	^{12}C, ^{16}O, ^{32}S
$\frac{1}{2}$	$-\frac{1}{2}, +\frac{1}{2}$	^{1}H, ^{3}H, ^{3}He, ^{13}C, ^{15}N, ^{19}F, ^{29}Si, ^{31}P
1	$-1, 0, +1$	^{2}H
$\frac{3}{2}$	$-\frac{3}{2}, -\frac{1}{2}, +\frac{1}{2}, +\frac{3}{2}$	^{7}Li, ^{11}B, ^{19}F, ^{23}Na

Table 19.2 Magnetic spin quantum number, m, of selected nuclei

where m is the magnetic spin quantum number given by $m = -I, -I + 1, \ldots +I$. Equation (19.2) is analogous to Eq. (3.15). The magnetic spin quantum number, m, of selected nuclei is given in Table 19.2.

19.2 Two-Level System

For an $I = 1/2$ nucleus, there are two m levels, $m = \pm 1/2$. In a magnetic field, say $B_o \hat{z}$, these levels are separated due to the Zeeman effect, as shown in Fig. 19.2. Note the differences between Fig. 19.2 for nuclear spin, and Fig. 14.3 for the electron spin. Unlike electrons, the magnetic moment and spin are in the same direction for nuclei. Therefore, the lowest energy level corresponds to spin up (magnetic moment and spin parallel to the B-field) and the highest energy level corresponds to spin down (magnetic moment and spin anti-parallel to the B-field). Like Eq. (14.37), the energy separation ΔE between spin up and spin down states is:

$$\Delta E = \hbar \omega_o = \hbar \gamma B_o \tag{19.3}$$

where γ is the gyromagnetic ratio (in units of rad/Ts) and ω_o is the Larmor frequency (angular frequency in rad/s) of the nuclear spin. The nuclear precession for positively charged nuclei is clockwise about the z axis, opposite that of the electron precession considered in Chap. 14. Table 19.3 lists the gyromagnetic ratio, γ, and the Larmor frequency, ν_o (in units of MHz, $\nu_o = \omega_o/2\pi$), for selected nuclei at a magnetic field of $B_o = 10$ T.

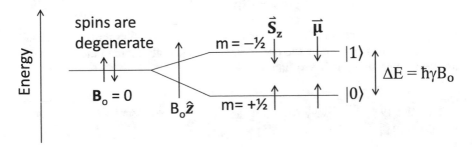

Fig. 19.2 Two-level quantum system for $I = 1/2$ nuclei

Table 19.3 Gyromagnetic ratio, γ, of selected nuclei

Nucleus	p	n	I	γ ($\times 10^7$ rad/Ts)	ν_0 (MHz) at $B_0 = 10$ T
^{1}H	1	0	$\frac{1}{2}$	26.8	427
^{2}H	1	1	1	4.1	65
^{3}H	1	2	$\frac{1}{2}$	28.5	454
^{13}C	6	7	$\frac{1}{2}$	6.7	107
^{15}N	7	8	$\frac{1}{2}$	2.7	43
^{17}O	8	9	$\frac{5}{2}$	3.6	57
^{19}F	9	10	$\frac{1}{2}$	25.2	401
^{23}Na	11	12	$\frac{3}{2}$	7.1	113
^{31}P	15	16	$\frac{1}{2}$	10.8	172

Note that ΔE is directly proportional to γ and B_o, and each element has a different γ. Therefore, ΔE and the Larmor frequency are unique to each element. Each atom in the molecule can act as a separate qubit that can be individually addressed by its unique Larmor frequency. In a molecule, local magnetic fields from the electron spins in the local molecular bonds may reduce B_o (this effect is known as "shielding"). This means that even identical atoms in a molecule can be addressed individually because they have different Larmor frequencies (this is known as the "chemical shift"). The observation of such magnetic resonance frequencies of the nuclei present in a molecule allows chemists to discover chemical and structural information about the molecule, making NMR one of the most important tools for chemical spectroscopy.

In the absence of an external magnetic field ($B_o = 0$), the nuclear spin energy levels are degenerate ($\Delta E = 0$). When $B_o > 0$, the nuclear spin energy levels for $I = 1/2$ are split and denoted as α and β. The lower energy level, α, is for nuclear spins and magnetic moments parallel to the B_o field, and corresponds to the $|0\rangle$ state. The upper energy level, β, is for nuclear spins and magnetic moments anti-parallel to the B_o field, and corresponds to the $|1\rangle$ state. Using the Boltzmann equation, the population distribution is given as:

$$N_\beta / N_\alpha = \exp(-\Delta E / k_B T) \tag{19.5}$$

Hence, the Boltzmann equation states that the population of spins will be distributed between the quantized lower and upper states with slightly more in the lower (spin up) than the upper (spin down) state.

Exercise 19.1 What is N_β / N_α for ^{1}H at 1.41 T and $T = 300$ K?

Suppose B_o is set along the z-axis. The z-axis is called the "longitudinal axis" and the (x, y)-plane is called the "transverse plane". Each magnetization vector of a given nuclei will be rotating (precessing) clockwise around the z-axis at the same frequency, ω_o (Fig. 19.3). The spin precession of many nuclei at equilibrium will trace a cone around the z-axis. In 1946, Felix Bloch (Fig. 19.4) proposed that the

Fig. 19.3 A net magnetization, M_o (thick black arrow), is produced by a static field, B_o, along the z-axis as a result of the collection of N_α spin up and N_β spin down nuclei

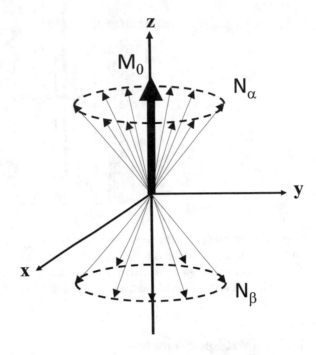

Fig. 19.4 Felix Bloch (1905–1993); Nobel Prize for Physics in 1952. *Credit* Wikimedia Commons [2]

Fig. 19.5 Cross-section of an NMR spectrometer. The superconducting magnet produces the static longitudinal field, $B_o\hat{z}$, while the probe coil produces the transverse RF field, $B_1 \cos(\omega_o t)\hat{x}$

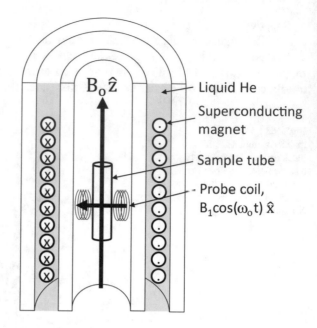

population of nuclear spins could be represented by a macroscopic magnetization vector which shows the net orientation and magnitude of magnetization of the entire population of spins. At equilibrium, the net magnetization is denoted M_o and will align with the orientation of the applied magnetic field, B_o (Fig. 19.3).

19.3 NMR Spectrometer

A typical NMR spectrometer is illustrated in Fig. 19.5. A superconducting magnet (cooled with liquid He) applies the longitudinal static field, $B_o\hat{z}(\sim 10–15$ T). A sample tube sits at room temperature (liquid He is used to cool the superconducting magnets, not the sample). The "probe", surrounding the sample, consists of coils for applying transverse radio frequency (RF) pulses, $B_1 \cos(\omega_o t)\hat{x}$, and for spin read-out discussed below.

19.4 Single-Qubit Gates

NMR is similar to electron spin resonance (ESR), where a field, $B(t) = B_o\hat{z} + B_1 \cos(\omega_o t)\hat{x}$, allows us to control the qubit state. In NMR, the qubit is a nuclear spin, instead of an electron spin. Like in ESR, the field applied along $\hat{z}$ $(B_o\hat{z} \sim 10–15$ T) results in the Zeeman effect, creating a two-level qubit system. This longitudinal field also results in clockwise precession of the nuclear spins about the $\hat{z}$ axis. For example, at a typical field of $B_o = 11.7$ T, the proton will precess at a frequency $v_o = \omega_o/2\pi = 500$ MHz. At equilibrium, the net magnetization along the z-axis is $M_z = M_o$, and the net magnetization in the x-y plane is zero ($M_x = 0$, $M_y = 0$).

In NMR, like in ESR, the transverse field $B_1 \cos(\omega_o t)\hat{x}$ results in single-qubit rotation. The transverse field, $B_1 \cos(\omega_o t)$, is implemented along the $+x$-axis for a certain duration of time and is therefore called an RF pulse. The RF pulse causes the magnetization vector to precess about the x-axis with a precession frequency $\Omega = \gamma B_1/2$ (Rabi frequency). The angle of rotation, θ, depends on the duration, t, that the B_1 field is applied. Since $\Omega = \gamma B_1/2$, it follows that $\theta = \Omega t = \gamma B_1 t/2$. The Rabi frequency is in the RF range, so we apply pulses with durations typically on the order of microseconds. By timing the duration and phase of RF pulses, we can tip the magnetization in any direction, as we saw in Chap. 15. The only nuclear spins that will be rotated are those whose Larmor frequency are equal to the RF pulse frequency (resonance). In this way, we can address individual qubits (different nuclei) that have different Larmor frequencies. For example, as illustrated in Fig. 19.6, a $\theta = 90°$ RF pulse can tip a spin into the x-y plane at the frequency Ω. This tipped spin magnetization will continue to precess about the $B_o\hat{z}$ magnetic field at frequency ω_o. As we saw in Chap. 17, two-qubit gates can be implemented by a sequence of single-qubit gates and J-coupling due to spin-spin interaction. Hence, NMR RF pulse sequences can be used to implement all the basic logic gates needed for quantum computation.

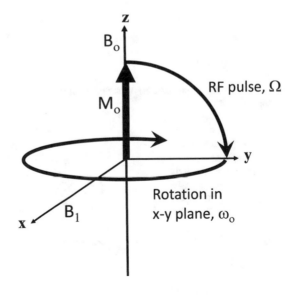

Fig. 19.6 A 90° RF pulse (B_1 field along the x-axis) rotates the net magnetization vector, M_o, into the x-y plane at the Rabi frequency, Ω. The B_o field rotates the magnetization vector in the x-y plane at the frequency ω_o

19.5   Read-Out

Based on Faraday's law of induction, a rotating (precessing) nuclear spin magnetization in the x-y plane will induce an oscillating voltage signal at frequency ω_o in
a probe coil positioned by convention along the y-axis (Fig. 19.7). To distinguish
between a $|0\rangle$ and $|1\rangle$ state, we can perform a 90° pulse and then measure the probe
signal. A 90° pulse applied to the $|0\rangle$ state results in a magnetization vector along +
y and will produce a positive signal in the probe coil, while a 90° pulse applied to the
$|1\rangle$ state will result in a magnetization vector in the $-y$ direction and will produce
a negative signal in the probe coil (Fig. 19.8). Due to the small value of the nuclear
magnetic moment, a large ($\sim 10^8$) number of molecules are required to produce a
measurable signal.

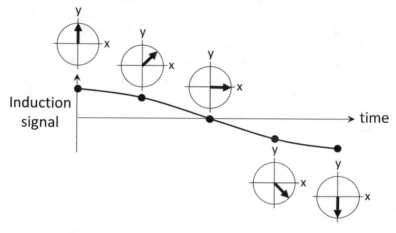

Fig. 19.7 Rotation (precession) of nuclear spin in the x-y plane, and the resulting induction signal
in the probe coil

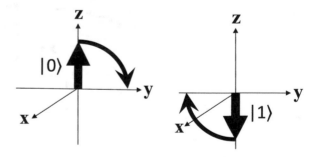

Fig. 19.8 90° RF pulse
applied to $|0\rangle$ or $|1\rangle$,
resulting in rotation in the
Bloch sphere and a positive
or negative signal amplitude,
respectively, in the probe coil

Fig. 19.9 Free induction
decay

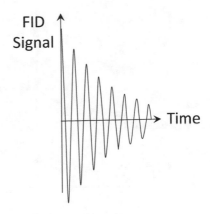

19.6   Free Induction Decay

Immediately after a 90° pulse, M is aligned along the x-y plane and $M_z = 0$. After removing B_1, M will relax back to its equilibrium position ($M_z = M_o$) in a characteristic duration of time, as the spin population returns to its equilibrium distribution. During this process, the oscillating signal collected by the detector coil will decay and is called the "free induction decay" (FID). The FID is a time-domain signal, illustrated in Fig. 19.9, analogous to the damping of vibrations.

19.7   Relaxation Time

The z-component of the magnetization will return to its equilibrium position, M_o, due to energy loss. This is known as the "longitudinal relaxation", and the length of time this takes is the "longitudinal relaxation time", T_1. The relaxation (or the return) of the z component towards its equilibrium value, M_o, is therefore given by:

$$M_z = M_o\left(1 - e^{-t/T_1}\right) \tag{19.6}$$

and is depicted in Fig. 19.10. T_1 necessarily involves energy loss as the nuclei return to their lower energy equilibrium orientation along B_o. This energy loss is transferred into nearby nuclei, atoms, and molecules through collisions, rotations, or electromagnetic interactions.

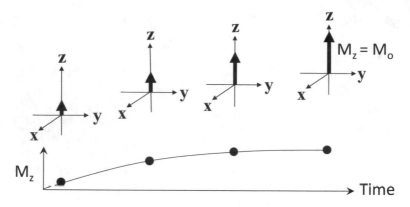

Fig. 19.10 Longitudinal relation

19.8   Decoherence

The loss of magnetization on the x-y plane, $M_x \to 0$ and $M_y \to 0$, is called "transverse relaxation", and is described by a time T_2 that is called the "transverse relaxation time":

$$M_x = M_o e^{-t/T_2} \qquad (19.7)$$

and

$$M_y = M_o e^{-t/T_2} \qquad (19.8)$$

T_2 is affected both by the motion of M in the x-y plane, as well as the return of M_z to M_o (that is, T_1). The motion of M in the x-y plane, called dephasing, is usually depicted as shown in Fig. 19.11. Dephasing is a result of the many nuclear spin rotations in the x-y plane falling out of phase with each other over time, so the net spin eventually cancels out. Contributions to dephasing include, for example, magnetic

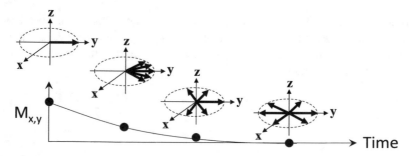

Fig. 19.11 Dephasing of a collection of nuclear spins

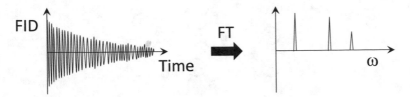

Fig. 19.12 Fourier transform (FT) of the time-domain FID gives the NMR frequency-domain spectrum

field inhomogeneity, so identical spin qubits on different molecules in the liquid sample have different Larmor frequencies resulting in different phase accumulations over time. Typically, $T_2 \leq T_1$; that is, the longitudinal relaxation time, T_1, is the upper limit of the relaxation process. In NMR, T_1 can be as long as hours or days, and T_2 can be as long as seconds. This gives enough time to perform hundreds or thousands of gate operations before the magnetized state (M_x, M_y) is destroyed. Since T_2 is typically much less than T_1, the FID follows an exponential decay which is proportional to e^{-t/T_2}. Chapter 25 discusses decoherence and characteristic times in more detail.

19.9 NMR Spectrum

Each atom in the molecule may resonate at a different frequency, ω_o. Thus, the FID is a superposition of all the different resonance frequencies from each atom of the molecule. Frequency analysis (Fourier transform) of the FID signal yields the component frequencies (Fig. 19.12). About 10^8 molecules are needed to generate a large enough signal.

19.10 Magnetic Resonance Imaging (MRI)

NMR is the basis for magnetic resonance imaging (MRI) used to image the human body, first done on humans in 1977. A 90° pulse tilts the spin vector along the x-y plane, and the FID decay signal is observed. Spatial resolution is achieved by varying the static magnetic field strength along the body. The magnitude of the FID signal is proportional to the proton density; i.e., the density of water, allowing a differentiation of various soft tissues. One can also generate a T_1 or T_2 weighted image to obtain further information. T_1 can be made intentionally shorter by introducing gadolinium (Gd) into the human body. Dipolar (spin-spin) interactions between water nuclei in tissue and electron spins in the Gd reduce T_1. For this reason, Gd is used as an MRI contrast agent.

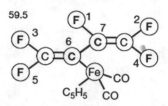

Fig. 19.13 Perfluorobutadienyl iron complex molecule used in NMR for implementation of Shor's algorithm. The 7 qubits are labelled. Qubits 1–3 form the first register for Shor's algorithm, and qubits 4–7 form the second register. Reproduced by permission from L.M.K. Vandersypen et al., Nature 414 (2001) 883, https://doi.org/10.1038/414883a, Copyright © 2001 [6]

19.11   Algorithms

NMR was the first implementation of a quantum computer [3]. NMR has been successfully used to demonstrate the Deutsch [3, 4], Grover [5], and Shor [6] quantum algorithms. In 2001, researchers at IBM famously reported the successful implementation of Shor's algorithm in a 7-qubit NMR quantum computer to factor $15 = 3 \times 5$ [6]. Shor's algorithm was implemented with ~300 RF pulse sequences for a total duration of ~ 1 s applied to $n = 7$ qubits in a perfluorobutadienyl iron complex molecule (Fig. 19.13). Figure 19.14 illustrates the RF pulse sequences used to implement Shor's algorithm.

19.12   Outlook

One of the difficulties of NMR for quantum computing is the Boltzmann distribution of states at room temperature, so it is difficult to initialize to $|0\rangle$. Also, scaling to many qubits would require molecules with a large number of atoms. Due to these difficulties, NMR is likely unfeasible for large-scale quantum computing.

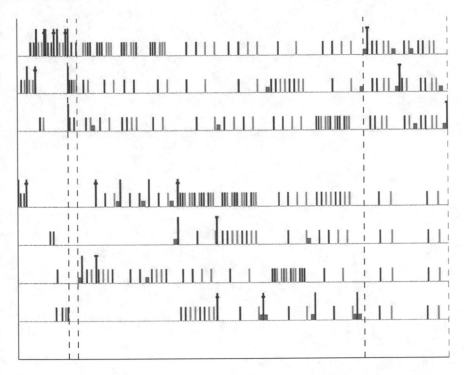

Fig. 19.14 NMR pulse sequences for Shor's algorithm. The tall red lines represent 90° pulses selectively acting on one of the seven qubits (horizontal rows) about positive *x* (no cross), negative *x* (lower cross) and positive *y* (top cross). The smaller blue lines denote 180° selective pulses used for refocusing about positive *x* (darker shade) and negative *x* (lighter shade). Refocusing is a technique used in NMR to reduce spin decoherence. Rotations about *z* are denoted by smaller and thicker green rectangles. Reproduced by permission from L. M. K. Vandersypen et al., Nature 414 (2001) 883, https://doi.org/10.1038/414883a, Copyright © 2001 [6]

References

1. J.A. Jones, Phys. Chem. Comm. **4**, 49 (2001)
2. Author: Stanford University/ Courtesy Stanford News Service. This file is licensed under the Creative Commons Attribution 3.0 Unported license (https://creativecommons.org/licenses/by/3.0/deed.en). File: Felix Bloch, Stanford University.jpg. (2020, September 27). *Wikimedia Commons, the free media repository.* Retrieved 16:43, December 7, 2020 from https://commons.wikimedia.org/w/index.php?title=File:Felix_Bloch,_Stanford_University.jpg&oldid=473329961
3. J.A. Jones, J. Chem. Phys. **109**, 1648 (1998)
4. I.L. Chuang et al., Nature **393**, 143 (1998)
5. I.L. Chuang, N. Gershenfeld, M. Kubinec, Phys. Rev. Lett. **80**, 3408 (1998)
6. L.M.K. Vandersypen et al., Nature **414**, 883 (2001)

Chapter 20
Solid-State Spin Qubits

Solid-state spin qubits are based on the electron spin in solid materials. We want to manufacture a spin system by confining/trapping a single electron in a quantum two-level system containing spin up or spin down, and then use electron spin resonance to control the spin. This chapter will discuss methods of trapping single electrons and implementing quantum computing in solid materials.

20.1 Electron Confinement

How do we trap a single electron in solid materials to form a spin qubit? We use quantum confinement in semiconductors, as we discussed in Chap. 2. Recall that a free particle (e.g., an electron) is described by a wavefunction:

$$\psi = \psi_o e^{i(kx-\omega t)} \tag{20.1}$$

where k is the wavevector ($k = 2\pi/\lambda$), λ is the de Broglie wavelength, and ω is the angular frequency. Quantum confinement occurs when the electron is confined along one or more directions with dimensions on the order of λ. The electron energy becomes quantized for motion along the confinement direction.

20.2 Quantum Wells, Wires, and Dots

Quantum wells (QWs) are structures (called 2-dimensional, 2-D) in which electron motion is confined along one direction, creating a so-called "two-dimensional electron gas" (2DEG). The electron motion is free along two directions (in the plane of the 2DEG), and its energy becomes quantized for motion in the perpendicular direction, as discussed in Chap. 2.

© The Author(s), under exclusive license to Springer Nature Switzerland AG 2021
R. LaPierre, *Introduction to Quantum Computing*, The Materials Research Society Series,
https://doi.org/10.1007/978-3-030-69318-3_20

Fig. 20.1 Semiconductor
heterostructure with its
corresponding energy band
diagram

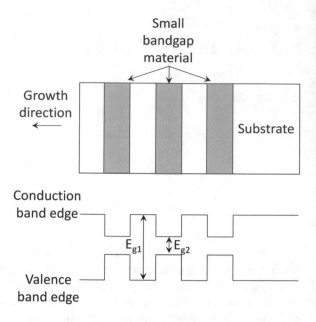

QWs are typically created using semiconductor "heterostructures", where the chemical composition of the semiconductor, and therefore the bandgap, changes with position (Fig. 20.1). For example, we might form a QW by sandwiching a small bandgap material such as gallium arsenide (GaAs) by a larger bandgap material such as aluminum gallium arsenide (AlGaAs). The 2000 Nobel Prize in Physics was awarded "for basic work on information and communication technology" with one half jointly to Zhores Alferov and Herbert Kroemer "for developing semiconductor heterostructures used in high-speed-electronics and opto-electronics" and the other half to Jack Kilby "for his part in the invention of the integrated circuit".

QWs were experimentally realized in 1974 [1] and are typically formed by "epitaxy". Epitaxy is a method of forming single crystal semiconductor thin films by impinging atomic or molecular beams onto a single crystal substrate that acts as a template for the film growth. Thus, the film has the same or related crystal structure as the substrate.

Quantum wires (QWrs) are structures (called one-dimensional, 1-D) in which electron motion is strongly confined in two dimensions, with electrons free to move in only one direction along the wire. QWrs can be produced by etching a semiconductor (called "top down" fabrication) or by "bottom-up" fabrication—for example, by using "selective-area epitaxy" methods. Figure 20.2 shows a scanning electron microscopy image of QWrs grown using bottom-up selective-area epitaxy.

Quantum dots (QDs) are structures (called zero-dimensional, 0-D) in which electron motion is strongly confined in three dimensions (with energy quantization in all directions). A QD will exhibit discrete energy levels like an atom. Today, QDs

Fig. 20.2 Scanning electron microscopy image of gallium arsenide quantum wires. The scale bar is 200 nm

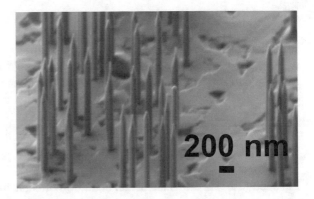

may be realized by a variety of different techniques and materials including (1) gate-defined QDs where electrostatic gates are used to confine charge, (2) pillars etched out of quantum well heterostructures, (3) self-assembled QDs induced by strain in the "Stranski–Krastanov" growth method, and (4) QDs formed in bottom-up grown nanowires (Fig. 20.3).

Exercise 20.1 Estimate the dimensions of a QD that gives a Larmor frequency of 10 GHz.

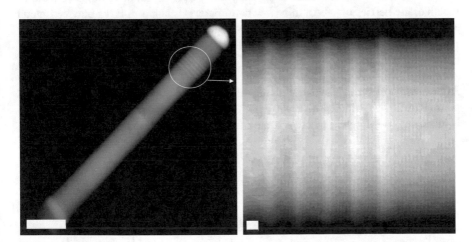

Fig. 20.3 Transmission electron microscopy image of a quantum wire, containing five quantum dots (bright contrast). The scale bar is 50 nm in the left image and 2 nm in the right image

20.3   Coulomb Blockade

A QD can be used to trap a single electron to form a spin or charge qubit. How do we move only one electron onto a QD? To understand that, we need to review some basics of capacitors. Suppose a potential, V, is applied to a capacitor, C, as shown in Fig. 20.4a. A charge Q will form on the capacitor that is proportional to the applied potential V:

$$Q = CV \qquad (20.2)$$

Equation (20.2) is the fundamental definition of capacitance, derived from electro-statics.

The energy required (work performed) to put charge Q on the capacitor is derived from electrostatics:

$$E = \frac{Q^2}{2C} \qquad (20.3)$$

This energy is stored in the electric field of the charge distribution. Suppose we allow one electron to cross the gap between the electrodes of a parallel plate capacitor from the $+Q$ side to the $-Q$ side. This process may occur by quantum tunneling if the capacitor gap is sufficiently thin. The capacitor is then called a tunnel junction and the gap is called a barrier. The symbol for a tunnel junction is shown in Fig. 20.4b, which looks like two electrodes or parallel plates of a capacitor in close contact. In practice, the barrier may be a thin (few nanometers) insulating layer between metal electrodes. The initial electrostatic energy of the capacitor is:

$$E_i = Q^2/2C \qquad (20.4)$$

After a single electron moves across the tunnel barrier, the positive charge on the left plate (Fig. 20.4a) is increased to $+(Q + e)$ and the negative charge on the right plate is increased to $-(Q + e)$ where e is the fundamental charge. The final energy is:

$$E_f = (Q + e)^2/2C \qquad (20.5)$$

Fig. 20.4   a A potential, V, applied to a capacitor, C. **b** A potential, V, applied to a tunnel junction

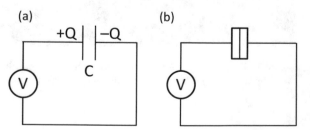

Fig. 20.5 Current–voltage
characteristic demonstrating
Coulomb blockade

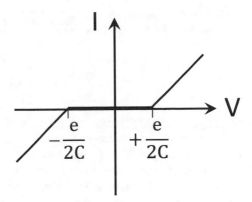

The change in energy is:

$$\Delta E = E_f - E_i = \frac{e}{C}\left(Q + \frac{e}{2}\right) \tag{20.6}$$

This process is favorable only if $\Delta E < 0$, meaning $Q < -e/2$. Thus, the required potential is $V = Q/C < -e/2C$. If the polarity is reversed, then we find $V > +e/2C$ is required. Considering both polarities, $V > |e/2C|$ is required for an electron to cross the tunnel barrier and produce a current flow (Fig. 20.5). If $V < |e/2C|$, then no current flows. The energy required to transfer the electron is $E_C = eV$, or:

$$E_C = e^2/2C \tag{20.7}$$

which is called the "charging energy". If $E < e^2/2C$, then there is insufficient energy from the source to transfer an electron from the $+Q$ electrode to the $-Q$ electrode due to the repulsion of the $-Q$ electrons. This is known as "Coulomb blockade". Coulomb blockade is a classical effect arising from electrostatics, but it is only observable in nanometer-scale objects where the magnitude of the charging energy becomes appreciable (see below). Note that Q can be continuous while e is discrete. Q can be continuous because it includes a polarization charge where the electrons are merely displaced relative to the background of positive ions.

What is the energy required to charge a sphere with one electron? The capacitance of a sphere in free space is $C = 4\pi\epsilon_o R$ where R is the radius of the sphere and ϵ_o is the permittivity of free space. Table 20.1 shows the charging energy of a sphere for difference radii. The charging energy increases from 0.07 to 70 meV as the radius of the sphere decreases from 10 to 0.01 μm. Small islands have small capacitance and therefore large charging energy. Hence, a QD is required to observe Coulomb blockade in semiconductors.

The charging energy is the energy that an electron needs to enter the QD. Where does the electron get this energy? It gets it from a source potential, V. If the energy provided by the source, eV, is less than the charging energy, then there is insufficient

Table 20.1 Capacitance and charging energy for a spherical island

Radius (μm)	$C = 4\pi\epsilon_o R$ (F)	$E_C = e^2/2C$(meV)
10	1.1×10^{-15}	0.07
1	1.1×10^{-16}	0.7
0.1	1.1×10^{-17}	7
0.01	1.1×10^{-18}	70

energy to move the electron onto the QD and we have Coulomb blockade. Thermal energy may also provide sufficient energy for the electron to be added to the QD. Thus, thermal energy is an unwanted source of energy (noise) that needs to be minimized for the observation of Coulomb blockade. Thermal energy is minimized by cooling the QD to very low temperature, typically mK, using liquid He in a dilution refrigerator. For example, at 30 mK, the thermal energy, kT, is only 0.0026 meV, far below the charging energy for a typical QD.

20.4 Single Electron Transistor

Consider the device shown schematically in Fig. 20.6a, which is called a single electron transistor. The charge island (QD) is separated from source (S) and drain (D) electrodes by tunnel barriers. The tunnel barriers allow us to confine electrons

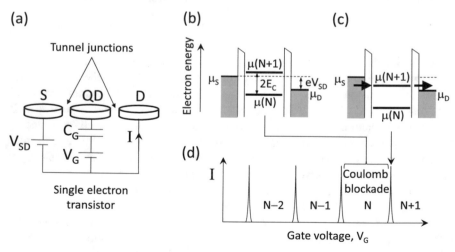

Fig. 20.6 a Schematic of single electron transistor with source (S), quantum dot (QD), drain (D), source-drain voltage (V_{SD}) and gate voltage (V_G). **b** Electron energy diagram for Coulomb blockade with N electrons on the QD. **c** Application of a gate voltage results in the lifting of Coulomb blockade, allowing current to flow across the QD from source to drain. **d** Current versus gate voltage, showing regions of Coulomb blockade and Coulomb oscillations (current spikes)

to the QD, while also allowing the addition or removal of electrons across the tunnel barriers. The QD and barriers may be formed in many different material systems such as GaAs/AlGaAs heterostructures (see Sect. 20.7). If the barriers are sufficiently thin (less than the de Broglie wavelength), then electrons can tunnel through the barriers to hop on or off the QD, providing certain conditions are met (see below).

The charging of the QD can be visualized in an electron energy diagram, as depicted in Fig. 20.6b. In an electron energy diagram, the electron energy increases vertically. On the left side of the energy diagram, we have a metal electrode (the source) used to inject electrons onto the QD through the tunnel barrier. In the metal electrodes, all energy levels are filled up to some energy level, μ_S, called the "Fermi energy", named after the physicist Enrico Fermi. Similarly, we have a metal electrode on the right (the drain) with a Fermi energy, μ_D, separated from the QD by a tunnel barrier. μ_S and μ_D are also referred to as the chemical potential, which is the work required to add or remove an electron. If a potential difference, V_{SD}, is applied between the source and drain electrodes, then an energy difference, $\mu_S - \mu_D = eV_{SD}$, forms between the two electrodes. This energy difference is called the bias window. When $\mu_S \neq \mu_D$, a current will try to flow to equilibrate the chemical potentials and achieve thermal equilibrium.

The electrostatic energy of the QD containing N excess electrons is:

$$E(N) = \frac{(-eN + C_G V_G)^2}{2C_\Sigma} \tag{20.8}$$

where N is the number of excess electrons (above the positive background charge), $C_G V_G$ is the continuous charge due to the gate voltage (V_G), $C_\Sigma = 2C + C_G$ is the total capacitance of the QD, C is the capacitance of each tunnel junction, and C_G is the gate capacitance. The chemical potential, $\mu(N)$, of a QD with N electrons is defined as the work or energy required to add the last electron to the QD:

$$\mu(N) = E(N) - E(N-1) = \frac{\left(N - \frac{1}{2}\right)e^2}{C_\Sigma} - e\frac{C_G}{C_\Sigma}V_G \tag{20.9}$$

If one electron is added to a QD at constant gate voltage, the chemical potential changes by an amount:

$$\Delta\mu = \frac{e^2}{C_\Sigma} = 2E_C \tag{20.10}$$

Thus, the separation of chemical potential levels of the QD is twice the charging energy, E_C. This is called the addition energy. Also, the separation of chemical potential levels in a QD are affected by the different quantized energy levels of electrons in the QD, which we neglect here.

Suppose N electrons occupy the QD up to the level, $\mu(N)$. We assume the bias window, eV_{SD}, is less than the addition energy, which is called the small bias approximation. Increasing the number of electrons on the QD to $N + 1$ requires an increase of chemical potential by $2E_C$, raising the chemical potential to $\mu(N + 1)$. However, as shown in Fig. 20.6b, μ_S is below $\mu(N + 1)$, meaning the source potential is insufficient to provide the energy that is required to add the electron. In addition, $\mu(N)$ is below μ_D, which inhibits electrons from hopping off the QD to the drain electrode. Hence, the number of electrons on the QD is fixed at N and no current can flow. This is Coulomb blockade.

Suppose a voltage, V_G, is applied to the QD through the gate electrode. If the gate voltage, V_G, is positive, then the chemical potential levels of the QD will move down by an amount qV_G since the charge q of the electron is negative, as indicated in Eq. (20.9). If $\mu(N + 1) \leq \mu_S$, then an electron can hop from the source electrode onto the QD into the level $\mu(N + 1)$. Similarly, if $\mu_D \leq \mu(N + 1)$, then the electron can hop off the QD to the drain electrode. Thus, if a potential V_{SD} is applied between the source and drain, and $\mu_D \leq \mu(N + 1) \leq \mu_S$, then current can flow across the QD from the source to drain electrode (Fig. 20.6c). In other words, for current to flow, a chemical potential of the QD must lie within the bias window of width eV_{SD}. By continuously tuning the gate voltage, we can move each chemical potential level into the bias window resulting in current spikes called "Coulomb oscillations" (Fig. 20.6d). The period of the Coulomb oscillations versus gate voltage is $2E_C/e = e/C_\Sigma$. Between the Coulomb oscillations, we have Coulomb blockade. At sufficiently low gate voltages, the Coulomb oscillations will end, meaning the QD has been depleted of all excess electrons. By counting the Coulomb oscillations, we can count the number of electrons on the QD. Thus, the charge on the QD can be controlled at the single electron level, forming a single electron transistor [2–4].

20.5 Spin Qubits

Once a single electron is trapped in a QD, you can use magnetic fields to control the spin state of the single electron using electron spin resonance (ESR). Using ESR, you can form any superposition of spin up and spin down; i.e., the single electron acts as a spin qubit. A spin-based quantum computer was first proposed by D. Loss and DiVincenzo in 1998 [5], and Rabi oscillations have been successfully observed in GaAs [6] and Si [7] spin qubits. Quantum computing based on spin has certain advantages, since spin is less sensitive to noise such as other charges, photons, phonons, etc.

Fig. 20.7 A spin filter

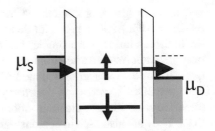

20.6 Spin Read-Out

The magnetic moment of a single-electron spin is equal to the Bohr magneton, $\mu_B = 9.27 \times 10^{-24}$ J/T. It is difficult to measure such a small magnetic moment. Instead, we can use spin-to-charge conversion. To convert spin to a charge signal, we use the ability of a single electron to tunnel into or out of a QD according to its spin state. If we apply a large static magnetic field, then the spin degenerate energy levels of the QD will be split by the Zeeman effect. By arranging the gate voltage (Fig. 20.7), only current comprised of spin up electrons will flow from source to drain while the spin down electrons are trapped. Thus, we can distinguish between $|0\rangle$ and $|1\rangle$ states. A spin filter or spin-to-charge convertor that can be used for spin readout has been built [8]. However, this method requires a large field to split the spin degenerate energy levels. An alternative method, called Pauli spin blockade, involves a double QD without the need for very strong magnetic fields [9].

20.7 Electrostatic Quantum Dots

A common method of carrier confinement, including the formation of QDs, is electrostatic gates. First, a two-dimensional electron gas (2DEG) can be formed at a heterointerface, for example between a large bandgap material like AlGaAs and a smaller bandgap material like GaAs. Figure 20.8 shows an AlGaAs/GaAs heterostructure

Fig. 20.8 AlGaAs/GaAs heterostructure and its corresponding conduction band energy diagram, showing electrons donated from the AlGaAs and forming a 2DEG near the heterointerface

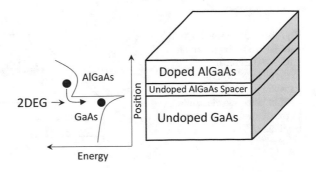

and its corresponding conduction band energy diagram. A portion of the wider bandgap AlGaAs is doped with donor impurities, typically Si. Electrons from the donor atoms fall into the smaller bandgap GaAs and become trapped in the GaAs near the AlGaAs/GaAs heterointerface, forming a 2DEG of electrons in a triangular quantum well. A spacer of undoped AlGaAs separates the ionized donor impurities from the 2DEG, increasing the electron mobility of the 2DEG. These 2DEGs are therefore used in high electron mobility transistors (HEMTs) for telecommunications. They are also used to demonstrate many interesting physical phenomena, such as the integer quantum Hall effect (IQHE) and the fractional quantum Hall effect (fQHE). Klaus von Klitzing won the 1985 Nobel Prize in Physics for the discovery of the IQHE. The 1998 Nobel Prize in Physics was awarded to Robert Laughlin, Horst Störmer and Daniel Tsui for their discovery of the fQHE. The IQHE and fQHE will be examined further in Chap. 26 when we consider topological quantum computing.

The 2DEG can be manipulated by depositing metal electrodes on top of the structure (Fig. 20.9). A negative potential applied to the electrodes will repel (deplete) electrons from the 2DEG located beneath the electrodes. In other words, the electrodes form an electrostatic barrier that can confine electrons. Electrons remain in the 2DEG in regions where electrodes on the surface are absent. Thus, electron islands or QDs can be formed in the 2DEG by an appropriate electrode geometry, as shown in Fig. 20.9. These types of QDs are called "electrostatic QDs" or gate-defined QDs. Two openings in the electrodes, as shown in Fig. 20.9, result in a lowering of the electrostatic barrier, allowing electrons to enter or leave the QD, forming a current, I. These openings are referred to as quantum point contacts (QPCs). Using Coulomb blockade, you can control the number of electrons on the QD at the single electron level. Once you have formed a QD and trapped a single electron, you can use magnetic fields to control the electron spin state using electron spin resonance (ESR), forming a spin qubit.

Figure 20.10 shows a scanning electron microscopy image of actual electrostatic QDs formed from several gates. In Fig. 20.10, nearby QPCs, connecting a source and drain reservoir of electrons, is used as a sensitive single-electron electrometer. A single electron in a QD is sufficient to raise the electrostatic potential of the nearby QPC. The opening of the QPC is reduced, which reduces the current that flows from source to drain. Hence, the QPC can observe single electrons jumping on or off the QD in real time [10].

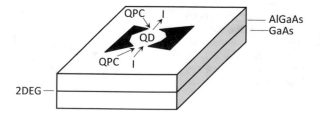

Fig. 20.9 QD formed by gate electrodes on top of an AlGaAs/GaAs 2DEG. Gaps in the electrodes form quantum point contacts (QPCs), allowing electrons to enter/exit the QD forming a current, I

Fig. 20.10 Two electrostatic QDs (trapped electrons are indicated by white circles). Current, I_{QPC}, flowing through two QPCs are used as charge detectors for each QD. The X in a box indicates contacts to the 2DEG. Reprinted with permission from *R*. Hanson et al., Rev. Mod. Phys. 79 (2007) 1217, https://doi.org/10.1103/Rev ModPhys.79.1217. Copyright © 2007 by the American Physical Society [2]

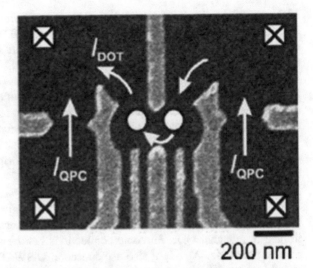

200 nm

For a complete set of universal quantum gates, we require not only single-qubit operations but also two-qubit operations. For example, you cannot create entanglement using only single-qubit operations. We need a double QD where the state of one QD influences the state of an adjacent QD, forming a CNOT gate. Figure 20.10 shows two electrostatic QDs with QPC charge detectors (I_{QPC}) for the left and right QD. The two thin gates below the QDs (called plunger gates) raise or lower the chemical potential of the left or right QD; i.e., these form gate voltages, V_{gL} and V_{gR}, used to control the charge on the QDs. Current, I_{DOT}, will flow only when the chemical potential of both QDs align and lie within the "bias window" between μ_S and μ_D, as controlled by V_{gL} and V_{gR} (Fig. 20.11a). Otherwise, current flow is blocked by Coulomb blockade (Fig. 20.11b). This can be used as the basis for a CNOT gate, where the state of the right QD depends on the state of the left QD.

The double QD has also been proposed as a charge qubit, with an electron in the left QD representing $|0\rangle$ and an electron in the right QD representing $|1\rangle$ [11].

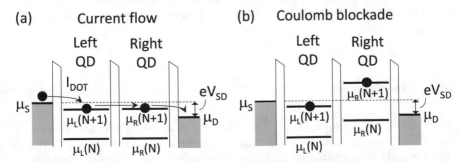

Fig. 20.11 **a** A double QD with aligned chemical potentials inside the bias window, allowing current to flow from source to drain. **b** A double QD under Coulomb blockade

If the tunnel barrier between the two QDs is sufficiently thin, then the two QDs become coupled. The resonant energy levels in each QD will split (anti-crossing), as discussed in Chap. 16, resulting in a symmetric and anti-symmetric wavefunction. A double QD is prepared with an electron in the right QD, labelled (0, 1). The barrier between the two QDs is tuned by a gate for a short time to allow electron tunneling between the two QDs. (0, 1) is no longer an eigenstate of the system due to the tunnel coupling. Rabi oscillations will occur between qubit states (0, 1) and (1, 0), meaning the electron charge will oscillate as a superposition between the left and right QD, forming a charge qubit. Spin qubits can also be developed based on the two-electron singlet–triplet state in a double QD, which was the original idea for a spin-based quantum computer [5].

One of the problems with solid-state qubits is that we need interactions between any two qubits, not just adjacent ones. One of the major challenges in using electrostatic QDs for quantum computation is that the interaction between electrons is limited to adjacent QDs. Photonic connections between distant QDs may help to resolve this issue. We might also use consecutive SWAP operations to move qubit states next to each other for their two-qubit interaction.

20.8   Other Spin Qubits

Besides QD qubits, there are other methods being developed for spin-based solid-state quantum computing. One of the most promising of these methods is the nitrogen-vacancy (NV) center in diamond [12, 13]. NV centers consist of a substitutional nitrogen atom replacing a carbon atom, and a neighbouring vacancy (missing carbon atom) in the diamond crystal lattice. This creates a point defect in the diamond lattice, containing an electron that can be used as a spin qubit. NV centers are randomly oriented within the diamond crystal lattice, but ion implantation techniques can be used to control their location [14]. NV fluorescence is spin-state dependent, allowing simple methods for optical readout of spin states. Quantum Diamond Technologies is a company developing NV-based spin qubits. There are many other proposals for forming solid-state spin qubits. For example, a single dopant atom, such as ^{31}P, in Si can also act as a spin qubit by using the electron or nuclear spin of the impurity atom [15].

20.9   Decoherence

As we discussed in Chap. 19, the Rabi oscillations in spin qubits are damped by "phase decoherence" (described by a characteristic time, T_2). Here, we consider these effects in solid-state spin qubits, where the phase decoherence occurs predominantly by "spin–orbit interaction" and "hyperfine interaction".

Fig. 20.12 Spin–orbit interaction where an electron moving at velocity **v** in an electric field **E** experiences a magnetic field **B**$_{so}$, resulting in precession of its spin **S**

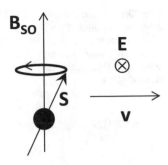

Spin–orbit interaction (SOI) is an example of a decoherence process that can occur in solid-state (spin) qubits. An internal electric field **E** may exist in the crystal unit cell of the host material, depending on its symmetry. As described by special relativity, an electron moving with velocity **v** in an electric field **E** will experience an effective magnetic field (spin–orbit field), **B**$_{so}$, given by:

$$\mathbf{B_{SO}} \propto (\mathbf{v} \times \mathbf{E}) \tag{20.11}$$

The electron spin will precess around **B**$_{SO}$ (Fig. 20.12). Spin randomization occurs when the electron follows random trajectories between scattering events. This is the dominant spin relaxation mechanism in semiconductors (decoherence of electron spins) and is known as the "Dresselhaus effect" [16].

The same effect can occur in an electric field arising from an asymmetric structure, such as the triangular quantum well at an AlGaAs/GaAs interface used to create a 2DEG (Fig. 20.8). In this case, the spin–orbit coupling is called the Rashba effect. In 1990, this effect was proposed for a field effect transistor based on electron spin, which launched the field of spintronics [17].

Another decoherence mechanism is "hyperfine coupling". The electron spin in a QD can be coupled to many ($\sim 10^6$) nuclear spins (e.g., Ga and As nuclei in gallium arsenide); i.e., the electron wavefunction overlaps with $\sim 10^6$ nuclei. Electrons experience a randomly fluctuating magnetic field from the nuclear spins (called the Overhauser field) on the order of mT (Fig. 20.13) [18–22]. This effect produces a T_2 time on the order of tens of nanoseconds in GaAs.

One way to eliminate nuclear spins is to use materials made of isotopes with no nuclear spin (e.g., isotopically pure silicon and germanium). Naturally available silicon, for example, is composed of 92.2% ^{28}Si ($I = 0$), 4.7% ^{29}Si ($I = \frac{1}{2}$) and 3.1% ^{30}Si ($I = 0$). Si can be purified to ^{28}Si by centrifuging SiF$_4$ gas. Saeedi et al. [23] observed a room-temperature coherence time of over 39 min using ionized donors in isotopically purified Si-28.

Fig. 20.13 Illustration of
hyperfine coupling. The
circle represents the extent of
the electron wavefunction,
which is overlapping many
nuclear spins

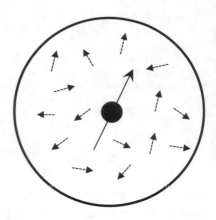

20.10   Outlook

A review article on solid-state spin qubits, with an emphasis on III-V materials, is
available in Ref. [2]. Spin qubits can also be manufactured in Si, leveraging the
huge Si electronics industry [15]. Thus, there are many companies developing solid-
state spin qubits including HRL Laboratories, Intel, Silicon Quantum Computing,
Quantum Motion, CEA-LETI, Hitachi and IMEC.

References

1. R. Dingle, W. Wiegmann, C.H. Henry, Phys. Rev. Lett. **33**, 827 (1974)
2. R. Hanson et al., Rev. Mod. Phys. **79**, 1217 (2007)
3. A.N. Korotkov, Int. J. Electron. **86**, 511 (1999)
4. L. Zhuang, L. Guo, S.Y. Chou, Appl. Phys. Lett. **72**, 1205 (1998)
5. D. Loss, D.P. DiVincenzo, Phys. Rev. A **57**, 120 (1998)
6. F.H.L. Koppens et al., Nature **17**, 766 (2006)
7. M. Veldhorst et al., Nat. Nanotechnol. **9**, 981 (2014)
8. R. Hanson et al., Phys. Rev. B **70**, 241304 (2004)
9. K. Ono et al., Science **297**, 1313 (2002)
10. S. Gustavsson et al., Appl. Phys. Lett. **85**, 4394 (2005)
11. K.D. Petersson et al., Phys. Rev. Lett. **105**, 246804 (2010)
12. L. Childless, R. Hanson, MRS Bull. **38**, 134 (2013)
13. M.W. Doherty et al., Phys. Rep. **528**, 1 (2013)
14. B. Naydenov et al., Appl. Phys. Lett. **95**, 181109 (2009)
15. B.E. Kane, Nature **393**, 133 (1998)
16. G. Dresselhaus, Phys. Rev. **100**, 580 (1955)
17. S. Datta, B. Das, Appl. Phys. Lett. **56**, 665 (1990)
18. I.A. Merkulov, Al.L. Efros, M. Rosen, Phys. Rev. B **65**, 205309 (2002)
19. F.H.L. Koppens et al., Science **309**, 1346 (2005)

20. A.C. Johnson et al., Nature **435**, 925 (2005)
21. S. Nadj-Perge et al., Phys. Rev. B **81**, 201305 (2010)
22. A. Pfund et al., Phys. Rev. Lett. **99**, 036801 (2007)
23. K. Saeedi et al., Science **342**, 830 (2013)

Chapter 21
Trapped Ion Quantum Computing

In trapped ion quantum computing (TIQC), ions are created and trapped to form a qubit register. Ions can be suspended and trapped in free space using electromagnetic fields. The qubits, defined by the electronic states of the ions, are manipulated and read out using lasers.

21.1 Paul Trap

The trapped ion quantum computer (Fig. 21.1) was first proposed in 1995 by Ignacio Cirac and Peter Zoller (Fig. 21.2) [1]. For TIQC, it is necessary to trap ions to form a qubit register. The ions are trapped in a "Paul trap", invented in the 1950s by Wolfgang Paul (Nobel Prize in Physics in 1945) [2]. According to Gauss's law, it is impossible to trap a single charge in free space by static electric fields alone (Earnshaw's theorem) since there can be no net inward force ($\nabla \cdot E = 0$) to constrain the motion of the ions. There will be at least one direction where ions can escape. We can, however, use time varying electromagnetic fields to trap charge.

Figure 21.3 illustrates the end view of a linear ion trap with the four electrodes extending perpendicular to the page. An alternating current (ac) potential is applied to the electrodes, resulting in a saddle-shaped potential that rotates at the ac frequency (typically radio frequencies). The alternating forces create a trap for ions. Coulomb repulsion of the ions distributes them in a linear chain (1D crystal) along the trap (perpendicular to the page in Fig. 21.3). The ions can remain trapped for hours, days, or even months. In 2016, more than 200 Be ions were trapped [5]. In 2018, a quantum register of 20 trapped ions were entangled [6]. Microfabricated 2D ion trap arrays are under development. RF and DC fields can be applied by planar electrodes to move the ions around on the chip (ion shuttling) like electrons in a CCD camera [7]. Paul traps are also used in mass spectrometers, Bose–Einstein condensates, and atomic clocks.

© The Author(s), under exclusive license to Springer Nature Switzerland AG 2021
R. LaPierre, *Introduction to Quantum Computing*, The Materials Research Society Series,
https://doi.org/10.1007/978-3-030-69318-3_21

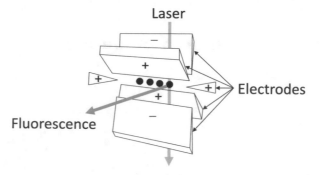

Fig. 21.1 Schematic of a trapped ion quantum computer. The ions are shown as black dots. The potential on the four larger electrodes alternates to create a "rotating saddle potential" that traps the ions. A laser is used to excite Rabi oscillations. Fluorescence from the ions can be read by a CCD camera or photodetector to determine the state of the ions

Fig. 21.2 Left: Ignacio Cirac [3]. Right: Peter Zoller [4]. *Credit* Wikimedia Commons [3, 4]

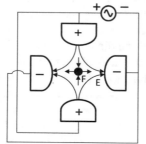

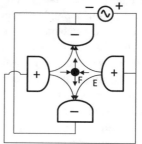

Fig. 21.3 A Paul trap illustrating the alternating potential at two different times (separated by a half-period of the ac potential), and the resulting electric field lines (E) and forces (F) on a positively charged ion. The resulting saddle potential rotates, resulting in an ion trap

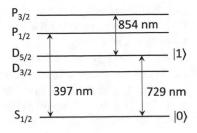

Fig. 21.4 Electronic structure for Ca$^+$. Transition wavelengths are indicated in nanometers

21.2 Ion Production

Ions are produced by evaporating a metal (e.g., Ca) to produce neutral atoms. These atoms can be ionized by an electron beam or laser beam that knock an electron off the atom. The ions then get trapped in the Paul trap. Be$^+$, Mg$^+$, Ca$^+$, Sr$^+$, Ba$^+$, Zn$^+$, Cd$^+$, Hg$^+$ and Yb$^+$ are commonly used, which have single valence electrons after ionization. Figure 21.4 shows the electronic structure for Ca$^+$. S, P and D refer to the orbital angular momentum state ($l = 0$, 1 and 2, respectively), and the subscript denotes the total (spin + orbital) angular momentum quantum number. $S_{1/2}$ is used as the $|0\rangle$ state and $D_{5/2}$ is used as the $|1\rangle$ state.

21.3 Qubits

Two types of qubit are possible with trapped ions: radio frequency (RF) qubits and optical qubits [8]. In RF qubits, qubits are encoded in the electronic ground state ($S_{1/2}$) of the ions which are split by the hyperfine interaction. Alternatively, an applied magnetic field (~0.01 T) can be used to create a Zeeman splitting of the ground state $S_{1/2}$ energy level. Addressing of individual qubits requires a magnetic field gradient to separate the transition frequencies of the ions. Since these energy transitions are in the radio frequency range, these qubits are referred to as RF qubits. Alternatively, in an optical qubit, $|0\rangle$ and $|1\rangle$ are defined by the electronic ground state and an excited electronic state (e.g., $S_{1/2}$ and $D_{5/2}$ in Ca$^+$). Since this energy difference is in the visible range, these are referred to as optical qubits. Separate lasers can be used to address individual qubits.

21.4 Ion Cooling

The ions are cooled to near absolute zero to supress unwanted electronic and vibrational transitions. The suppression of ion motion is described by the Lamb-Dicke criterion where the amplitude of the ion motion in the trap should be much less than the wavelength of the incident light. The cooling is performed using laser cooling or "Doppler cooling". Steven Chu, Claude Cohen-Tannoudji and William Phillips (Fig. 21.5) were awarded the 1997 Nobel Prize in Physics for their work on laser cooling and atom trapping.

As its name suggests, Doppler cooling involves the Doppler effect. A laser of frequency v is detuned from an electronic energy transition of the ion, meaning its frequency is slightly lower than the transition. At atom approaching the laser source will observe a higher photon frequency due to the Doppler effect:

$$v_{\text{ion}} = \sqrt{\frac{1 + v/c}{1 - v/c}} \tag{21.1}$$

The Doppler-shifted frequency is resonant with the electronic transition of the approaching ion, meaning the atom will absorb the photon. The momentum of the photon is transferred to the ion, giving it a "kick" in a direction that is opposite to the ion momentum. After photon absorption, the ion will move back into its ground state by spontaneous emission of the absorbed photon. However, this process occurs in a random direction. Hence, after repeated absorption and emission events, the emission recoil momentum averages to zero but the absorption recoil momentum does not. As a result, the ion momentum and kinetic energy are reduced, cooling the ion to lower temperatures (Fig. 21.6). For Ca^+ ions, the laser cooling takes place between the $S_{1/2}$ and $P_{1/2}$ levels at 397 nm with a detuning of ~10 MHz. Further cooling is then accomplished using a related technique, called "sideband" cooling. With these

Fig. 21.5 Left to right: Steven Chu [9], Claude Cohen-Tannoudji [10] and William Phillips [11]; Nobel Prize in Physics in 1997. *Credit* Wikimedia Commons [9–11]

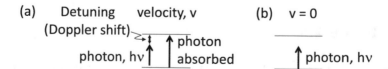

Fig. 21.6 Principle behind Doppler cooling. **a** Ion motion toward the photon results in a Doppler shift of photon frequency, resulting in resonant absorption. **b** Other ion motion, such as a stationary ion, results in no photon absorption

methods, the ions can be cooled such that the thermal energy is much less than the vibrational energy of the ions.

> **Exercise 21.1** Estimate a typical ion speed at 10 mK and calculate the corresponding Doppler frequency shift for the 397 nm transition in Ca^+.

21.5   Initialization

In the ground state of Ca^+, the single valence electron is in the 4s orbital with orbital angular momentum quantum number $l = 0$. An electron in the excited 3d state has $l = 2$. Hence, an excited electron can only decay to the ground state by emitting two quanta of angular momentum. These quadrupole transitions are forbidden in the dipole approximation, resulting in long lifetimes of the excited state. A method is needed to rapidly initialize the electrons to the ground state.

The qubits can be initialized to the ground state ($|0\rangle$) by a technique called "optical pumping". Optical pumping means that the qubit is excited by a laser from the excited state ($|1\rangle$) to a metastable state, which then rapidly decays by spontaneous emission to the ground state ($|0\rangle$). For Ca^+, we can pump from the D to the P states using 854 and 866 nm. The state rapidly decays from the metastable P state to the $|0\rangle$ state ($S_{1/2}$). Any qubit in the ground state is unaffected.

21.6   Single-Qubit Operations

Individual ions (qubits) can be addressed by separate lasers focussed on each ion (Fig. 21.1). To change any individual qubit, we simply illuminate the relevant ion with a pulse of light, creating Rabi oscillations between the two quantum levels. This is possible because the ions are separated by tens of microns, and we can focus a laser beam down to a width of a few microns to address individual ion qubits. The lifetime (T_2) is on the order of seconds, relatively long compared to other techniques since the individual ions are well separated and isolated from the environment.

21.7   Two-Qubit Operations

For universal quantum gates, we also need two-qubit operations. The CNOT gate, together with single-qubit rotations, form a universal set of gates. A control-Z gate (also called a controlled-phase gate) sandwiched between two Hadamard gates

Fig. 21.7 CNOT gate
formed from Hadamard and
control-Z gates

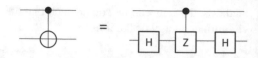

produces a CNOT gate (Fig. 21.7). As we saw in Chap. 14, the Hadamard gate can
be formed by combinations of single-qubit rotations. Thus, all we need to do is find
a way to create a control-Z gate.

David Wineland (Fig. 21.8) developed the methods used in the CNOT gate, for
which he won the Nobel Prize in Physics in 2012 with Serge Haroche. The ions
form a linear crystal and oscillate in collective vibrational motional modes around
their equilibrium positions, mediated by the Coulomb interaction between ions, like
an array of pendulums connected by springs. The ion motion is described using a
parabolic potential well. The quantized vibrational energies in the potential well are
called "phonons", with an equal energy separation, $\hbar\omega$ (Fig. 21.9).

Using the phonon vibrational modes of the ions, we can produce a control-Z gate
between any two ions as follows [1]. Referring to Fig. 21.10, ion A is the control
qubit, while ion B is the target qubit. Ion A and B do not have to be neighbors. A
π pulse ($R_x(\pi)$) on ion A is detuned (red-shifted) from the electronic ground (g)
to excited (e_o) state by an amount equal to the first excited phonon mode. If ion A
is in the ground state, nothing happens. If ion A is in the excited state, stimulated
emission to the ground state in the 1st phonon mode occurs. This is indicated as step
(1) in Fig. 21.10. The phonon mode is transmitted via the strong ion-ion electrical

Fig. 21.8 David Wineland;
Nobel Prize in physics in
2012. *Credit* Wikimedia
Commons [12]

Fig. 21.9 Quantized energy
levels of ions in a harmonic
potential well for motion
along x, y or z

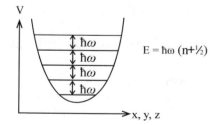

$E = \hbar\omega (n+\frac{1}{2})$

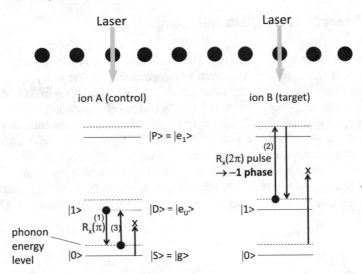

Fig. 21.10 Implementation of a control-Z gate

repulsion to all the other ions. The ion vibrations act as a "data bus". Thus, ion B also gets excited to the phonon mode. A 2π pulse about the x-axis at ion B induces an excitation from the e_o excited state to the higher e_1 excited state and back, inducing a -1 phase shift for the $|1\rangle$ state (you can verify that $R_x(2\pi) = -I$). If ion B is in the electronic ground state ($|0\rangle$), nothing happens. This is indicated as step (2) in Fig. 21.10. Another π pulse on ion A returns it to the excited state and cancels the phonon mode. This is indicated as step (3) in Fig. 21.10. The effect of these interactions is to change the sign of the state of ion B only when both ions are initially excited in the $|1\rangle$ state. We have successfully implemented a control-Z gate.

21.8 Read-Out

We can tune a laser, so it is resonant with a transition from the ground state ($|0\rangle$), but not with the excited state ($|1\rangle$); e.g., the 397 nm transition for Ca^+. At 397 nm, only the ground state will be excited to the higher energy P state and then decay back to the ground state. The emitted photon fluorescence can be detected by an ordinary CCD camera or photodetector. Thus, the $|0\rangle$ state lights up. Nothing happens with the excited state, so $|1\rangle$ appears dark.

21.9   Neutral Atom Quantum Computing

A related approach to TIQC is to use the electronic states in cold neutral atoms as qubits [13]. In this approach, atoms can be trapped using the standing light waves of an "optical lattice" created by a network of crossed laser beams. QuEra, ColdQuanta, and Atom Computing are companies working on the cold atom approach.

Rydberg atoms are also being explored for atom-based quantum computing. Rydberg states are atoms that have an electron excited to a state with a high principal quantum number. Rydberg atoms have a large electric dipole moment, making them easier to manipulate with lasers. A Rydberg atom also has a large dipole–dipole interaction with other nearby Rydberg atoms, making two-qubit gates easier to implement. A review on quantum computing by Rydberg atoms is available in Ref. [14].

21.10   Outlook

The Deutsch-Jozsa [15], Grover [16], and Shor algorithm [17] have been implemented by TIQC. Shor's algorithm on a 5-ion system was used to factor the number 15 [17]. However, trapping and cooling ions is a complicated process, and it may be challenging to scale this approach to many thousands of ions required for some quantum algorithms.

TIQC offers a major advantage compared to other methods: the ions are all identical and well isolated from the environment, resulting in long coherence times. As a result, despite the difficulties of ion trapping and cooling, the ion approach to quantum computing has attracted some industrial interest. IonQ, AQT (Alpine Quantum Technologies), NextGenQ, Honeywell, and Universal Quantum are companies developing TIQC. A review of TIQC is available in Ref. [7].

References

1. J.I. Cirac, P. Zoller, Phys. Rev. Lett. **74**, 4091 (1995)
2. W. Paul, Rev. Mod. Phys. **62**, 531 (1990)
3. This file is licensed under the Creative Commons Attribution-Share Alike 3.0 Unported license (https://creativecommons.org/licenses/by-sa/3.0/deed.en). File: Prof Cirac office klein (cropped).jpg. (2020, October 4). *Wikimedia Commons, the free media repository*. Retrieved 18:56, December 7, 2020 from https://commons.wikimedia.org/w/index.php?title=File:Prof_C irac_office_klein_(cropped).jpg&oldid=480274796
4. Author: Hans Weber. This file is licensed under the Creative Commons Attribution-Share Alike 3.0 Unported, 2.5 Generic, 2.0 Generic and 1.0 Generic license (https://creativecommons.org/licenses/by-sa/3.0/deed.en). File: Zoller Peter.jpg. (2020, September 5). *Wikimedia Commons, the free media repository*. Retrieved 19:02, December 7, 2020 from https://commons.wikime dia.org/w/index.php?title=File:Zoller_Peter.jpg&oldid=448828794.
5. J.G. Bohnet et al., Science **352**, 1297 (2016)

6. N. Friis et al., Phys. Rev. X **8**, 021012 (2018)
7. C.D. Bruzewicz et al., Appl. Phys. Rev. **6**, 021314 (2019)
8. H. Häffner, C.F. Roos, R. Blatt, Phys. Rep. **469**, 155 (2008)
9. Attribution: The Royal Society. This file is licensed under the Creative Commons Attribution-Share Alike 3.0 Unported license (https://creativecommons.org/licenses/by-sa/3.0/deed.en). File: Professor Steven Chu ForMemRS headshot.jpg. (2020, October 30). *Wikimedia Commons, the free media repository.* Retrieved 18:38, December 8, 2020 from https://commons.wikimedia.org/w/index.php?title=File:Professor_Steven_Chu_ForMemRS_headshot.jpg&oldid=507616885
10. Author: Amir Bernat. This file is licensed under the Creative Commons Attribution-Share Alike 4.0 International, 3.0 Unported, 2.5 Generic, 2.0 Generic and 1.0 Generic license (https://creativecommons.org/licenses/by-sa/4.0/). File:Claude Cohen-Tannoudji.JPG. (2020, August 28). *Wikimedia Commons, the free media repository.* Retrieved 18:06, December 8, 2020 from https://commons.wikimedia.org/w/index.php?title=File:Claude_Cohen-Tannoudji.JPG&oldid=444509221
11. Author: Markus Pössel. This file is licensed under the Creative Commons Attribution-Share Alike 3.0 Unported license (https://creativecommons.org/licenses/by-sa/3.0/deed.en). File: William D. Phillips.jpg. (2020, October 26). *Wikimedia Commons, the free media repository.* Retrieved 18:08, December 8, 2020 from https://commons.wikimedia.org/w/index.php?title=File:William_D._Phillips.jpg&oldid=502669574
12. File: David Wineland 2008.jpg. (2020, September 19). *Wikimedia Commons, the free media repository.* Retrieved 19:12, December 7, 2020 from https://commons.wikimedia.org/w/index.php?title=File:David_Wineland_2008.jpg&oldid=464547801
13. L. Henriet et al., Quantum **4**, 327 (2020)
14. M. Saffman, T.G. Walker, K. Mølmer, Rev. Mod. Phys. **82**, 2313 (2010)
15. S. Gulde et al., Nature **421**, 48 (2003)
16. K.-A. Brickman et al., Phys. Rev. A **72**, 050306 (2005)
17. T. Monz et al., Science **351**, 1068 (2016)

Chapter 22
Superconducting Qubits

Thus far, we have examined quantum computing based on single particle states in atoms, ions and semiconductor structures. In this chapter, we will examine quantum states in superconductors and their application as qubits. This chapter is particularly extensive due to the large variety of possible superconducting quantum circuits. We introduce superconductivity and examine the three main types of superconducting qubit: the flux qubit, the charge qubit and the phase qubit. We will also examine the transmon qubit and circuit quantum electrodynamics.

22.1 Superconductivity

A quantum computer cannot be made of regular electronic circuits due to electrical resistance. The electron scattering (i.e., the resistance) causes energy dissipation and destroys the phase coherence of the electron waves. In contrast, current in a superconductor flows without dissipation (i.e., without resistance). Therefore, superconducting electronic circuits can be used to form qubits.

In 1908, Heike Kamerlingh Onnes (Fig. 22.1) succeeded to liquefy He at temperatures below 4.22 K. Subsequently, in 1911, Onnes used his low temperature apparatus to measure the resistance of Hg and demonstrated the superconducting state below 4.2 K [1]. Below a critical temperature (also called the superconducting transition temperature), T_c, superconductors exhibit zero resistance. How do we know the resistance is truly zero, and not just a very low value? The zero-resistance state can be confirmed by inducing a superconducting current, called a supercurrent or persistent current, in a loop of superconducting wire cooled below its transition temperature. The supercurrent flows without dissipation, meaning that electrons do not lose energy by scattering processes—thus, the current does not decay. Measurements of the persistent current in a superconducting loop indicate that it will dissipate after 10^{92} years, indicating that the resistance is effectively zero! [2].

© The Author(s), under exclusive license to Springer Nature Switzerland AG 2021
R. LaPierre, *Introduction to Quantum Computing*, The Materials Research Society Series,
https://doi.org/10.1007/978-3-030-69318-3_22

(a)   (b)

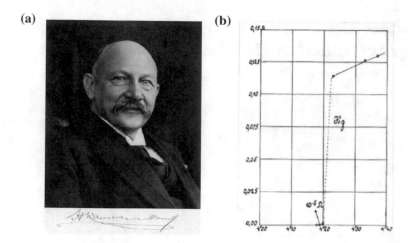

Fig. 22.1 a Signed photo of Heike Kamerlingh Onnes (1853–1926); Nobel Prize in Physics in 1913. *Credit* Wikimedia Commons [4]. **b** Onnes' measurement of the resistance of Hg. *Credit* Ref. [1]

Al ($T_c = 1.2$ K) or Nb ($T_c = 9.26$ K) are often used for superconducting quantum computing circuits. Qubit frequencies are usually 5–10 GHz which corresponds to 0.25–0.5 K. Thus, the circuits are operated in dilution refrigerators, typically at temperatures of 10–30 mK, to minimize thermal excitation. After many decades of research, a room temperature superconductor has been achieved [3], although a complete understanding of superconductivity is still elusive.

22.2 BCS Theory

Since the discovery of superconductivity in 1911, nearly 50 years passed before a microscopic theory was developed in 1957 by Bardeen, Cooper and Schrieffer (Fig. 22.2), known as the BCS theory [5]. In a normal metal, all electrons obey Fermi-Dirac statistics and fill an energy band up to the Fermi energy. In the BCS theory, a fraction of the electrons near the Fermi energy combine into "Cooper" pairs having opposite spin and momentum (Fig. 22.3). In type I superconductors, this attractive force is attributed to an electron-phonon interaction that overcomes the Coulomb repulsion of electrons (the microscopic mechanism in type II superconductors is still under debate). This attractive force is manifest at temperatures below the superconducting transition temperature, T_c. A superconducting energy gap ($\Delta \sim kT_c$) forms at the Fermi energy and protects the Cooper pairs from dissociation (Fig. 22.3). The energy band below the gap is full and contains the normal electrons, while the energy band above the gap is empty. Normal electrons have insufficient energy to cross the gap at low temperatures. Thus, conduction is only by Cooper pairs at the Fermi energy.

Fig. 22.2 From left: John Bardeen (1908–1991) [6], Leon Cooper (1930–) [7], and John Schrieffer (1931–2019) [8]; Nobel Prize in Physics in 1972. *Credit* Wikimedia Commons [6–8]

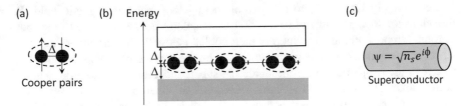

Fig. 22.3 **a** Cooper pairs of opposite spin with binding energy Δ. **b** Superconducting energy gap Δ with Cooper pairs at the Fermi energy. All states above the gap are empty and all states below the gap are full of normal electrons. **c** Quantum state for the ensemble of Cooper pairs in a superconductor

The radius of the Cooper pairs is on the order of 10–1000 nm, so that the Cooper pairs significantly overlap with other pairs and their motion is highly correlated. Hence, the Cooper pairs behave coherently, like photons in a laser, and can be described by a single wave function:

$$\psi = \sqrt{n_s} e^{i\phi} \tag{22.1}$$

where n_s is the density of Cooper pairs and ϕ is the superconducting phase. In a normal (non-superconducting) metal, all the electrons obey Fermi-Dirac statistics and occupy different (spin degenerate) energy levels. The phases differ for each electron in a normal metal and average out in any macroscopic quantity. Conversely, the net spin of a Cooper pair is zero (Cooper pairs form a spin singlet). Hence, Cooper pairs are bosons and obey Bose-Einstein statistics, rather than the Fermi-Dirac statistics of normal electrons. This means that Cooper pairs do not obey the Pauli exclusion principle and can condense to the same quantum state in a single energy level at the Fermi energy. This process is called "Bose-Einstein condensation" with all Cooper pairs acting as a single quantum mechanical object. Superconductivity is therefore a form of macroscopic quantum behaviour; i.e., it is observable even in macroscopic-sized superconductors.

22.3   Flux Quantization

In the presence of a magnetic field, quantum mechanics tells us that the superconducting phase accumulation along a path length from point 1–2 is given by the path integral:

$$\phi_{1\rightarrow 2} = \frac{2\pi}{\Phi_o} \int_1^2 \mathbf{A} \cdot \mathbf{dl} \tag{22.2}$$

where $\mathbf{dl}$ is a differential line element along the current path, $\Phi_o = h/2e \sim 2 \times 10^{-15}$ Vs is called the flux quantum, and $\mathbf{A}$ is the magnetic vector potential defined by $\mathbf{B} = \nabla \times \mathbf{A}$. The flux quantum is the ratio of Planck's constant to the charge of the Cooper pair ($2e$).

Let us apply Eq. (22.2) to a superconducting loop with a magnetic field, $\mathbf{B}$, threading the loop as shown in Fig. 22.4. A magnetic field is applied to the loop at room temperature, and the loop is then cooled below the superconducting transition temperature. A supercurrent will form on the surface of the superconductor to screen the magnetic field from inside the superconductor (i.e., the loop exhibits perfect diamagnetism, which is another hallmark of superconductivity). When the external magnetic field is switched off, a flux Φ is maintained through the loop because of the persistent surface current. In the following, we will relate this flux to the phase.

Using Eq. (22.2), the phase change of the wavefunction around a closed loop (closed path) due to the magnetic field is:

$$\phi = \frac{2\pi}{\Phi_o} \int_{\text{closed path}} \mathbf{A} \cdot \mathbf{dl} \tag{22.3}$$

Using Stokes' theorem and $\mathbf{B} = \nabla \times \mathbf{A}$, the integral on the right side of Eq. (22.3) can be written as:

Fig. 22.4 Magnetic field, **B**, threading a superconducting loop. The magnetic flux, Φ, threading the loop is quantized in units of the flux quantum, $\Phi_o = h/2e$

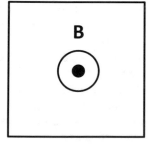

$$\int_{\text{closed path}} \mathbf{A} \cdot d\mathbf{l} = \int_{\text{enclosed area}} (\nabla \times \mathbf{A}) \cdot d\mathbf{a} = \int_{\text{enclosed area}} \mathbf{B} \cdot d\mathbf{a} = \Phi \qquad (22.4)$$

where Φ is the total magnetic flux enclosed by the area of the loop, and $d\mathbf{a}$ is a differential area of the loop. Substituting Eq. (22.4) into (22.3) gives:

$$\phi = 2\pi \frac{\Phi}{\Phi_o} \qquad (22.5)$$

Much like Bohr's atomic orbits, the superconducting wavefunction must be single-valued around the loop. After traversing a complete loop, the phase must return to an integer multiple of 2π. Hence,

$$\phi = 2\pi \frac{\Phi}{\Phi_o} = 2\pi n \qquad (22.6)$$

or

$$\Phi = n\Phi_o \qquad (22.7)$$

Hence, the flux Φ is an integer multiple of the flux quantum, $\Phi_o = h/2e$, due to the requirement of a single-valued wavefunction. Equation (22.7) is called the flux quantization condition. This is an example of macroscopic quantization since the superconducting loop can have macroscopic dimensions. Note that the externally applied magnetic flux need not be quantized. Rather, it is the total flux, comprised of the external flux and the current-induced flux, which is quantized. The current-induced flux is given by LI where L is the inductance of the loop and I is the current flowing in the loop. Flux quantization was experimentally verified in 1961, using a torsion balance (Fig. 22.5) [9, 10].

22.4 Quantum Harmonic Oscillator

Consider a loop made of superconducting material cooled below its transition temperature and containing a capacitor, C, as shown in Fig. 22.6a. Suppose the loop has some self-inductance, L, and magnetic flux, Φ, threading the loop due to a persistent current, I, in the loop. This is equivalent to an LC circuit with lumped elements, L and C, as shown in Fig. 22.6b, with the magnetic flux threading the inductor. This is a realization of the idealized LC circuit without dissipation (zero resistance). From Eq. (22.5), there is a phase change, ϕ, around the loop due to the magnetic flux threading the loop. Equivalently, ϕ is the phase difference across the lumped inductor element containing the flux Φ.

An oscillating supercurrent is sustained in the LC circuit as the energy oscillates indefinitely between electrostatic energy stored in the electric field of the capacitor,

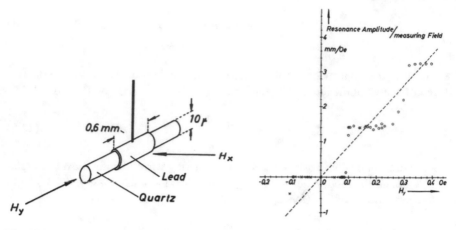

Fig. 22.5 Flux quantization experiment using a torsion balance. Reprinted with permission from R. Doll and M. Näbauer, Phys. Rev. Lett. 7 (1961) 51, https://doi.org/10.1103/PhysRevLett.7.51. Copyright © 1961 by the American Physical Society [10]

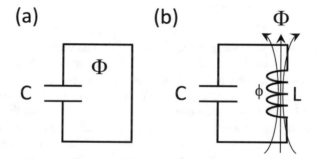

Fig. 22.6 **a** A loop with self-inductance, L, capacitance, C, and magnetic flux, Φ, threading the loop. **b** Equivalent LC circuit with lumped elements, C and L. ϕ is the superconducting phase change around the closed loop in (**a**) due to the magnetic flux threading the loop, which is equal to the phase difference across the lumped inductor element

U_C, and magnetic energy stored in the magnetic field of the inductor, U_L. A standard result from elementary electrodynamics gives:

$$U_C = \frac{Q^2}{2C} \tag{22.8}$$

and

$$U_L = \frac{LI^2}{2} \tag{22.9}$$

where Q is the charge on the capacitor, and I is the current flowing through the inductor. Also, the magnetic flux through the inductor is given by:

$$\Phi = LI \tag{22.10}$$

Hence, we can write U_L in terms of the magnetic flux:

$$U_L = \frac{\Phi^2}{2L} \tag{22.11}$$

Combining Eqs. (22.8) and (22.11) gives the total energy of the LC circuit:

$$U_{\text{tot}} = U_C + U_L = \frac{Q^2}{2C} + \frac{\Phi^2}{2L} \tag{22.12}$$

Using this classical result, the Hamiltonian for the quantum version of the LC circuit is:

$$\widehat{H} = \frac{\widehat{Q}^2}{2C} + \frac{\widehat{\Phi}^2}{2L} \tag{22.13}$$

where $\widehat{Q}$ and $\widehat{\Phi}$ are quantum operators.

We are interested in finding the quantized energy levels for the quantum LC circuit. The two lowest quantized energy levels (ground state and excited state) may be used to represent the qubit states, $|0\rangle$ and $|1\rangle$. To find the quantized energy levels, we must solve the Schrodinger equation:

$$\widehat{H}\psi = E\psi \tag{22.14}$$

where $\widehat{H}$ is given by Eq. (22.13). Hence,

$$\left(\frac{\widehat{Q}^2}{2C} + \frac{\widehat{\Phi}^2}{2L} \right)\psi = E\psi \tag{22.15}$$

To solve Eq. (22.15), an analogy can be made to the familiar problem in quantum mechanics of a particle of mass m in a harmonic potential (i.e., a mass on a spring obeying Hooke's law):

$$\widehat{H} = \frac{\widehat{p}^2}{2m} + \frac{1}{2}k\widehat{x}^2 \tag{22.16}$$

where $\widehat{p}$ is the linear momentum operator, $\widehat{x}$ is the position operator, and k is the force constant. $\widehat{H}$ can also be written in the form:

Table 22.1 Analogy between SHO and *LC* circuit

SHO	*LC* circuit
$\widehat{H} = \frac{\hat{p}^2}{2m} + \frac{1}{2}k\hat{x}^2$	
	$\widehat{H} = \frac{\widehat{Q}^2}{2C} + \frac{\widehat{\Phi}^2}{2L}$
$\hat{p}$	$\widehat{Q}$
$\hat{x}$	$\widehat{\Phi}$
m	C
k	$1/L$
$\hat{p} = -i\hbar\frac{\partial}{\partial x}$	$\widehat{Q} = -i\hbar\frac{\partial}{\partial \Phi}$
$\omega_o = \sqrt{k/m}$	$\omega_o = \frac{1}{\sqrt{LC}}$

$$\widehat{H} = \frac{\hat{p}^2}{2m} + \frac{1}{2}m\omega_o^2\hat{x}^2 \tag{22.17}$$

where $\omega_o = \sqrt{k/m}$ is the resonant frequency. The first term of Eq. (22.17) is the kinetic energy and the second term is the potential energy, $U(x)$. Equation (22.17) describes a simple harmonic oscillator (SHO). Equation (22.17) is also called a "linear" harmonic oscillator because the force on the particle obeys Hooke's law, $F = -kx$; i.e., the force is linearly dependent on the displacement.

Comparing Eqs. (22.13) and (22.16), we see that they are mathematically identical if the substitutions in Table 22.1 are made. $\widehat{Q}$ is equivalent to $\hat{p}$, $\widehat{\Phi}$ is equivalent to $\hat{x}$, m is equivalent to C, and k is equivalent to $1/L$. In the position representation, the linear momentum operator is expressed as $\hat{p} = -i\hbar\frac{\partial}{\partial x}$. Using our analogy, the charge operator may be expressed as:

$$\widehat{Q} = -i\hbar\frac{\partial}{\partial \Phi} \tag{22.18}$$

Hence, the Schrodinger equation for the quantum *LC* circuit may be written in terms of Φ as:

$$\left(-\frac{\hbar^2}{2C}\frac{\partial^2}{\partial \Phi^2} + \frac{\Phi^2}{2L}\right)\psi = E\psi \tag{22.19}$$

By analogy, the first term of Eq. (22.19) is called the kinetic energy and the second term is called the potential energy. Hence, the potential energy is:

$$U(\Phi) = \frac{\Phi^2}{2L} \tag{22.20}$$

as depicted in Fig. 22.7.

The solution to Eq. (22.19) will give a wavefunction, ψ, that is a function of Φ. We already know from elementary quantum mechanics that the solution of Schrodinger's

Fig. 22.7 Quantized energy levels superimposed on the harmonic potential well of a quantum LC circuit, which is equivalent to a simple harmonic oscillator (SHO). The potential energy is proportional to ϕ^2 or Φ^2. The energy is quantized as $E_n = (n + 1/2)\hbar\omega_o$

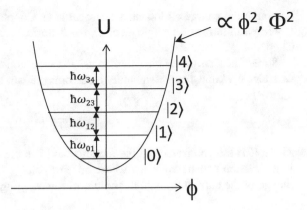

equation for the SHO gives eigenvalues with quantized energy levels, E_n:

$$E_n = (n + 1/2)\hbar\omega_o, \quad n = 0, 1, 2, \ldots \tag{22.21}$$

where

$$\omega_o = \sqrt{k/m} \tag{22.22}$$

By analogy, the solution to the quantum LC circuit, Eq. (22.19), gives the identical energy quantization as Eq. (22.21). Substituting $m \equiv C$ and $k \equiv 1/L$ into Eq. (22.22) gives the resonant frequency for the quantum LC circuit:

$$\omega_o = \frac{1}{\sqrt{LC}} \tag{22.23}$$

Equation (22.23) is identical to the classical result for an LC circuit.

The Hamiltonian and Schrodinger's equation for the quantum LC circuit can be written in several alternative formats. We can write U_C in the alternative form:

$$U_C = 4E_C N^2 \tag{22.24}$$

where E_C is the single electron charging energy introduced in Chap. 20:

$$E_C = e^2/2C \tag{22.25}$$

N is the number of Cooper pairs on the capacitor (also sometimes called the "reduced charge"):

$$N = Q/2e \tag{22.26}$$

Although the charge on the capacitor occurs as Cooper pairs, the charging energy is written in terms of a single electron for historical reasons (and consistency with Chap. 20).

Sometimes it is more convenient to use the superconducting phase as the quantum variable. We know from elementary electromagnetism that:

$$V = L\frac{dI}{dt} \tag{22.27}$$

where V is the voltage across the inductor and I is the current. Equation (22.27) is an expression of Faraday's law where an emf voltage is proportional to the rate of change of the current through the inductor. Rearranging:

$$\int V dt = \int L dI = LI = \Phi \tag{22.28}$$

where Φ is the flux threading the inductor. Note that this is the derivation of Eq. (22.10). Later we show that the relationship between superconducting phase and voltage is given by $V = \frac{\Phi_o}{2\pi}\frac{d\phi}{dt}$. This is known as the second Josephson relation or the voltage-phase relation. Hence, starting from Eq. (22.28):

$$\Phi = \int V dt = \frac{\Phi_o}{2\pi}\int \frac{d\phi}{dt}dt = \frac{\Phi_o}{2\pi}\int d\phi = \frac{\Phi_o}{2\pi}\phi \tag{22.29}$$

Note that this is the same as Eq. (22.5). Equation (22.29) can be used to express the Hamiltonian in terms of phase, ϕ, instead of flux, Φ.

Using Eq. (22.29), we can express the number of Cooper pairs, N, as an operator in the same way as Q:

$$\widehat{Q} = -i\hbar\frac{\partial}{\partial\Phi} \rightarrow \widehat{N} = -i\frac{\partial}{\partial\phi} \tag{22.30}$$

The first term (kinetic term) of the Hamiltonian in Eq. (22.19) becomes:

$$U_C = -\frac{\hbar^2}{2C}\frac{\partial^2}{\partial\Phi^2} \rightarrow -4E_C\frac{\partial^2}{\partial\phi^2} \tag{22.31}$$

Similarly, substituting Eq. (22.29) into Eq. (22.20), we can also write U_L in terms of ϕ:

$$U_L = \left(\frac{\Phi_o}{2\pi}\right)^2\frac{\phi^2}{2L} \tag{22.32}$$

or

$$U_L = \frac{1}{2}E_L\phi^2 \tag{22.33}$$

where E_L is called the inductive energy:

$$E_L = \left(\frac{\Phi_o}{2\pi}\right)/L \tag{22.34}$$

Combining the previous equations, the Hamiltonian may be written as:

$$\hat{H} = -4E_C\frac{\partial^2}{\partial\phi^2} + U(\phi) \tag{22.35}$$

where $U(\phi) = \frac{1}{2}E_L\phi^2$. The resonance frequency in Eq. (22.23) can also be written as:

$$\omega_o = \sqrt{8E_C E_L}/\hbar \tag{22.36}$$

The quantized energy levels for the quantum LC circuit, equivalent to a SHO, are shown in Fig. 22.7. The ground state ($n = 0$) and first excited state ($n = 1$) of the SHO could represent $|0\rangle$ and $|1\rangle$, respectively. For typical values of E_C and E_L, the resonant frequency, ω_o, is typically in the microwave region (~10 GHz). 10 GHz corresponds to a temperature of 500 mK, which is much higher than dilution refrigerator temperatures (~mK). Hence, we can initialize the LC qubit to the ground state simply by cooling the qubit and waiting for a time t much greater than the relaxation time T_1. We could excite Rabi oscillations between $|0\rangle$ and $|1\rangle$ by applying a microwave signal to the LC circuit at the resonant frequency, ω_o. This signal can be applied to the LC circuit by capacitive coupling (gate capacitance, C_g) of a voltage oscillator, V_g, as shown in Fig. 22.8.

The energy separation, $\hbar\omega_{01}$, between $|0\rangle$ and $|1\rangle$ is identical to all the other energy separations in the SHO ($\hbar\omega_{12}$, $\hbar\omega_{23}$, ...). The energy level spacings are degenerate.

Fig. 22.8 Circuit for inducing Rabi oscillations in a quantum LC circuit

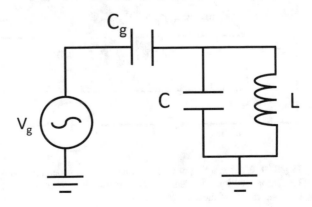

This means that we cannot select $|0\rangle$ and $|1\rangle$ for Rabi oscillations without avoiding excitations between $|1\rangle$ and $|2\rangle$, between $|2\rangle$ and $|3\rangle$, etc. This is called "leakage" and represents a serious problem when using a SHO for quantum computing. What we need instead is a "non-linear" or anharmonic potential with unequal separations between energy levels, so that we can individually address $|0\rangle$ and $|1\rangle$ without leakage. The solution to this problem is to use a "non-linear" circuit element; i.e., one with a potential energy that is different than the SHO. This element is the Josephson junction, considered in the next section. Note that leakage is not a problem in atomic or ionic qubits because their potentials are naturally anharmonic.

22.5 Josephson Junction

The Josephson effect (Fig. 22.9) is the quantum tunneling of Cooper pairs without resistance from one superconductor to another through a thin insulating barrier (~1 nm thick), predicted in 1962 by Brian Josephson (Fig. 22.10), who was awarded the Nobel Prize in Physics in 1973 together with Leo Esaki and Ivar Giaever. Like the flux quantization discussed earlier, the Josephson effect is another manifestation of the superconducting state. In the Josephson effect, a superconducting current (super-current) flows between two superconductors separated by the tunnel barrier (known as a Josephson junction) when a current bias is applied.

The Josephson junction is fabricated by lithographic techniques using Nb or Al films deposited on Si or sapphire substrate with an AlO_x tunnel barrier. The Josephson junction is the basic building block of all superconducting qubits. Typical dimensions

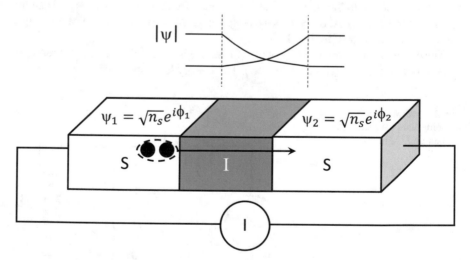

Fig. 22.9 The tunneling of Cooper pairs across a superconductor-insulator-superconductor (SIS) Josephson junction. The decay of the wavefunction, representing Cooper pair tunneling, is shown from either side of the superconductor

Fig. 22.10 Brian Josephson (1940–); Nobel Prize in Physics in 1973. *Credit* Wikimedia Commons [11]

of superconducting circuits are on the scale of micrometers. Thus, it is relatively easy to fabricate large numbers of qubits using conventional microelectronic processing techniques.

Before Josephson, physicists thought that the probability of Cooper pair tunneling was very low. They believed that if the probability of single electron tunneling was P, then the probability of a Cooper pair tunneling would be P^2. Josephson showed that this reasoning is incorrect. Instead, it is the single macroscopic wavefunction, Eq. (22.1), that is tunneling through the barrier, as shown in Fig. 22.9.

While Josephson was a Ph.D. student, he predicted the relationship between current, voltage and phase of the junction, called the Josephson relations. Josephson predicted that a phase difference occurs across the junction, $\phi = \phi_2 - \phi_1$ (Fig. 22.9), which we simply call the phase. In the first Josephson relation, the supercurrent across the junction, I, is related to the phase:

$$I = I_o \sin \phi \tag{22.37}$$

where I_o is the maximum supercurrent across the junction before the onset of a resistive state. The critical current is dependent on the material (the superconducting gap), and the junction resistance (area and thickness). Equation (22.37) is also called the current-phase relation. The state described by Eq. (22.37) is called the zero-voltage state. This state is depicted by a vertical line in a current-voltage characteristic, as shown in Fig. 22.11a, since the supercurrent flows without dissipation (zero voltage).

At finite temperature ($0 < T < T_c$), there is a finite probability that Cooper pairs can be broken up, producing electrons in the band above the superconducting gap. Electrons may also be created by thermal excitation across the superconducting gap. In the zero-voltage state, these electrons do not contribute to the current. However, as soon as the current exceeds the critical current, the normal electrons can contribute to the current. However, unlike Cooper pairs, these electrons can undergo scattering, so a voltage drop appears across the junction. This current contribution is called the normal current and obeys Ohm's law as shown in Fig. 22.11a. This resistive state

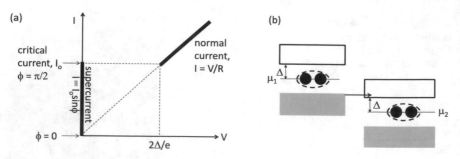

Fig. 22.11 a Current-voltage (I-V) characteristic of a Josephson junction. **b** Energy band diagram showing onset of the resistive state at $V = 2\Delta/e$

becomes significant when $V = (\mu_1 - \mu_2)/e > 2\Delta/e$. At these voltages, electron states in the filled band of the left superconductor align with the empty states of the right superconductor. In this situation, electrons can tunnel from filled electron states on the left to empty electron states on the right, as depicted in Fig. 22.11b. Note that the I-V characteristic in Fig. 22.11a is usually hysteretic [12], which is omitted from the present discussion.

In the second Josephson relation, an alternating phase results in a voltage drop across the junction:

$$(2e)V = \hbar\frac{d\phi}{dt} \tag{22.38}$$

where $2e$ is the Cooper pair charge. Alternatively,

$$(2e)V = \hbar\omega \tag{22.39}$$

where $\omega = \frac{d\phi}{dt}$ is the angular frequency at which the Cooper pairs oscillate back and forth across the Josephson junction. The second Josephson relation is often written in the alternative form:

$$\frac{d\phi}{dt} = \frac{2\pi}{\Phi_o}V \tag{22.40}$$

where Φ_o is the flux quantum. The second Josephson relation is also called the voltage-phase relation. Both Josephson relations, Eqs. (22.37) and (22.38), can be derived by solving the Schrodinger equation with an overlap of the wavefunctions on the two sides of the junction, as depicted in Fig. 22.9 [13].

If a constant voltage is applied to the junction, then Eq. (22.38) can be solved:

$$\phi = \phi_o + \frac{2e}{\hbar}Vt \tag{22.41}$$

Substituting Eq. (22.41) into (22.37) results in an alternating supercurrent with a voltage-tunable angular frequency, $\omega = 2\,eV/\hbar$, or frequency, $\nu = \omega/2\pi = 2\,eV/h$, or 483.6 GHz/mV, which is called the Josephson frequency. We see that the Josephson junction can produce a voltage-controlled oscillator at a very high frequency. Due to the difficulty of detecting this high frequency supercurrent, I-V characteristics like Fig. 22.11 show only the time-averaged (dc) current obtained from a typical measurement. Besides superconducting qubits, the Josephson junction has many other applications in superconducting electronics (RSFQ logic), sensors, and metrology (as a voltage standard). In qubits, the typical voltage across a Josephson junction is a few μV with operation frequencies of a few GHz (microwave region).

The Josephson junction is like a parallel plate capacitor and exhibits a capacitance:

$$C_J = \frac{\kappa \epsilon_o A}{d} \tag{22.42}$$

where κ is the dielectric constant (relative permittivity) of the junction dielectric (e.g., Al_2O_3), ϵ_o is the permittivity of free space (8.85×10^{-12} F/m), A is the junction area and d is the junction thickness. For example, with typical values of $d = 1$ nm, $A = 1\,\mu m \times 1\,\mu m$ and $\kappa = 9$, we obtain $C_J \sim 0.1$ pF.

The Josephson relations indicate that any change in phase and therefore Josephson current will result in a finite voltage across the Josephson junction. Therefore, the Josephson junction exhibits an inductance as follows. Combining Eqs. (22.37) and (22.38), we obtain:

$$\frac{dI}{dt} = I_o \cos\phi \frac{d\phi}{dt} = \frac{2\,eV}{\hbar} I_o \cos\phi \tag{22.43}$$

Rearranging gives:

$$V = \frac{\hbar}{2eI_o \cos\phi} \frac{dI}{dt} \tag{22.44}$$

Comparing Eq. (22.44) to the fundamental formula for inductance, $V = L\frac{dI}{dt}$, gives the inductance for a Josephson junction:

$$L_J = \frac{\hbar}{2eI_o \cos\phi} = \frac{L_{J_o}}{\cos\phi} \tag{22.45}$$

where $L_{J_o} = \frac{\Phi_o}{2\pi I_o}$. Hence, the Josephson junction can be described by an equivalent circuit containing a parallel arrangement of C_J and L_J, like Fig. 22.6b. The circuit symbol for a Josephson junction is an "x", or an "x" in a box when its capacitance is included, as shown in Fig. 22.12.

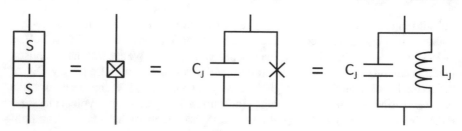

Fig. 22.12 Equivalent circuit models of a Josephson junction

22.6 Potential Energy of a Josephson Junction

If a current (supercurrent), I, flows through the junction with a potential drop, V, then the energy stored in the junction, U_J, is obtained by integrating the power, IV, with respect to time:

$$U_J = \int_0^t IV\, dt \tag{22.46}$$

Substituting the first and second Josephson relations, Eqs. (22.37) and (22.40), we get:

$$U_J = \int_0^\phi (I_o \sin\phi)\left(\frac{\Phi_o}{2\pi}\frac{d\phi}{dt}\right)dt \tag{22.47}$$

Solving Eq. (22.47) gives:

$$U_J = E_J(1 - \cos\phi) \tag{22.48}$$

where E_J is called the Josephson energy or Josephson coupling energy:

$$E_J = \frac{I_o \Phi_o}{2\pi} \tag{22.49}$$

E_J arises due to the overlap or coupling of the wavefunction on either side of the junction. The energy offset, E_J, in Eq. (22.48) simply shifts the energy scale, so it is often dropped giving:

$$U_J = -E_J \cos\phi \tag{22.50}$$

Now consider the circuit of Fig. 22.13. A Josephson junction with capacitance C_J and inductance L_J is shunted by a capacitance C. The Hamiltonian becomes:

Fig. 22.13 Josephson
junction shunted by a
capacitance, C

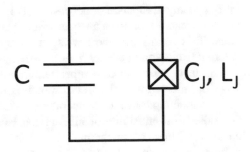

$$\widehat{H} = U_C + U_J = -4E_C\frac{\partial^2}{\partial\phi^2} - E_J\cos\phi \qquad (22.51)$$

where U_C is obtained from Eq. (22.31) with $E_C = e^2/2C_\Sigma$, and $C_\Sigma = C + C_J$ is the total parallel capacitance. Schrodinger's equation with this Hamiltonian is called the Mathieu equation, named after the French mathematician, Émile Léonard Mathieu. Solving the Schrodinger equation with the Hamiltonian of Eq. (22.51) gives the quantized energy levels illustrated in Fig. 22.14, which are superimposed on the $-\cos\phi$ dependence of the potential energy, U_J. Due to the $\cos\phi$ dependence of U_J, the potential energy is anharmonic and the energy level splittings are unequal, unlike the equal energy level splittings of the SHO in Fig. 22.7. The separation between consecutive energy levels in the anharmonic potential decreases as the quantum number increases. Hence, we call the Josephson junction a "non-linear" element (the potential is not quadratic; i.e., Hooke's linear force law is not applicable). Alternatively, the non-linearity can be attributed to a "non-linear" inductance given by Eq. (22.45); i.e., the inductance is non-linear in ϕ.

The inductance of a Josephson junction is different from the usual inductance that would be formed, for example, by a coil of superconducting wire. As described earlier, a superconducting LC circuit would not work as a qubit, because (like any harmonic oscillator) it has infinitely many evenly spaced energy levels (Fig. 22.7). Resonant driving by ESR does not let us isolate two levels to use as $|0\rangle$ and $|1\rangle$. The Josephson junction is the only known non-linear circuit element that does not

Fig. 22.14 Quantized
energy levels superimposed
on the anharmonic potential
well of a Josephson junction

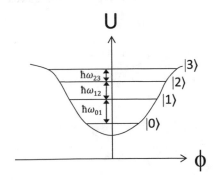

introduce any dissipation or dephasing. This non-linearity results in an anharmonic potential, and therefore an unequal spacing between the energy levels of the Cooper pairs. This means that the lowest levels can be individually addressed by using external excitation of energy $\hbar\omega_{01}$ between $|0\rangle$ and $|1\rangle$.

The level separations in superconducting qubits are typically in the range 5–20 GHz (microwave region), corresponding to a temperature of 250–1000 mK. The superconducting transition temperatures ($\sim\Delta/k$) are typically 1–10 K, and the qubits are cooled to a temperature of milliKelvins. Hence, $kT < \hbar\omega_{01} < \Delta$. The first inequality ensures that there are no thermal excitations from $|0\rangle$ to $|1\rangle$ and we can initialize the qubit to the ground state ($|0\rangle$). The second inequality ensures no excitations across the superconducting gap.

22.7   Flux Quantization in a Josephson Junction

Now let us consider the effect of a magnetic flux, Φ, threading a superconducting loop with a Josephson junction, as shown in Fig. 22.15. Suppose the current flows counter-clockwise in the loop. Starting at point a in Fig. 22.15, the phase winding along a counter-clockwise contour line of the loop is given by:

$$\frac{2\pi}{\Phi_o}\int_a^b \mathbf{A}\cdot d\mathbf{l} + \phi + \frac{2\pi}{\Phi_o}\int_b^a \mathbf{A}\cdot d\mathbf{l} = \frac{2\pi}{\Phi_o}\int_{\text{closed path}} \mathbf{A}\cdot d\mathbf{l} + \phi = 2\pi n \qquad (22.52)$$

where $\phi = \phi_a - \phi_b$ is the phase difference across the Josephson junction. Similar to Eqs. (22.5) and (22.29), Eq. (22.52) gives the phase difference across the Josephson junction within an arbitrary phase difference of 2π:

$$\phi = 2\pi\frac{\Phi}{\Phi_o} \qquad (22.53)$$

Hence, the phase difference across the Josephson junction, ϕ, and the magnetic flux, Φ, are related and can be used interchangeably.

Fig. 22.15 Magnetic field threading a superconducting loop with a Josephson junction

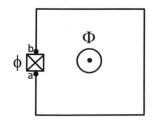

22.8 DC-SQUID (Tunable Josephson Junction)

A superconducting quantum interference device (SQUID) consists of a parallel arrangement of two Josephson junctions, as shown in Fig. 22.16. This is called a DC-SQUID for historical reasons.

The total current is equal to the sum of the two branch currents:

$$I = I_1 + I_2 = I_{o1} \sin \phi_1 + I_{o2} \sin \phi_2 \tag{22.54}$$

where the critical currents and phase across the two junctions may be different. In a similar fashion as Eq. (22.53), we obtain:

$$\phi_1 - \phi_2 = 2\pi \frac{\Phi}{\Phi_o} \tag{22.55}$$

where ϕ_1 is the phase difference across the first Josephson junction and ϕ_2 is the phase difference across the second Josephson junction. Let us assume that the critical currents are identical ($I_{o1} = I_{o2} = I_o$). Using the trigonometric identity $\sin(x) + \sin(y) = 2 \cos\left(\frac{x-y}{2}\right) \sin\left(\frac{x+y}{2}\right)$, Eq. (22.54) becomes:

$$I = 2I_o \cos\left(\frac{\phi_1 - \phi_2}{2}\right) \sin\left(\phi_2 + \frac{\phi_1 - \phi_2}{2}\right) \tag{22.56}$$

Substituting Eq. (22.55) gives:

$$I = 2I_o \cos\left(\pi \frac{\Phi_e}{\Phi_o}\right) \sin\left(\phi_2 + \pi \frac{\Phi_e}{\Phi_o}\right) \tag{22.57}$$

where we assume the total flux Φ through the loop is dominated by an externally applied flux, Φ_e (negligible inductance of the loop). Note that there is no flux quantization in this case since there is no circulating supercurrent. The maximum current in Eq. (22.57) is obtained with an appropriate value of ϕ_2 such that the sine term is ± 1. We see from Eq. (22.57) that the maximum (critical) current of the DC-SQUID

Fig. 22.16 DC-SQUID

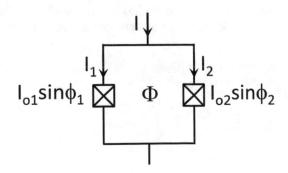

Fig. 22.17 Maximum
current, I_{max}, as a function of
flux in a DC-SQUID

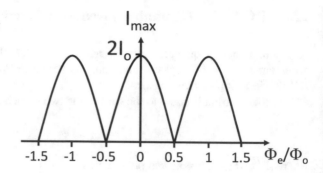

becomes $I_{max} = 2I_o \left| \cos\left(\pi \frac{\Phi_e}{\Phi_o}\right) \right|$, as shown in Fig. 22.17. Hence, the Josephson junction energy E_J is replaced by $2E_J \cos\left(\pi \frac{\Phi_e}{\Phi_o}\right)$. An external flux, Φ_e, allows control of the junction energy between 0 and $2E_J$. A SQUID device is also used as a sensitive magnetometer, capable of measuring a fraction of the flux quantum Φ_o. This is used to read-out flux qubits, as discussed later, in addition to many other applications in biomagnetism and geophysics, for example.

The pattern in Fig. 22.17 is analogous to that obtained from the interference of two light waves in a double-slit experiment. In the case of the DC-SQUID, interference takes place between the two currents, I_1 and I_2. The outcome of the interference is either destructive or constructive depending on the difference of the Josephson phases, ϕ_1 and ϕ_2, which in turn depends on the magnetic field via Eq. (22.55).

22.9   General Superconducting Circuit

We can generalize the circuits considered thus far by considering a parallel arrangement of a Josephson junction, a shunt capacitor C and a shunt inductor L, with corresponding energies E_J, E_C and E_L (Fig. 22.18). An externally applied flux, Φ_e, may be applied to the circuit. If the circuit has some inductance, then a current-induced flux, LI, also penetrates the loop. This gives a total flux, Φ, through the loop. There are three types of qubit circuit, all using Josephson junctions: the phase qubit, the charge qubit, and the flux qubit depending on how the circuit is biased (current bias, voltage bias, or flux bias) and the relative magnitude of E_C, E_J and E_L.

Fig. 22.18 General circuit
for superconducting qubits

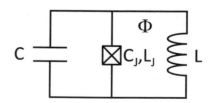

22.10 Phase Qubit

The phase qubit is a current-biased Josephson junction (Fig. 22.19) without any shunt capacitor or inductor (negligible self-inductance). The current source resides at room temperature, while the Josephson junction resides in a low temperature cryostat. The circuit can be modelled by a current source, I, applied to a junction capacitance C_J in parallel with a supercurrent $I_o \sin \phi$. This model is called the resistively and capacitively shunted junction (RCSJ) model, developed by Stewart and McCumber [14, 15]. The resistance, R, of the junction in the superconducting state ($I < I_o$) is due to normal electrons that are thermally excited at finite temperature and can cross the tunnel barrier, or due to shunt paths created by defects. R is very large ($G\Omega$), so we will ignore it in this analysis.

From Kirchoff's current law, we can write the total current, I, as the sum of the currents in the two parallel branches of the equivalent circuit:

$$I = I_o \sin \phi + C_J \frac{dV}{dt} \tag{22.58}$$

where $I_o \sin \phi$ is the supercurrent flowing across the Josephson junction, and $I = \frac{dQ}{dt} = \frac{d(C_J V)}{dt} = C_J \frac{dV}{dt}$ is the displacement current across the capacitor. Using the second Josephson relation, we get:

$$I = I_o \sin \phi + C_J \frac{\hbar}{2e} \frac{d^2\phi}{dt^2} \tag{22.59}$$

Rearranging and multiplying by $\hbar/2e$ gives:

$$C_J \left(\frac{\hbar}{2e}\right)^2 \frac{d^2\phi}{dt^2} + \frac{\hbar}{2e}(I_o \sin \phi - I) = 0 \tag{22.60}$$

Equation (22.60) is analogous to the equation of motion for a driven simple pendulum, with ϕ equivalent to the angle of the pendulum. Using this analogy, the first term in

Fig. 22.19 Equivalent circuit for the phase qubit

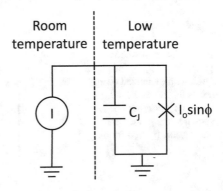

Eq. (22.60) is an inertial term (with acceleration, $\frac{d^2\phi}{dt^2}$) and the second term is a force term $\left(F = -\frac{dU(\phi)}{d\phi} = -\frac{\hbar}{2e}(I_o \sin\phi - I) \right)$. Hence, the potential energy is:

$$U(\phi) = \int \frac{\hbar}{2e}(I_o \sin\phi - I)d\phi \qquad (22.61)$$

Integrating gives:

$$U(\phi) = -\frac{\hbar I_o}{2e}(\cos\phi + I\phi/I_o) \qquad (22.62)$$

or

$$U(\phi) = -E_J(\cos\phi + I\phi/I_o) \qquad (22.63)$$

where $E_J = \frac{\hbar I_o}{2e} = \frac{I_o \Phi_o}{2\pi}$ is the Josephson energy (Eq. 22.49). Equation (22.63) looks like the potential energy of the Josephson junction ($U_J = -E_J\cos\phi$) but tilted due to the second term that is proportional to the current bias, I. The resulting anharmonic potential, $U(\phi)$, illustrated in Fig. 22.20a, is called a "washboard" potential due to its resemblance to an old-fashioned laundry washboard. For small bias current, the depth of the potential well is $\sim E_J$.

For the phase qubit, $E_J \gg E_C$. Recall that $E_J = \frac{I_o \Phi_o}{2\pi}$ where the critical current is proportional to the junction area, A. On the other hand, $E_C = \frac{e^2}{2C_J}$ where $C_J = \frac{\kappa \epsilon_o A}{d}$. Hence, E_J is proportional to A, E_C is inversely proportional to A, and E_J/E_C is proportional to A^2. Therefore, the condition $E_J \gg E_C$, used for phase qubits, can be achieved with large junction areas.

The energy levels in the potential well can be obtained in the usual manner by solving the Schrodinger equation. Since $E_J \gg E_C$, we can ignore the charging energy (kinetic energy) and the Hamiltonian is $\hat{H} = U(\phi)$ with $U(\phi)$ given by Eq. (22.63). The resulting energy levels of the anharmonic potential well are illustrated in Fig. 22.20a. The bias current, I, can be adjusted to tilt the washboard potential to the point where only two localized wavefunctions (energy levels) are inside the well, as shown in Fig. 22.21a. If the bias current, I, is sufficiently small ($I < I_o$),

Fig. 22.20 a Washboard potential, $U(\phi)$, of the phase qubit with quantized energy levels for the case $I < I_o$. A fictitious phase particle is shown in the ground state energy, $|0\rangle$. **b** $U(\phi)$ for the case $I > I_o$

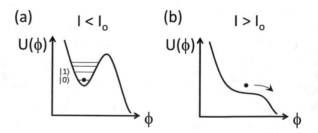

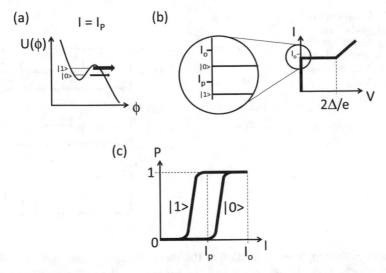

Fig. 22.21 **a** Macroscopic quantum tunneling (MQT) at an applied current $I = I_P$, where the tunneling probability of the $|1\rangle$ state is much higher than that for the $|0\rangle$ state. **b** Current-voltage characteristic showing the onset of the voltage-state at a current $I_P < I_o$ due to MQT. At the current, I_P, only the $|1\rangle$ state contributes to the MQT and to the onset of the resistive state. **c** Switching curves, giving the probability, P, of switching to the resistive state from states $|0\rangle$ and $|1\rangle$

then the energy levels are close to the bottom of the potential well and the wavefunction is sharply peaked. In this case, the phase ϕ is well defined. Hence, the phase is a convenient variable to describe the dynamics of the phase qubit (ϕ is a "good quantum number"). We can think of the qubit as a fictitious "phase particle" sitting in the energy levels of the potential well.

Exercise 22.1 Show that the potential minimum in Fig. 22.20a disappears when $I > I_o$.

A microwave ac current signal can be added to the applied bias current (a microwave transmission line capacitively coupled to the circuit) to resonantly drive Rabi oscillations. The phase particle oscillates between the $|0\rangle$ and $|1\rangle$ state of the potential well, forming a qubit.

When the applied current exceeds the critical supercurrent, I_o, the washboard potential tilts to such an extent that there is no longer any potential energy minimum to confine the fictitious phase particle. The phase particle "rolls down the potential hill" (Fig. 22.20b). Thus, the phase will continuously increase ($d\phi/dt \neq 0$), and the system switches to the resistive state resulting in a voltage, $V = \frac{\Phi_o}{2\pi} \frac{d\phi}{dt}$. Before the current, I, reaches the critical supercurrent, I_o, the phase particle can escape the potential well by tunneling through the anharmonic potential barrier, as shown in Fig. 22.21a. This is known as "macroscopic quantum tunneling" (MQT), since it

Fig. 22.22 Sequence for measuring Rabi oscillations, showing applied bias current, applied microwave (μW) signal, and measured output voltage

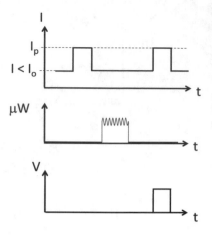

is another macroscopic manifestation of the superconducting state. The probability of tunneling is different for the $|0\rangle$ and $|1\rangle$ state due to the different widths of the potential barrier at the different heights of the $|0\rangle$ and $|1\rangle$ energy level in the well. As shown in Fig. 22.21b, the $|1\rangle$ state will undergo MQT and switch to the resistive state with high probability at a bias current slightly below I_p, while MQT of the $|0\rangle$ state is negligible at I_P. If we measure a voltage pulse after applying a bias current I_p, then the phase qubit was in the $|1\rangle$ state with high probability; otherwise, if no voltage pulse is obtained, then the qubit was in the $|0\rangle$ state with high probability. These switching probabilities are described by "switching curves", shown in Fig. 22.21c.

Figure 22.22 shows a sequence of current pulses for measuring Rabi oscillations. A phase qubit is prepared in the ground state, $|0\rangle$, by starting with a current $I < I_o$. A first current pulse I_p is applied to measure the state of the qubit. No voltage pulse is produced, verifying that the qubit is in the $|0\rangle$ state. A microwave signal tuned to frequency ω_{01} is applied for a certain duration, t, causing Rabi oscillations between the $|0\rangle$ and $|1\rangle$ states. A second current pulse, I_p, is used to read-out the qubit state after Rabi oscillations. A voltage pulse indicates the $|1\rangle$ state, while no voltage indicates the $|0\rangle$ state. The experiment is repeated many times for each duration, t, so the probability of measuring $|0\rangle$ or $|1\rangle$ can be plotted versus time. The resulting probability versus time is oscillatory, indicating Rabi oscillations.

22.11 Flux Qubit

The flux qubit is based on "persistent currents" induced by an external magnetic field applied to a superconducting loop containing a Josephson junction (Fig. 22.23) [16, 17]. For historical reasons, a superconducting loop containing a single Josephson junction is called a radio-frequency superconducting quantum interference device (RF-SQUID). Today, most flux qubits contain three or more Josephson junctions.

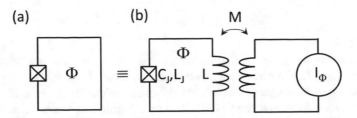

Fig. 22.23 a A flux qubit containing a single Josephson junction (RF-SQUID) and total flux, Φ.
b Equivalent circuit with loop inductance L, and a flux-bias circuit coupled to the qubit circuit by
mutual inductance, M

One of the junctions is made with smaller area than the others and forms the qubit,
while two larger junctions are used to introduce inductance into the loop without
the need for large loop areas, thereby reducing the sensitivity to flux noise. In the
following, we will show only a single junction to understand the basic operating
principles of the flux qubit.

The flux qubit is "flux biased", meaning an external magnetic flux, Φ_e, is applied
to the superconducting loop. The total flux in the loop, Φ, is the externally applied
flux (Φ_e) plus a self-induced flux (LI) due to induced supercurrent (I) in the loop.
The flux is produced by the flux-bias circuit on the right of Fig. 22.23b. The qubit
circuit and the flux-bias circuit are coupled inductively by a mutual inductance, M,
with the equivalent circuit of the qubit containing an inductance L.

For the flux qubit, $E_J > E_C$, similar to the phase qubit. However, unlike the phase
qubit, there is an inductance in the circuit. Since the charging energy is negligible in
comparison to the Josephson energy, $\widehat{H}$ is given only by the Josephson energy and
the magnetic energy from the loop inductance:

$$\widehat{H} = -E_J \cos\phi + \frac{1}{2}LI^2 \tag{22.64}$$

For the flux qubit, it is more convenient to use the flux, Φ, as the variable rather than
the phase ϕ. The total flux threading the superconducting loop is:

$$\Phi = \Phi_e + LI \tag{22.65}$$

where Φ_e is the externally applied flux and LI is the current-induced flux. Substituting
LI from Eqs. (22.65) into (22.64) gives:

$$\widehat{H} = -E_J \cos\phi + \frac{(\Phi - \Phi_e)^2}{2L} \tag{22.66}$$

Substituting the relationship between ϕ and Φ (Eq. 22.55) gives:

$$\widehat{H} = -E_J \cos\left(2\pi \frac{\Phi}{\Phi_o}\right) + \frac{(\Phi - \Phi_e)^2}{2L} \qquad (22.67)$$

Since the charging energy (equal to the kinetic energy term) is negligible in the flux qubit, Eq. (22.67) gives the potential energy, $U(\Phi)$. Inspection of Eq. (22.67) indicates that $U(\Phi)$ is a parabola modulated by a sinusoidal, as shown in Fig. 22.24a. This produces a double potential well with a tilt controlled by the externally applied flux, Φ_e. Only the lowest energy level in each well is important at low temperatures.

The control knob for the flux quit is the externally applied magnetic flux, Φ_e. When an external flux is applied to the loop, a circulating supercurrent is formed to push the total flux through the loop to the quantized values of the flux quantum, 0 or Φ_o, corresponding to the potential minima of the double well potential. When $\Phi_e > \Phi_o/2$, as depicted in Fig. 22.24a, then the right potential well has lower energy than the left well, favoring a counter-clockwise circulating supercurrent that pushes the total flux up to Φ_o. This counter-clockwise circulating supercurrent is equivalent to a magnetic flux pointing up ($|\uparrow\rangle$). Conversely, when $\Phi_e < \Phi_o/2$, then the left potential well is lower, favoring a clockwise circulating supercurrent that pushes the total flux to 0. This clockwise circulating supercurrent is equivalent to a magnetic flux pointing down ($|\downarrow\rangle$). Hence, in a sense, the flux qubit is equivalent to manufactured spins. When $\Phi_e = \Phi_o/2$, the energy levels of the $|\uparrow\rangle$ and $|\downarrow\rangle$ states are resonant, and their corresponding probabilities become equal, as shown in Fig. 22.24b.

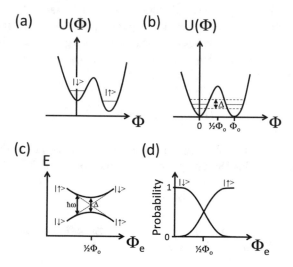

Fig. 22.24 a Potential energy, $U(\Phi)$, versus total flux Φ for the flux qubit when $\Phi_e > \Phi_o/2$. **b** $U(\Phi)$ when $\Phi_e = \Phi_o/2$. The two energy levels corresponding to $|\uparrow\rangle$ and $|\downarrow\rangle$ states (solid black lines) become resonant at $\Phi_e = \Phi_o/2$, causing an energy level splitting, Δ, shown as dashed lines. **c** Energy levels for $|\uparrow\rangle$ and $|\downarrow\rangle$ states as a function of Φ_e, showing the avoided crossing at $\Phi_o/2$ due to E_J. The dashed lines correspond to the energies without Josephson coupling ($E_J = 0$). **d** Probability of $|\uparrow\rangle$ and $|\downarrow\rangle$ states as a function of Φ_e

This corresponds to a superposition of clockwise and counter-clockwise circulating supercurrent flowing simultaneously in the circuit.

The degenerate energy levels at $\Phi_e = \Phi_o/2$ become coupled by quantum tunneling through the potential barrier between the two states. This potential barrier is proportional to E_J as specified in the first term of the Hamiltonian, Eq. (22.67). As a result of this coupling, the degenerate energy levels are split (level repulsion) by an amount Δ, as shown in Fig. 22.24b, c, which depends on E_J (do not confuse Δ with the superconducting gap, which uses the same symbol). These energy levels correspond to the new eigenstates of symmetrical $\left(|-\rangle = \frac{1}{\sqrt{2}}(|\uparrow\rangle + |\downarrow\rangle)\right)$ and anti-symmetrical $\left(|+\rangle = \frac{1}{\sqrt{2}}(|\uparrow\rangle - |\downarrow\rangle)\right)$ superposition states in the lower and upper energy level, respectively, as discussed in Chap. 16. In other words, a superposition of clockwise and counter-clockwise supercurrent arises in the circuit. $\Delta(E_J)$ is equivalent to B_x ($\hat{\sigma}_x$ operator) using the spin analogue [18].

The qubit can be initialized in the $|\downarrow\rangle$ state with high probability by applying a flux far away from the degeneracy point (for example, $\Phi_e \ll \Phi_o/2$), and waiting longer than the relaxation time, T_1. A flux Φ_e can then be applied to tune the probability of the $|\uparrow\rangle$ and $|\downarrow\rangle$ states, as shown in Fig. 22.24d. Φ_e is equivalent to B_z ($\hat{\sigma}_z$ operator) using the spin analogue [18]. Near the degeneracy point, the initial $|\downarrow\rangle$ state is no longer an eigenstate of the system, so coherent Rabi oscillations will occur between the new eigenstates ($|+\rangle, |-\rangle$) at the resonance frequency, $\omega = \Delta/\hbar$. An ac signal can be added to the applied flux (a microwave transmission line inductively coupled to the qubit circuit) to resonantly drive Rabi oscillations between these states. The flux can be applied for a certain duration, t, allowing any single qubit unitary operation.

The resonance frequency at the degeneracy point is $\Delta/\hbar$. At a point away from the degeneracy point, the oscillation frequency is $\omega > \Delta/\hbar$, corresponding to the energy separation, $\hbar\omega$, between the states as shown in Fig. 22.24c. This frequency is dependent on E_J (the tunnel coupling Δ depends on the potential barrier E_J). The single Josephson junction can be replaced with two junctions to form a DC-SQUID. The energy gap E_J can then be tuned by an external flux (Fig. 22.17). Readout of the qubit can be performed by induction in another loop, usually using a DC-SQUID as a sensitive magnetometer. Figure 22.25 shows a micrograph of a flux qubit.

22.12 Charge Qubit

The charge qubit consists of a dc voltage source and gate capacitor connected to a Josephson junction (Fig. 22.26), forming a superconducting island (also known as a "Cooper pair box") [20]. Inductance is irrelevant in this circuit. The qubit is formed by Cooper pairs hopping (tunneling) on and off the island, forming a superposition of N and $N + 1$ Cooper pairs on the island.

Cooper pairs with excess charge, $Q = (2e)N$, reside on the island where N is a discrete variable representing the number of Cooper pairs. An additional charge, $C_g V_g$, is induced on the island by the gate capacitance, corresponding to a Cooper pair charge of:

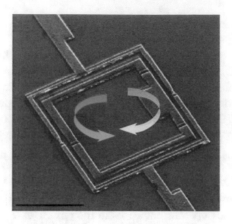

Fig. 22.25 Scanning electron micrograph of a flux qubit, showing the inner loop containing a small junction (forming the qubit) and two larger qubits (to increase the inductance of the loop). The arrows indicate the superposition of clockwise and counter-clockwise circulating supercurrent. The outer loop, containing two junctions, is a DC-SQUID for readout of the qubit state. Scale bar is 3 μm. Reprinted by permission from Springer Nature: J. Clarke and F.K. Wilhelm, Nature 453 (2008) 1031, https://doi.org/10.1038/nature07128, Copyright © 2008 [19]

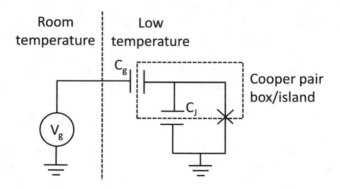

Fig. 22.26 Equivalent circuit for charge qubit

$$N_g = C_g V_g / 2e \tag{22.68}$$

Although N is discrete, N_g is a continuous variable since it corresponds to a polarization charge on the capacitor C_g. Hence, the Hamiltonian becomes:

$$\widehat{H} = 4E_C \left(\widehat{N} - N_g\right)^2 - E_J \cos \phi \tag{22.69}$$

where $E_c = e^2/2C_\Sigma$ is the charging energy with $C_\Sigma = C_J + C_g$. Using our earlier analogy, $\widehat{N} = -i\frac{\partial}{\partial \phi}$, gives:

$$\widehat{H} = 4E_C\left(-i\frac{\partial}{\partial\phi} - N_g\right)^2 - E_J\cos\phi \tag{22.70}$$

The first term in Eq. (22.70) is the energy required to add one Cooper pair to the island and the second term is the potential energy of the Josephson junction. Note that Eq. (22.70) is similar to the Hamiltonian for the non-linear harmonic oscillator discussed earlier (Eq. 22.51). However, Eq. (22.70) includes the variable, N_g, which can be tuned continuously via the gate voltage, V_g (Eq. 22.68).

For the charge qubit, $E_C \gg E_J$. This requires a small junction area with small capacitance C_Σ (~1 fF). First, suppose $E_J = 0$, so the total energy consists only of the charging energy, which is parabolic with N_g. The charging energy, $4E_C(N - N_g)^2$, consists of a series of parabolas as shown in Fig. 22.27a for different numbers of Cooper pairs, N, on the island. In this case, the charge states on the island ($N = \ldots-2, -1, 0, 1, 2, \ldots$) are completely independent.

For the case $E_J \neq 0$, the Schrodinger equation is again the Mathieu equation. Solving this differential equation gives the energy levels for the charge qubit. This differential equation is familiar from the problem that arises when considering the energy bands in a solid due to a periodic potential. The Josephson junction couples the charge states at the degeneracy points, causing an energy level splitting (anti-crossing). We find that gaps of magnitude ~E_J open in the energy spectrum at the crossings of the parabolas (degeneracy points), as shown in Fig. 22.27b, like the formation of a bandgap in a periodic potential. Only the two lowest lying energy bands are shown; all other charge states have a much higher energy and can therefore be ignored. "Sweet spots" occur at $N_g = n + 1/2$ ($n = 0, 1, 2, \ldots$), corresponding to a superposition of the states $|n\rangle$ and $|n + 1\rangle$; i.e., a superposition of n and $n + 1$ Cooper

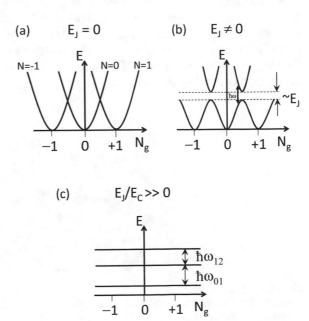

Fig. 22.27 a Energy spectrum of charge qubit in the case $E_J = 0$, and **b** $E_J \neq 0$

pairs on the island. At the sweet spot, the energy levels have zero slope $\left(\frac{\partial E}{\partial N_g} = 0\right)$, so the energy splitting (resonance frequency) becomes insensitive (to first order) to fluctuations in N_g caused by charge noise.

The charge qubit can be initialized by setting V_g far from the degeneracy point, say $V_g = 0$ (corresponding to $N_g = 0, N = 0$, and the $|0\rangle$ state with high probability). If V_g is pulsed to the "sweet spot" at $N_g = 0.5$, then $|0\rangle$ is no longer an eigenstate of the system. Rabi oscillations will occur between the new eigenstates of the system, which are the symmetric state $\left(|-\rangle = \frac{1}{\sqrt{2}}(|0\rangle + |1\rangle)\right)$ and anti-symmetric state $\left(|+\rangle = \frac{1}{\sqrt{2}}(|0\rangle - |1\rangle)\right)$ separated by the energy E_J. Coherent Rabi oscillations will occur between these two states at the frequency $\omega = E_J/\hbar$ at the degeneracy point, or at a higher frequency (dependent on E_J) away from the degeneracy point as shown in Fig. 22.27b. The single Josephson junction can be replaced with two junctions to form a DC-SQUID. The energy gap E_J can then be tuned by an external flux, Φ_e (Fig. 22.17).

The charge qubit can be represented using either the $|0\rangle$ and $|1\rangle$ basis, or the $|+\rangle$ and $|-\rangle$ basis. The probability of the $|0\rangle$ and $|1\rangle$ charge state depends on V_g. The relative phase between $|0\rangle$ and $|1\rangle$ will oscillate at the frequency, ω, which is dependent on E_J. Hence, the charge qubit is similar to the flux qubit but with the flux variable replaced by the charge variable. E_J is equivalent to B_x ($\hat{\sigma}_x$ operator) and V_g is equivalent to B_z ($\hat{\sigma}_z$ operator) using the spin analogue [18]. Hence, any single qubit gate (any qubit superposition state) can be achieved by pulsing V_g near the degeneracy point for some time, t. Resonant driving at the frequency ω can also be applied to the qubit by capacitively coupling an ac drive voltage to the dc voltage V_g.

Recall that these Rabi oscillations correspond to $N = 0$ or 1 Cooper pair on the island, or a superposition of these two states. E_J controls the frequency ($\omega = E_J/\hbar$) at which Cooper pairs enter and leave the island. Charge qubits can be read-out by a sensitive charge detector (electrometer) capacitively coupled to the qubit, such as the single-electron transistor (SET) discussed in Chap. 20. Figure 22.28 shows a micrograph of a charge qubit.

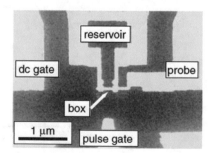

Fig. 22.28 Scanning electron micrograph of a charge qubit (Cooper pair box) and a probe that reads out the charge state. Scale bar is 1 μm. Reprinted by permission from Springer Nature: Y. Nakamura, Y.A. Pashkin and J.S. Tsai, Nature 398 (1999) 786, https://doi.org/10.1038/19718, Copyright © 1999 [20]

22.13 Complementary Variables

Recall that $\hat{x}$ and $\hat{p}$ are complementary variables (also called conjugate variables) satisfying the Heisenberg uncertainty relation:

$$\Delta x \Delta p \geq \hbar/2 \tag{22.71}$$

This means that $\hat{x}$ and $\hat{p}$ cannot be simultaneously defined with absolute precision. Similarly, using our prior analogy (Table 22.1), $\widehat{Q}$ and $\widehat{\Phi}$ are complementary variables satisfying an uncertainty relation:

$$\Delta Q \Delta \Phi \geq \hbar/2 \tag{22.72}$$

Equation (22.72) can also be expressed in terms of the number of Cooper pairs and the phase:

$$\Delta N \Delta \phi \geq 1 \tag{22.73}$$

Equation (22.73) is called the number-phase uncertainty relation.

The number-phase uncertainty relation can be used to further define the types of superconducting circuits. Recall that for the phase qubit $E_C \ll E_J$. Due to the small charging energy, Cooper pairs are easily added to the junction electrodes; i.e., charge readily fluctuates back and forth resulting in a large uncertainty in N. On the other hand, since E_J is large, ϕ is well defined, meaning ϕ is a good quantum number. Thus, we can describe the dynamics of the phase qubit using ϕ. Conversely, in the charge qubit, $E_C \gg E_J$. Here, N is a good quantum number while ϕ has a large quantum uncertainty.

22.14 Decoherence

The sources of decoherence in superconducting qubits are not completely understood. The qubits likely couple to defects (charge or other atomic-scale two-level systems) in the substrate, insulating layers, or elsewhere in the circuit [21, 22]. To reduce sensitivity to the environment (decoherence), many other types of superconducting qubits have been developed based on variations of the phase, charge and flux qubits. Today, the transmon qubit, described in the next section, has been the most successful with qubit lifetimes (T_2) on the order of 100 μs [23]. Unfortunately, this decoherence time is still relatively short compared to other systems, such as isolated atoms or ions.

22.15   Transmon Qubit

One of the problems with the chare qubit is its sensitivity to charge noise. The charge on the gate capacitor, $N_g = C_g V_g/2e$, can be tuned by the gate voltage. However, N_g can also change due to stray charge (effectively changing V_g). The transmon qubit is a charge qubit shunted by a large capacitance, C, as shown in Fig. 22.29a. The large capacitance reduces the charging energy, E_C, so that $E_J > E_C$ [24]. This makes the charge qubit less sensitive to charge noise. In practice, the large shunt capacitance necessary for the transmon qubit is achieved by an interdigitated electrode geometry as shown in Fig. 22.29b. In addition, the transmon qubit is comprised of two parallel Josephson junctions, forming a DC-SQUID. The DC-SQUID allows control of the Josephson energy by an external magnetic flux, as discussed previously.

The energy gap increases, and the dispersion flattens in Fig. 22.27b as E_J/E_C increases, as illustrated in Fig. 22.30. The resulting energy level splittings remain unequal by a few percent ($\hbar\omega_{12} - \hbar\omega_{01} \sim -200$ MHz), so that a microwave signal, $\hbar\omega_{01}$, can selectively address the two lowest energy levels for Rabi oscillations. As

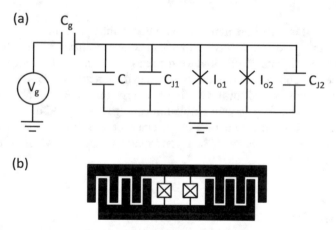

(a)

(b)

Fig. 22.29 **a** Equivalent circuit for the transmon qubit with two Josephson junctions (DC-SQUID) shunted by a large capacitance C. **b** Interdigitated electrode geometry for large shunt capacitance, C

Fig. 22.30 Energy spectrum of the transmon qubit

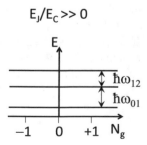

seen in Fig. 22.30, the energy splittings become insensitive to N_g, so the charge qubit can be made less sensitive to charge noise and it becomes unnecessary to tune the circuit to a sweet spot. For this reason, the transmon qubit has become the most popular variety of superconducting qubit.

22.16 Two-Qubit Operations

Superconducting qubit circuits are typically microns in size. Therefore, it is straightforward to couple nearby qubits using capacitive or inductive coupling. Figure 22.31a shows two anharmonic oscillators coupled by capacitance, C_{12}. Figure 22.31b shows two anharmonic oscillators coupled by their mutual inductance, M_{12}. A disadvantage of this scheme, however, is that the coupling is always turned on. A more flexible scheme uses a tunable coupler. As shown in Fig. 22.31c, this is achieved by a loop with a Josephson junction, whose inductance can be tuned by magnetic flux:

$$L_J = \frac{\hbar}{2eI_o \cos \phi} = \frac{\hbar}{2eI_o \cos\left(2\pi \frac{\Phi_c}{\Phi_o}\right)} \tag{22.74}$$

Fig. 22.31 a Direct capacitive coupling, **b** direct inductive coupling, and **c** tunable inductive coupling between two qubits

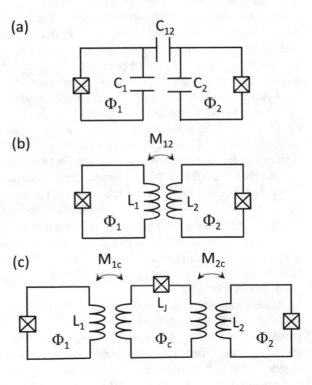

Distant qubits may be coupled by enclosing the two qubits inside a larger coupling loop, like a flux transformer. One qubit induces a flux in the coupling loop, which then induces a flux in the other qubit.

Like any two coupled LC circuits, energy is exchanged between the two resonators, analogous to two coupled pendulums. The resonance frequency of each qubit circuit is shifted due to the coupling (this is anti-crossing, as discussed in Chap. 16). Hence, the energy levels corresponding to $|00\rangle$, $|01\rangle$, $|10\rangle$ and $|11\rangle$ become unequal. This means that the transition between $|10\rangle$ and $|11\rangle$ can be selectively addressed, creating a CNOT gate, as discussed in Chap. 17.

22.17   Circuit QED

Quantum electrodynamics (QED) describes the quantum mechanical treatment of the interaction between light (photons) and matter (atoms). Experimentally, light-matter interaction can be studied by trapping a single atom in an optical cavity. The optical cavity, also known as an etalon or Fabry-Perot cavity, is formed from two mirrors. The cavity increases the amplitude of the electromagnetic field as compared to that in vacuum by bouncing the light back and forth between the mirrors. The electric field scales inversely with the volume of the cavity.

A photon, bouncing back and forth many times between the mirrors, has multiple opportunities to be absorbed by the atom. Hence, the interaction between the light and the atom is strongly enhanced by the cavity. An atom in an excited state can also spontaneously decay and emit a photon back into the cavity, which can be trapped and absorbed again later. This absorption and emission cycle is known as "vacuum Rabi oscillations". The cavity turns the absorption and spontaneous emission into a coherent process. The study of these processes is called cavity QED. Cavity QED exhibits many interesting phenomena. For example, due to the strong light-matter interaction, cavity QED can enable non-linear optics at the single photon level (the Kerr effect, for example, which can be used in two-qubit optical gates in Chap. 24).

Like QED, the interaction between a microwave signal and a qubit can be enhanced by placing the qubit inside a microwave cavity. The study of the interaction between the electromagnetic field and the qubit is called circuit QED, or cQED. The cavity can be formed in a coplanar geometry, illustrated in Fig. 22.32a, using Nb or Al on a sapphire substrate. The coplanar waveguide is similar to the cross-section of a coaxial waveguide or transmission line. A microwave signal is applied to the centre conductor with the two outer conductors grounded. The electromagnetic field propagates in the gap between conductors.

A microwave cavity is formed by introducing breaks in the centre conductor, which introduces a reflection of the electromagnetic field, effectively forming a mirror. In addition to forming the mirrors, the gaps also provide a capacitive link to the input (C_{in}) and output (C_{out}) microwave signal. If the cavity length is an integer number (n) of half wavelengths, $L = n\lambda/2$, then the field is resonant with the cavity and a standing wave (normal mode) is formed in the cavity. A resonant frequency

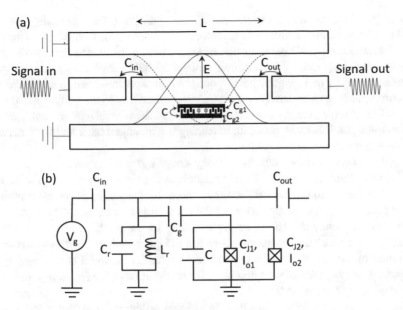

Fig. 22.32 **a** cQED with cavity length L containing a transmon qubit with two Josephson junctions (DC-SQUID). **b** Equivalent circuit for transmon cQED

of 10 GHz corresponds to a vacuum wavelength of 30 mm. Hence, the size of the cavity is macroscopic and can be constructed by standard microelectronic fabrication techniques. These transmission line constructions are also used as filters in microwave engineering.

The cavity volume in cQED is defined by the cavity length, L (a few centimetres) and the two transverse directions which are the gap between the signal and the ground lines (a few microns) and the thickness of the metal electrodes (a few hundred nanometers). Due to the small transverse directions, a large electric field amplitude can be produced in the cavity. The superconducting qubit circuit is placed at the anti-node of the electric field, where the electric field amplitude is greatest. The large electric field enables strong photon-qubit coupling.

The cavity can be modeled as an LC resonator (L_r, C_r) in parallel with the qubit circuit. The equivalent circuit is shown in Fig. 22.32b for a transmon qubit inside a cavity. The capacitance between the interdigitated electrodes and the transmission line electrodes (C_{g1}, C_{g2}) forms the total capacitance C_g that couples the transmon circuit to the resonator. The interdigitated electrodes of the transmon form the large shunt capacitance, C. The gaps in the transmission line, forming the cavity mirrors, introduce the capacitances C_{in} and C_{out}. A transmon qubit will interact with the electric field of the cavity. If, instead, we put a flux qubit in the cavity, then the qubit interacts with the magnetic field of the cavity. In the latter case, the coupling between the cavity and the flux qubit is described by inductive coupling (L_g) instead of the capacitive coupling, C_g.

The Hamiltonian describing the cavity, the qubit, and their coupling is called the Jaynes-Cummings Hamiltonian. However, the circuit of Fig. 22.32b allows us to understand the basic operating principles without getting into the details of the Hamiltonian. Inspection of the circuit in Fig. 22.32b indicates that the coupling between the cavity and the qubit corresponds to two capacitively coupled LC circuits, one corresponding to the cavity and another corresponding to the qubit. Like any two coupled LC circuits, energy is exchanged between the two resonators, analogous to two coupled pendulums. A quantum of energy is exchanged back and forth between the resonator cavity and the qubit.

cQED allows us to achieve the "strong coupling" regime where the period of the Rabi oscillations is shorter than the decoherence time (T_2) of the qubit and the lifetime of the photon in the cavity. When the qubit frequency is resonant with the cavity frequency, then the frequency of the transmitted signal is split into an upper and lower frequency due to the strong coupling of the qubit and the cavity (anti-crossing). As a result of the coupling between the qubit and the cavity, the frequency and phase of the cavity output is dependent on whether the qubit is in the $|0\rangle$ or the $|1\rangle$ state. Thus, the state of the qubit is easily read-out from the frequency and phase output of the transmission line.

We can couple multiple qubits by using higher order standing wave modes in the cavity and placing the qubits at different anti-nodes of the standing wave. The cavity acts as a quantum bus, like the phonon mode in the trapped ion quantum computer. Each qubit can be tuned to a different ω_o transition frequency by tuning $E_J(\Phi)$ of the DC-SQUID of each transmon. Hence, a qubit can emit a photon to the cavity and another qubit tuned to absorb it, producing a two-qubit gate between distant qubits. Figure 22.33 shows a transmon qubit in a cavity.

One can also imagine placing different qubit systems in the cavity, rather than superconducting qubits. These are known as hybrid systems [25]. For example, a spin qubit could couple to the magnetic field component of the cavity mode. Hybrid systems are an active field of research.

22.18 Outlook

Due to the ease of superconducting circuit fabrication, many companies are developing superconducting qubits, including Google, IBM, Intel, Rigetti, Quantum Circuits, Oxford Quantum Circuits, and Origin Quantum. The flexibility in design of superconducting circuits enables us to design the Hamiltonian, unlike atomic qubits. On the other hand, superconductivity requires very low temperature to produce the superconducting state. To circumvent this difficulty, companies are offering access to their superconducting quantum computers via cloud computing and new programming languages [27].

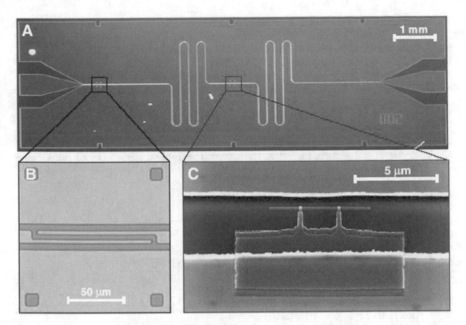

Fig. 22.33 **a** Optical micrograph showing cQED circuit. The transmission line "wiggles" to increase its cavity length. **b** The cavity mirror formed by a gap in the transmission line. The gap is extended to increase the capacitance. **c** The transmon qubit with two Josephson junctions (DC-SQUID). Reprinted by permission from Springer Nature: A. Wallraff et al., https://doi.org/10.1038/nature 02851, Nature 431 (2004) 162, Copyright © 2004 [26]

References

1. H. Kamerlingh Onnes, Commun. Phys. Lab. Univ. Leiden. Suppl. **29** (Nov 1911)
2. Y.B. Kim, C.F. Hempstead, A.R. Strnad, Phys. Rev. Lett. **9**, 306 (1962)
3. E. Snider et al., Nature **586**, 373 (2020)
4. Author: Nobel Foundation. File: Kamerlingh Onnes signed.jpg. (2018, January 17). *Wikimedia Commons, the free media repository*. Retrieved 14:26, December 8, 2020 from https://commons.wikimedia.org/w/index.php?title=File:Kamerlingh_Onnes_signed.jpg&oldid=280373897
5. J. Bardeen, L.N. Cooper, J.R. Schrieffer, Phys. Rev. **108**, 1175 (1957)
6. Author: Nobel Foundation. File: Bardeen.jpg. (2020, September 18). *Wikimedia Commons, the free media repository*. Retrieved 14:43, December 8, 2020 from https://commons.wikimedia.org/w/index.php?title=File:Bardeen.jpg&oldid=463602019
7. Author: Kenneth C. Zirkel. This file is licensed under the Creative Commons Attribution-Share Alike 3.0 Unported license (https://creativecommons.org/licenses/by-sa/3.0/deed.en). File: Nobel Laureate Leon Cooper in 2007.jpg. (2020, August 20). *Wikimedia Commons, the free media repository*. Retrieved 14:47, December 8, 2020 from https://commons.wikimedia.org/w/index.php?title=File:Nobel_Laureate_Leon_Cooper_in_2007.jpg&oldid=441386996
8. This is an image from the Nationaal Archief, the Dutch National Archives, and Spaarnestad Photo. This file is licensed under the Creative Commons Attribution-Share Alike 3.0 Netherlands license (https://creativecommons.org/licenses/by-sa/3.0/nl/deed.en). File: John Robert Schrieffer 1972.jpg. (2020, September 18). *Wikimedia Commons, the free media repository*.

Retrieved 14:51, December 8, 2020 from https://commons.wikimedia.org/w/index.php?title=
File:John_Robert_Schrieffer_1972.jpg&oldid=463418503

9. B.S. Deaver Jr., W.M. Fairbank, Phys. Rev. Lett. **7**, 43 (1961)
10. R. Doll, M. Näbauer, Phys. Rev. Lett. **7**, 51 (1961)
11. Author: Brian Josephson. This file is licensed under the Creative Commons Attribution 3.0 Unported license (https://creativecommons.org/licenses/by/3.0/deed.en). File: Brian Josephson, March 2004.jpg. (2020, October 4). *Wikimedia Commons, the free media repository*. Retrieved 15:13, December 8, 2020 from https://commons.wikimedia.org/w/index.php?title=File:Brian_Josephson,_March_2004.jpg&oldid=480177322
12. Y.C. Chen, J. Appl. Phys. **64**, 3119 (1988)
13. R. Feynman, R. Leighton, M. Sands, The feynman lectures on physics (1966). https://www.feynmanlectures.caltech.edu/
14. W.C. Stewart, Appl. Phys. Lett. **12**, 277 (1968)
15. D.E. McCumber, J. Appl. Phys. **39**, 3113 (1968)
16. J.R. Friedman et al., Nature **406**, 43 (2000)
17. C.H. van der Wal, Science **290**, 773 (2000)
18. Y. Makhlin et al., Rev. Mod. Phys. **73**, 357 (2001)
19. J. Clarke, F.K. Wilhelm, Nature **453**, 1031 (2008)
20. Y. Nakamura, Y.A. Pashkin, J.S. Tsai, Nature **398**, 786 (1999)
21. G. Ithier et al., Phys. Rev. B **72**, (2005)
22. W.D. Oliver, P.B. Welander, MRS Bull. **38**, 816 (2013)
23. H. Paik et al., Phys. Rev. Lett. **107**, (2011)
24. J. Koch et al., Phys. Rev. A **76**, (2007)
25. Z.-L. Xiang et al., Rev. Mod. Phys. **85**, 623 (2013)
26. A. Wallraff et al., Nature **431**, 162 (2004)
27. R. LaRose, Quantum **3**, 130 (2019)

Chapter 23
Adiabatic Quantum Computing

Adiabatic quantum computing (AQC) is an approach to quantum computing based on the adiabatic theorem. The adiabatic theorem states that if a Hamiltonian, $\widehat{H}$, changes slowly in time, then the state remains in the ground state. By starting with some simple Hamiltonian and evolving it to some $\widehat{H}$ representing the problem to be solved, the lowest energy configuration becomes the solution to the problem.

23.1 The Adiabatic Theorem

In the adiabatic approach, a system with a simple Hamiltonian, $\widehat{H}_i$, is prepared and initialized to the ground state. Next, a Hamiltonian, $\widehat{H}_f$, is found whose ground state describes the solution to the problem of interest. The system starts in the ground state of the simple Hamiltonian and then changes gradually to $\widehat{H}_f$. For example, we could ramp down $\widehat{H}_i$ and ramp up $\widehat{H}_f$ according to $\widehat{H}(t) = (1 - t)\widehat{H}_i + t\widehat{H}_f$ where t is a normalized time. $\widehat{H}_i$ is adiabatically evolved (no transfer of heat) to the desired complicated Hamiltonian $\widehat{H}_f$. The wavefunction $|\psi\rangle$ evolves according to the time-dependent Schrodinger equation, $\widehat{H}|\psi\rangle = i\hbar\frac{\partial|\psi\rangle}{\partial t}$. The adiabatic theorem states that if $\widehat{H}$ changes slowly, then $|\psi\rangle$ remains in the ground state (lowest energy configuration), which becomes the solution to the problem.

23.2 Ising Model

Consider the system shown in Fig. 23.1, called a "connection graph". Each S_i is called a "node" and represents a superconducting current loop (a flux qubit). $S_i = \pm 1$ represents the current direction of each superconducting loop. In a sense, each superconducting loop acts as an artificial spin. Suppose h_j represents the effect of

© The Author(s), under exclusive license to Springer Nature Switzerland AG 2021   323
R. LaPierre, *Introduction to Quantum Computing*, The Materials Research Society Series,
https://doi.org/10.1007/978-3-030-69318-3_23

Fig. 23.1 A connection graph

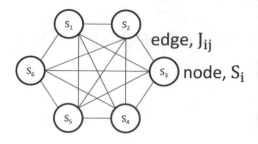

an external magnetic field on each qubit, and J_{ij} represents the coupling between qubits. h_j and J_{ij} are user programmable, defining the problem to be solved. The Hamiltonian for this system is:

$$\widehat{H}_f = \sum_{j=1}^{n} h_j \widehat{S}_j + \sum_{i \neq j} J_{ij} \widehat{S}_i \widehat{S}_j \tag{23.1}$$

This is called the Ising spin model [1], which we introduced in Chap. 17 for two-qubit coupling. There are 2^n possible configurations for n nodes. In computational complexity theory, finding the lowest energy configuration for this system is considered a hard problem. For n greater than about 50, this problem becomes intractable for classical computers. This is known as the many-body problem.

In adiabatic quantum computing (AQC), we start with each S_i in a superposition of $+1$ and -1 (superposition of clockwise and counter-clockwise current). This initial state represents $\widehat{H}_i$. Then, we slowly (on the order of milliseconds) change to $\widehat{H}_f$, representing the problem to be solved (with its specific values of h_j and J_{ij}). According to the adiabatic theorem, the system will settle into the configuration with the minimum energy out of the 2^n possible configurations (Fig. 23.2). This minimum energy configuration represents the solution to the problem. During AQC, quantum tunneling to the lowest energy state can occur in addition to thermal processes. This is known as "quantum annealing" [2].

AQC is different than a "universal quantum computer" where you have individual gate control on each qubit. AQC is designed to solve one class of problem: to find the ground state of a "generalized Ising model" on a particular class of "partially-connected graphs". AQC is generally used to solve optimization problems where you need to find some minimum, including *NP* problems, such as the traveling salesperson problem, routing air traffic, protein folding, and supply chain optimization. AQC can also be used to simulate quantum systems, especially many-body problems in physics such as spin systems [3, 4]. This is a realization of Feynman's dream of using controlled quantum systems to simulate other quantum systems, known as "quantum simulation" [5].

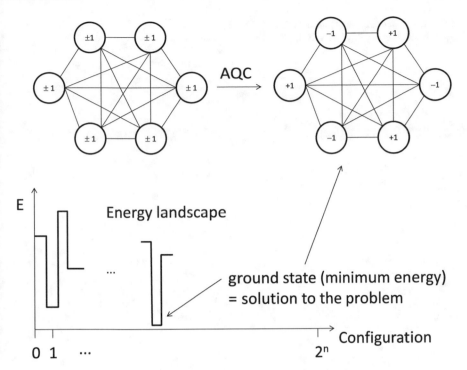

Fig. 23.2 Node configurations before and after adiabatic evolution, and the energy landscape

23.3 Outlook

The company, D-Wave Systems, based in British Columbia, Canada, uses AQC. As of 2020, the D-Wave system uses 2000 flux qubits of superconducting Nb loops. D-Wave Systems was the world's first quantum computing company. Systems have been sold to Google, Lockheed Martin, and others. The company now offers remote cloud-based access to source development tools and free, real-time access to a D-Wave quantum computer. D-Wave computers have been shown to exhibit quantum behavior, including quantum tunneling, superposition and entanglement [2, 6, 7]. D-Wave's system can solve certain optimization problems, but further work is needed to show if any speedup exists as compared to the best classical algorithms.

References

1. J. Brooke et al., Science **284**, 779 (1999)
2. M.W. Johnson et al., Nature **473**, 194 (2011)
3. R. Harris et al., Science **361**, 162 (2018)
4. A.D. King et al., Nature **560**, 456 (2018)
5. R.P. Feynman, Int. J. Theor. Phys. **21**, 467 (1982)

6. N. Dickson et al., Nat. Commun. **4**, 1903 (2013)
7. T. Lanting et al., Phys. Rev. X **4**, (2014)

Chapter 24
Optical Quantum Computing

Optical (also known as photonic or bosonic) quantum computing uses photons, so-called "flying qubits", to encode $|0\rangle$ or $|1\rangle$. Several different methods exist for encoding $|0\rangle$ and $|1\rangle$, including the photon polarization or the spin angular momentum state (polarization encoding), different optical paths in an optical system (path encoding), the spatial distribution of light such as different optical modes in an optical waveguide (orbital angular momentum encoding), and time or frequency encoding. We have already examined the case of polarization encoding for quantum communication in Chap. 6. Let us consider the key optical elements used for optical quantum computing, using path encoding as an example.

24.1 Phase Shifter

A phase shifter is a dielectric material of thickness L (Fig. 24.1). A classical light wave passing through a phase shifter will have a phase shift, ϕ, added to the wave, given by:

$$\phi = kL = \frac{2\pi}{\lambda}L = \frac{2\pi n}{\lambda_o}L \qquad (24.1)$$

were k is the wavevector, L is the length of the phase shifter, λ is the wavelength in the dielectric medium, λ_o is the wavelength in vacuum, and n is the dielectric refractive index. In terms of the photon picture, the phase ϕ is applied to the probability amplitude of a single photon (Fig. 24.1). Thus, an initial probability amplitude, α, becomes $\alpha e^{i\phi}$ after traversing the phase shifter. Since $\left|\alpha e^{i\phi}\right|^2 = |\alpha|^2$, there is no change in the probability due to the phase shifter. However, the phase shift can be important for interference, as we saw in the double-slit experiment of Chap. 1.

© The Author(s), under exclusive license to Springer Nature Switzerland AG 2021
R. LaPierre, *Introduction to Quantum Computing*, The Materials Research Society Series,
https://doi.org/10.1007/978-3-030-69318-3_24

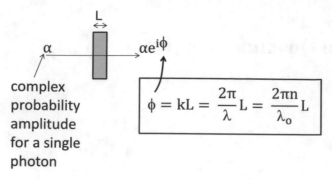

Fig. 24.1 A phase shifter

24.2   Path Encoding

Consider two optical paths with the upper path containing a phase shifter, as shown in Fig. 24.2. A photon in the upper path could represent $|0\rangle$, and a photon in the lower path could represent $|1\rangle$. This is known as "path encoding" or the "dual rail representation". The input probability amplitudes are α and β for the upper and lower path, respectively, which defines the input qubit state vector, $\begin{pmatrix} \alpha \\ \beta \end{pmatrix}$. The output probability amplitudes are α' and β', defining the output qubit state vector, $\begin{pmatrix} \alpha' \\ \beta' \end{pmatrix}$. $\beta' = \beta$ for the lower path, and $\alpha' = \alpha e^{i\phi}$ for the upper path due to the phase shifter. This optical system can be represented by the transformation matrix:

$$P_u = \begin{pmatrix} e^{i\phi} & 0 \\ 0 & 1 \end{pmatrix} \tag{24.2}$$

Fig. 24.2 Two optical paths with the upper path containing a phase shifter

$\alpha \longrightarrow \boxed{} \longrightarrow \alpha' = \alpha e^{i\phi}$

$\beta \xrightarrow{\hspace{3cm}} \beta' = \beta$

$$P_u = \begin{pmatrix} e^{i\phi} & 0 \\ 0 & 1 \end{pmatrix}$$

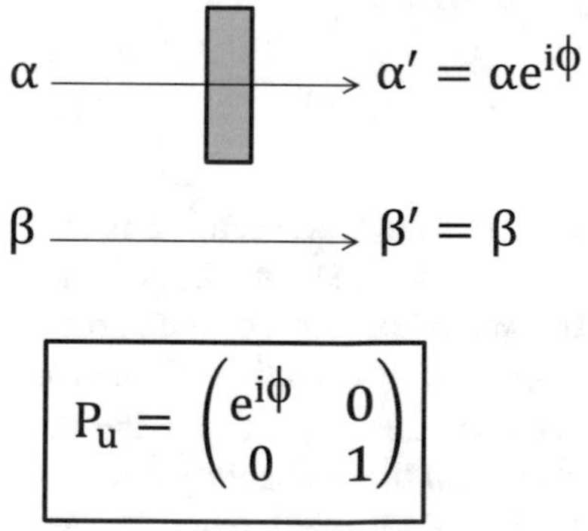

Fig. 24.3 Two optical paths with the lower path containing a phase shifter

$$\alpha \longrightarrow \alpha' = \alpha$$

$$\beta \longrightarrow \beta' = \beta e^{i\phi}$$

$$P_1 = \begin{pmatrix} 1 & 0 \\ 0 & e^{i\phi} \end{pmatrix}$$

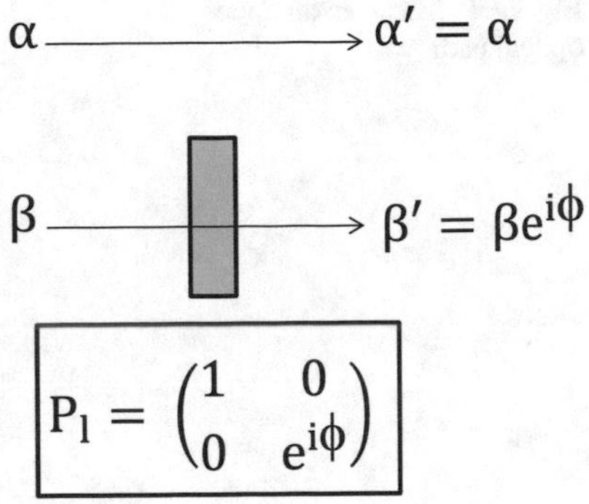

where the subscript "u" indicates that the phase shifter is present in the upper path. P_u transforms $\begin{pmatrix} \alpha \\ \beta \end{pmatrix}$ into $\begin{pmatrix} \alpha' \\ \beta' \end{pmatrix}$; i.e.,

$$\begin{pmatrix} \alpha' \\ \beta' \end{pmatrix} = \begin{pmatrix} e^{i\phi} & 0 \\ 0 & 1 \end{pmatrix} \begin{pmatrix} \alpha \\ \beta \end{pmatrix} = \begin{pmatrix} \alpha e^{i\phi} \\ \beta \end{pmatrix} \tag{24.3}$$

Now consider two optical paths with the lower path containing a phase shifter, as shown in Fig. 24.3. In this case, the transformation matrix is:

$$P_l = \begin{pmatrix} 1 & 0 \\ 0 & e^{i\phi} \end{pmatrix} \tag{24.4}$$

where the subscript "l" indicates that the phase shifter is in the lower path. The transformation is:

$$\begin{pmatrix} \alpha' \\ \beta' \end{pmatrix} = \begin{pmatrix} 1 & 0 \\ 0 & e^{i\phi} \end{pmatrix} \begin{pmatrix} \alpha \\ \beta \end{pmatrix} = \begin{pmatrix} \alpha \\ \beta e^{i\phi} \end{pmatrix} \tag{24.5}$$

This case is equivalent to the phase gate in Eq. (7.38), or the rotation operator in Eq. (14.4); i.e., it represents a rotation about the z-axis in the Bloch sphere, $\hat{R}_z(\phi)$.

24.3 Mirror

The mirror is another important optical element. Consider a mirror in the upper optical path, as shown in Fig. 24.4. The Fresnel equations, familiar from optics, tell us that a π phase shift ($e^{i\pi} = -1$) results when light is reflected from a mirror (a metallic surface). The transformation matrix corresponding to Fig. 24.4 is:

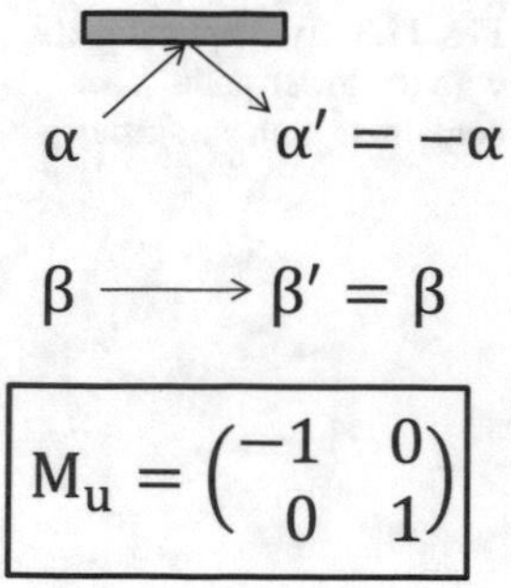

Fig. 24.4 Mirror in an upper
optical path

$$\alpha \qquad \alpha' = -\alpha$$

$$\beta \longrightarrow \beta' = \beta$$

$$M_u = \begin{pmatrix} -1 & 0 \\ 0 & 1 \end{pmatrix}$$

$$M_u = \begin{pmatrix} -1 & 0 \\ 0 & 1 \end{pmatrix} \tag{24.6}$$

such that:

$$\begin{pmatrix} \alpha' \\ \beta' \end{pmatrix} = \begin{pmatrix} -1 & 0 \\ 0 & 1 \end{pmatrix} \begin{pmatrix} \alpha \\ \beta \end{pmatrix} = \begin{pmatrix} -\alpha \\ \beta \end{pmatrix} \tag{24.7}$$

If the mirror is contained in the lower optical path, as shown in Fig. 24.5, then the transformation matrix is:

$$M_l = \begin{pmatrix} 1 & 0 \\ 0 & -1 \end{pmatrix} \tag{24.8}$$

such that:

$$\begin{pmatrix} \alpha' \\ \beta' \end{pmatrix} = \begin{pmatrix} 1 & 0 \\ 0 & -1 \end{pmatrix} \begin{pmatrix} \alpha \\ \beta \end{pmatrix} = \begin{pmatrix} \alpha \\ -\beta \end{pmatrix} \tag{24.9}$$

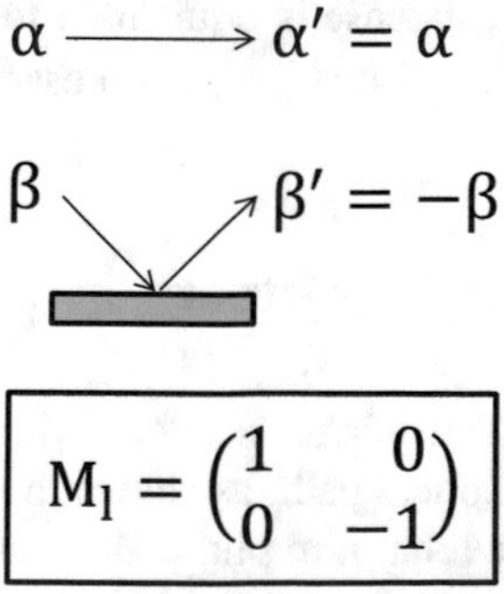

Fig. 24.5 Mirror in a lower
optical path

$$\alpha \longrightarrow \alpha' = \alpha$$

$$\beta \qquad \beta' = -\beta$$

$$M_l = \begin{pmatrix} 1 & 0 \\ 0 & -1 \end{pmatrix}$$

24.4 Beam Splitter

A 50:50 beam splitter is an optical element where 50% of incident light is reflected and 50% is transmitted. If single photons are incident on the beam splitter, this means that each photon has a 50% probability of either being reflected or transmitted. A beam splitter may be comprised of a semi-transparent metal film (usually Al with the real part of the refractive index equal to ~1.2 in the visible region of the light spectrum) deposited on a glass plate (with refractive index ~1.5). The Fresnel equations from optics tell us that light incident on a metal film from air (from a lower to a higher refractive index) will undergo a π phase shift in the reflected light. However, if light is incident on the metal from within the glass (from a higher to a lower refractive index), then this results in no phase shift of the reflected light. Therefore, when considering the effect of a beam splitter on incident light, we need to consider whether the light was incident on the metal film from the air or from within the dielectric. We will adopt a convention similar to Ref. [1] by indicating the side with the metal film by a dot, as shown in Fig. 24.6. Different phase conventions exist in the literature, depending on the type of beam splitter.

If we consider the 50:50 beam splitter oriented as shown on the left-hand side of Fig. 24.6 (with the metal on the lower side), then the transformation matrix is:

$$B_l = \frac{1}{\sqrt{2}} \begin{pmatrix} 1 & 1 \\ 1 & -1 \end{pmatrix} \tag{24.10}$$

Otherwise, if the metal is facing up, then the transformation matrix is:

$$B_u = \frac{1}{\sqrt{2}} \begin{pmatrix} -1 & 1 \\ 1 & 1 \end{pmatrix} \tag{24.11}$$

In addition to a 50:50 splitting of incident light, Eq. (24.10) also results in a π phase shift applied to the lower (β) input only, while Eq. (24.11) results in a π phase shift applied to the upper (α) input only.

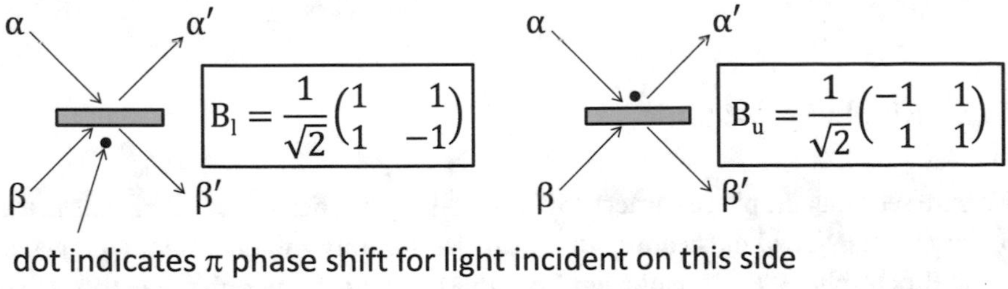

dot indicates π phase shift for light incident on this side

Fig. 24.6 50:50 beam splitter

We recognize B_u as the Hadamard gate. If we have a photon input in the upper path only, then the input qubit vector is $\begin{pmatrix} 1 \\ 0 \end{pmatrix}$. The output qubit vector is then:

$$\begin{pmatrix} \alpha' \\ \beta' \end{pmatrix} = \frac{1}{\sqrt{2}} \begin{pmatrix} 1 & 1 \\ 1 & -1 \end{pmatrix} \begin{pmatrix} 1 \\ 0 \end{pmatrix} = \frac{1}{\sqrt{2}} \begin{pmatrix} 1 \\ 1 \end{pmatrix} = \frac{1}{\sqrt{2}} (|0\rangle + |1\rangle) \tag{24.12}$$

The result is a superposition of photons in the upper and lower path, i.e., of $|0\rangle$ and $|1\rangle$, as expected for the Hadamard gate. If we have a photon input in the lower path only, then the input qubit vector is $\begin{pmatrix} 0 \\ 1 \end{pmatrix}$. The output qubit vector is then:

$$\begin{pmatrix} \alpha' \\ \beta' \end{pmatrix} = \frac{1}{\sqrt{2}} \begin{pmatrix} 1 & 1 \\ 1 & -1 \end{pmatrix} \begin{pmatrix} 0 \\ 1 \end{pmatrix} = \frac{1}{\sqrt{2}} \begin{pmatrix} 1 \\ -1 \end{pmatrix} = \frac{1}{\sqrt{2}} (|0\rangle - |1\rangle) \tag{24.12}$$

The result is a superposition of $|0\rangle$ and $|1\rangle$ with a phase shift of -1, as expected for the Hadamard gate. The 50:50 beam splitter creates a superposition of probability amplitudes for a single photon in the two paths.

The beam splitter can be designed to give an arbitrary reflectance/transmittance ratio different than 50:50. The resulting transformation matrix is:

$$\begin{pmatrix} \alpha' \\ \beta' \end{pmatrix} = \begin{pmatrix} r & t \\ t & -r \end{pmatrix} \begin{pmatrix} \alpha \\ \beta \end{pmatrix} \tag{24.13}$$

where r is called the "reflection coefficient" and t is the "transmission coefficient". We can rewrite Eq. (24.13) using θ as a parameter:

$$\begin{pmatrix} \alpha' \\ \beta' \end{pmatrix} = \begin{pmatrix} \cos\frac{\theta}{2} & \sin\frac{\theta}{2} \\ \sin\frac{\theta}{2} & -\cos\frac{\theta}{2} \end{pmatrix} \begin{pmatrix} \alpha \\ \beta \end{pmatrix} \tag{24.14}$$

The resulting matrix is equivalent to arbitrary rotations about the y-axis, $\hat{R}_y(\theta)$. $\theta = 90°$ results in the 50:50 beam splitter.

24.5 Single-Qubit Gate

We have seen that the phase shifter provides arbitrary rotations about the z-axis in the Bloch sphere, $\hat{R}_z$, and the beam splitter provides arbitrary rotations about the y-axis in the Bloch sphere, $\hat{R}_y$. Recall from Eq. (15.8) that all single-qubit operations can be achieved by rotation about any two axis in the Bloch sphere:

$$\hat{R}_n(\theta) = \hat{R}_y(\alpha)\hat{R}_z(\beta)\hat{R}_y(\delta) \tag{24.15}$$

Fig. 24.7 Dual rail representation of NOT gate

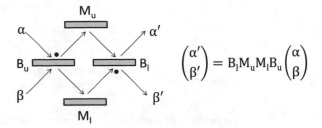

$$\begin{pmatrix} \alpha' \\ \beta' \end{pmatrix} = B_l M_u M_l B_u \begin{pmatrix} \alpha \\ \beta \end{pmatrix}$$

for some angles α, β, δ. Therefore, any single qubit operation can be generated by a combination of phase shifters and beam splitters.

24.6 NOT Gate

Consider the optical circuit shown in Fig. 24.7. The transformation is:

$$\begin{pmatrix} \alpha' \\ \beta' \end{pmatrix} = B_l M_u M_l B_u \begin{pmatrix} \alpha \\ \beta \end{pmatrix} \tag{24.16}$$

or

$$\begin{pmatrix} \alpha' \\ \beta' \end{pmatrix} = \frac{1}{\sqrt{2}} \begin{pmatrix} 1 & 1 \\ 1 & -1 \end{pmatrix} \begin{pmatrix} -1 & 0 \\ 0 & 1 \end{pmatrix} \begin{pmatrix} 1 & 0 \\ 0 & -1 \end{pmatrix} \frac{1}{\sqrt{2}} \begin{pmatrix} -1 & 1 \\ 1 & 1 \end{pmatrix} \begin{pmatrix} \alpha \\ \beta \end{pmatrix} \tag{24.17}$$

$$= \begin{pmatrix} 0 & -1 \\ 1 & 0 \end{pmatrix} \begin{pmatrix} \alpha \\ \beta \end{pmatrix} = \begin{pmatrix} -\beta \\ \alpha \end{pmatrix} \tag{24.18}$$

Note that we have created a NOT gate within an arbitrary phase factor; i.e., an input of $\begin{pmatrix} 1 \\ 0 \end{pmatrix}$ gives $\begin{pmatrix} 0 \\ 1 \end{pmatrix}$, and an input of $\begin{pmatrix} 0 \\ 1 \end{pmatrix}$ gives $-\begin{pmatrix} 1 \\ 0 \end{pmatrix}$.

Exercise 24.1 Show that the order of matrix multiplication for the mirrors does not matter: $M_u M_l = M_l M_u$

24.7 Two-Qubit Gate

Recall that universal quantum gates also require a two-qubit operation, such as a CNOT gate. As we showed previously in Chap. 21, the CNOT gate is equivalent to

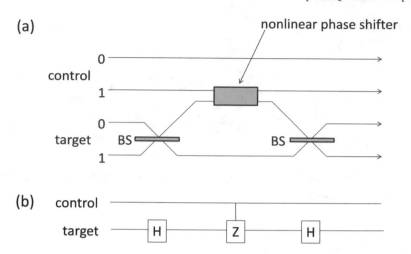

Fig. 24.8 **a** CNOT gate with dual rail representation. **b** Equivalent gate model

a control-Z gate sandwiched between two Hadamard gates as shown in Fig. 24.8b. A dual rail representation of a CNOT gate is shown in Fig. 24.8a. We recognize the two beam splitters as the Hadamard gates. The control-Z gate is more difficult to implement. It relies on non-linear optics.

Non-linear optics is based on the Kerr effect, discovered in 1875 by John Kerr, a Scottish physicist. The Kerr effect describes the non-linear dependence of the polarizability of a material, P, on the electric field, E, of incident light:

$$\text{Polarizability:} \ P = \ \underbrace{\epsilon_o \chi E}_{\text{linear optics}} + \underbrace{\epsilon_o \chi^2 E^2 + \epsilon_o \chi^3 E^3 + \cdots}_{\text{nonlinear optics}} \tag{24.19}$$

where ϵ_o is the permittivity of free space and χ is the electric susceptibility. The first term is linear in E, while the higher order terms describe a non-linear dependence of P on E. The non-linear terms give a refractive index, n, that is proportional to the light intensity, I. This is known as the Kerr effect. Analogous to Eq. (24.1), after passing through a nonlinear material, incident light will experience an extra phase shift due to the Kerr effect of:

$$\phi = 2\pi \, \Delta n L / \lambda_o \tag{24.20}$$

where L is the thickness of nonlinear material, λ_o is the incident wavelength, and Δn is the refractive index change that is proportional to light intensity. The thickness L can be chosen to incur a π phase shift. Thus, a $|1\rangle$ state in the control qubit, combined with a $|1\rangle$ in the target qubit, provides two photons with sufficient intensity to induce a π phase shift (a control Z gate), while $|0\rangle$ in the control qubit does nothing. Unfortunately, the Kerr effect is a very weak effect in bulk material, making two-qubit gates difficult to construct. An alternative approach is circuit quantum electrodynamics (cQED), considered in Chap. 22, where non-linear interactions can be very strong. Alternatively, in 2001, it was shown that a CNOT gate could be theoretically implemented using entirely linear optics [2], and this was demonstrated experimentally in 2003 [3]. However, linear optical quantum computing is probabilistic and requires many more gates.

24.8 Outlook

We can build an optical quantum computer using optical components (phase shifters, mirrors, beam splitters). Optical quantum computers have several advantages including a lack of interaction with the environment and hence a long coherence time, room temperature operation, and the possibility of a compact design using photonic integrated circuits. The disadvantage is that photons do not interact, making two-qubit gates difficult to implement. Optical quantum computing also requires single photon sources and detectors, which are not trivial to build [4]. Several companies, such as Xanadu, QuiX, PsiQuantum, and TundraSystems are pursuing photonic-based quantum computers.

References

1. M.A. Nielsen, I.L. Chuang, in *Quantum Computation and Quantum Information* (Cambridge University Press, 2017)
2. E. Knill, R. Laflamme, G.J. Milburn, Nature **409**, 46 (2001)
3. J.L. O'Brien et al., Nature **426**, 264 (2003)
4. F. Flamini, N. Spagnolo, F. Sciarrino, Rep. Prog. Phys. **82**, (2019)

Chapter 25
Quantum Error Correction

DiVincenzo criterion #5 from Chap. 18 states that the qubit lifetimes should be long compared to the duration of the algorithm. However, quantum systems are fragile. Unwanted external perturbations from the environment (e.g., electromagnetic waves, thermal excitations, charge, etc.) can interact with the quantum system, such that the system becomes entangled with the environment. Quantum computing presents two conflicting requirements: qubits should be isolated from the environment for long qubit lifetimes (DiVincenzo criterion #5), while also possessing good addressability (DiVincenzo criterion #2 and #4). We need to be able to address qubits without letting the noisy environment leak into the system. Due to the difficulty of this challenge, present day quantum algorithms have been demonstrated with only a few 10s of qubits. We need a way of correcting errors, known as quantum error correction (QEC).

25.1 Leakage

There may be other energy levels in the system where qubits may transition (Fig. 25.1). This problem is called leakage. For example, as we saw in Chap. 22 with superconducting qubits, a harmonic oscillator would not be appropriate for a qubit because the energy level separations are all identical. This problem can be solved by using an anharmonic potential.

25.2 Longitudinal Relaxation

Many qubit systems are composed of a ground state and an excited state. The excited state can decay to the ground state due to unintentional spontaneous or stimulated emission, thereby introducing errors. This is called "longitudinal relaxation" or

© The Author(s), under exclusive license to Springer Nature Switzerland AG 2021

R. LaPierre, *Introduction to Quantum Computing*, The Materials Research Society Series, https://doi.org/10.1007/978-3-030-69318-3_25

Fig. 25.1 Undesired
precession causing leakage

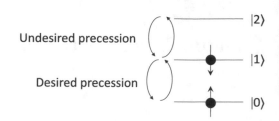

Fig. 25.2 Energy level
diagram and Bloch sphere
representation of
longitudinal relaxation with
characteristic time, T_1

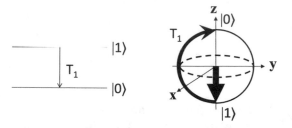

simply "relaxation" and occurs in a time T_1, called the "longitudinal coherence time" or simply the "relaxation time", as shown in Fig. 25.2. T_1 measures the loss of energy in the system due to radiative or non-radiative processes. As we saw in Chap. 19, this introduces a damping to the Rabi oscillations. Of course, a long T_1 time is desired—it is difficult to observe Rabi oscillations if T_1 is less than the Rabi oscillation period.

How do we measure T_1? We assume $|0\rangle$ represents the ground state and $|1\rangle$ represents the excited state. As shown in Fig. 25.3, we initialize the qubit to the ground state, $|0\rangle$. We apply a π pulse to put the qubit into the excited state, $|1\rangle$. We then wait for a time t, and then measure the probability of being in the $|1\rangle$ state. We repeat this procedure for different wait times, t. We expect an exponential decay of the probability, e^{-t/T_1}, allowing us to determine T_1. The value of T_1 depends on the type of qubit system and the energy loss mechanisms. T_1 ranges from microseconds in some solid-state systems to hundreds of seconds in NMR.

25.3   Pure Dephasing

Other interactions, such as spin-orbit interaction or hyperfine interaction (discussed in Chap. 20), can change the phase of a qubit; i.e., the relative phase between $|0\rangle$ and $|1\rangle$ in a superposition (Fig. 25.4). This effect is called "pure dephasing" or "decoherence", and is described by a time T_ϕ, called the "pure dephasing time", "decoherence time" or "phase coherence time". Pure dephasing is represented by a fanout of the Bloch vector in the x–y plane, as shown in Fig. 25.4.

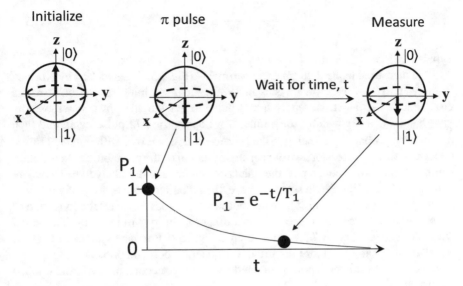

Fig. 25.3 Procedure for measurement of T_1

Fig. 25.4 Pure dephasing on the Bloch sphere

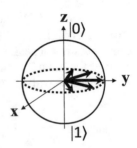

25.4 Transverse Relaxation

A combination of T_1 and T_ϕ leads to the "transverse relaxation time", denoted T_2^*, leading to a loss of the superposition state as shown in Fig. 25.5. It can be shown by the theory of relaxation that [1, 2]:

Fig. 25.5 Transverse relaxation in the Bloch sphere resulting from pure dephasing and longitudinal relaxation

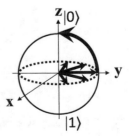

$$\frac{1}{T_2^*} = \frac{1}{2T_1} + \frac{1}{T_\phi} \tag{25.1}$$

Typically, $T_1 \gg T_\phi$, so that $T_2^* \sim T_\phi$.

T_2^* is measured using a "Ramsey experiment" (Fig. 25.6), developed by Norman Ramsey Jr. (Nobel Prize in Physics in 1989). First, we initialize the qubit to the ground state $|0\rangle$. Next, we apply a $\pi/2$ pulse to get the state in a superposition, $\frac{1}{\sqrt{2}}(|0\rangle + |1\rangle)$. After waiting for a time, t, we apply the $\pi/2$ pulse again and then measure the probability of being in the $|1\rangle$ state. We repeat with different wait times, t. Due to Larmor precession, we expect an oscillation of the probability. On the other hand, due to phase docoherence, the spin is eventually equally likely to be anywhere in the x–y plane of the Bloch sphere. Hence, the second $\pi/2$ pulse is equally likely to put the state back into $|0\rangle$ or $|1\rangle$, so the measurement goes to 1/2 after a long time as we slowly lose phase coherence. The resulting probability will be an oscillation but with a decay described by T_2^*, as illustrated in Fig. 25.6. For most systems, $T_2^* \ll T_1$, which means that T_2^* is more important for quantum computation. T_2^* ranges from nanoseconds in semiconductor quantum dots to many seconds in NMR and trapped ion qubits [3].

Why do we use a superscript "*" in T_2^*? T_2^* applies to an ensemble measurement; i.e., measurement of a large number of physical qubits like in NMR, or repeated measurements on a single qubit. A related time, T_2, applies to a single measurement on a single qubit. $T_2^* < T_2$ due to low frequency (slowly time varying) noise that may cause variations in ensemble measurements. T_2 can be measured from an ensemble by using the Hahn Echo technique. A Hahn Echo experiment is the same as a Ramsey experiment but with a $\hat{\sigma}_y$ π-pulse applied in the middle of the wait time. This refocuses

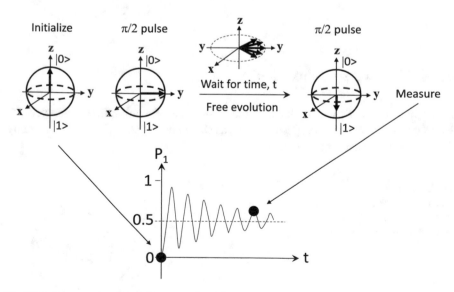

Fig. 25.6 Ramsey experiment for measuring T_2^*

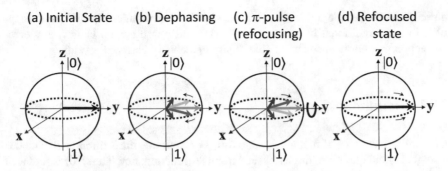

Fig. 25.7 Refocusing used in Hahn Echo measurement of T_2. **a** An initial state with Bloch vector along the y-axis. **b** Dephasing of Bloch vector. **c** A π-pulse rotates the Bloch vectors about the y-axis, so that the diverging vectors in (**b**) are now converging. **d** After some wait time, the Bloch vectors are refocused back to the original state

the diverging Bloch vector in the x-y plane, as illustrated in Fig. 25.7. This is known as "refocusing" and eliminates low frequency noise. Hence, $T_2 > T_2^*$. Refocusing originated as a commonly used technique in NMR.

25.5 Noise Mitigation Strategies

There are three approaches to dealing with noise:

1. Complete isolation of the quantum system from its environment; i.e., careful elimination of noise. In most quantum computing systems, extrinsic noise has been reduced to below the level of intrinsic effects. Thus, the qubits are now limited by the materials technologies. An example is the transmon qubit where a large shunt capacitance reduces the sensitivity to charge noise.
2. Use protocols for quantum error correction.
3. Use topological quantum computation.

Here, we will examine quantum error correction, while topological quantum computation is examined in Chap. 26.

25.6 Quantum Error Correction

The classical approach to error correction is to make multiple copies of the information. This is not possible with quantum error correction (QEC) due to the no-cloning theorem. Also, direct measurement cannot be used in QEC, since this will act to destroy any quantum superposition that is being used for computation. Error correction protocols must therefore be employed that can detect and correct errors without

directly measuring the qubit state. Unlike classical information, qubits are suscep-
tible to both traditional bit errors ($|0\rangle \rightarrow |1\rangle$)) and phase errors. Hence, any error
correction procedure needs to be able to simultaneously correct for both.

25.7   Redundant Encoding

QEC utilizes the idea of redundant encoding. For example, the 3-qubit code encodes
a single logical qubit into three physical qubits for redundancy. Thus, $|0\rangle$ is replaced
with $|000\rangle$, and $|1\rangle$ is replaced with $|111\rangle$:

$$|0\rangle \rightarrow |000\rangle \tag{25.2}$$

$$|1\rangle \rightarrow |111\rangle \tag{25.3}$$

The superposition, $\alpha|0\rangle + \beta|1\rangle$, becomes $\alpha|000\rangle + \beta|111\rangle$, which is an entangled
state:

$$\alpha|0\rangle + \beta|1\rangle \rightarrow \alpha|000\rangle + \beta|111\rangle \tag{25.4}$$

Redundant encoding works on the premise that an error on two or more qubits
is highly unlikely. For example, an error may occur that results in one of the three
qubits flipping, say from 0 to 1. However, it becomes increasingly unlikely for 2 or
all 3 of the qubits to flip. If only a single qubit is flipped, then we can still recognize
the original state. For example, if $|000\rangle$ becomes $|100\rangle$, we can still recognize an
error in the first qubit and recover the original state.

The circuit for implementing the 3-qubit code was presented in Fig. 7.14 and is
shown again in Fig. 25.8. Note that $\alpha|000\rangle + \beta|111\rangle \neq (\alpha|0\rangle + \beta|1\rangle)^{\otimes 3}$; i.e., we are
not copying the system; therefore, the no-cloning theorem is not violated.

Exercise 25.1 Check that the circuit works; i.e., $|0\rangle \rightarrow |000\rangle$, $|1\rangle \rightarrow |111\rangle$,
and $\alpha|0\rangle + \beta|1\rangle \rightarrow \alpha|000\rangle + \beta|111\rangle$

Fig. 25.8 Quantum circuit
for implementing the 3-qubit
code

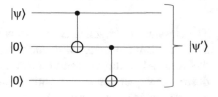

Exercise 25.2 If the probability of a bit flip error occurring is p, show that the probability of two or more errors occurring in a 3-qubit encoding is $3p^2 - 2p^3$.

As shown in Exercise 25.2, the probability of an error occurring in a single physical qubit is p. With a three-qubit encoding, two qubits would need to be flipped to produce an unrecoverable error; i.e., if $|000\rangle$ is flipped to $|110\rangle$, we cannot know if the original state was $|000\rangle$ with the first two qubits flipped or $|111\rangle$ with only the last qubit flipped. The probability of two qubits being flipped in a three-qubit encoding is $3p^2 - 2p^3$. This error is smaller than p, providing $p < 1/2$.

25.8 Correcting Bit Flips

Suppose an error occurs which flips only one of the 3 qubits. The circuit that detects and corrects this single bit flip is shown in Fig. 25.9. The measurements produce the "syndrome" shown in Table 25.1, which indicates which of the 3 qubits were flipped.

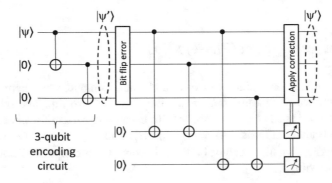

Fig. 25.9 QEC circuit that detects and corrects a single bit flip error on a 3-qubit encoding

Table 25.1 Syndrome used to detect a single bit flip

State	Syndrome
$|000\rangle$	00
$|001\rangle$	01
$|010\rangle$	10
$|011\rangle$	11
$|100\rangle$	11
$|101\rangle$	10
$|110\rangle$	01
$|111\rangle$	00

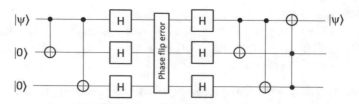

Fig. 25.10 QEC circuit that detects and corrects a sign flip error

If 00 is measured, then there is nothing to correct. If 01 is measured, then the bit flip error occurred on the 3rd qubit. If 10 is measured, then bit flip error occurred on the 2nd qubit. If 11 is measured, then the bit flip error occurred on the 1st qubit. The error can therefore be corrected with a controlled X gate (a NOT gate) applied to the flipped qubit. At no point during correction do we gain any information regarding the coefficients of $|\psi'\rangle$; hence, superposition will remain intact during correction.

Exercise 25.3 Verify the syndromes in Table 25.1 from Fig. 25.9.

25.9   Correcting Sign (Phase) Flips

Flipped bits are the only kind of error in a classical computer, but there is another possible error with quantum computers—the sign or phase flip. The relative sign between $|0\rangle$ and $|1\rangle$ can become inverted. To detect and correct for phase flips, we can use the Hadamard basis: $|+\rangle = \frac{1}{\sqrt{2}}(|0\rangle+|1\rangle)$ and $|-\rangle = \frac{1}{\sqrt{2}}(|0\rangle - |1\rangle)$. In the Hadamard basis, sign flips become bit flips. To correct phase flips, we implement the following encoding:

$$|\psi\rangle = \alpha|0\rangle + \beta|1\rangle \rightarrow |\psi'\rangle = \alpha|+++\rangle + \beta|---\rangle \qquad (25.5)$$

The circuit which can correct a single sign flip is shown in Fig. 25.10. The first half of the circuit (before the phase flip error) implements the code in Eq. (25.5). The other half corrects the phase flip.

25.10   Correcting Bit Flips and Phase Flips

In 1995, Peter Shor developed a 9-qubit encoding and algorithm able to correct for both bit flips and sign flips on any one of the 9 physical qubits [4]. Remarkably, this code can correct for any arbitrary single qubit error! In 1996, Steane [5] developed a 7-qubit code that performed as well as the Shor code. In 1996, Laflamme et al. [6] developed a 5-qubit code. 5 qubits turn out to be the least number of physical qubits needed to detect and correct errors on logical qubits. Another approach to QEC is to build a circuit that simply detects a possible error of an encoded qubit without correcting for it. If an error is detected, then the computation is simply started again. These are known as quantum error detection circuits. Many other codes and circuits have been developed that protect against various errors and sources of noise [7]. "Surface codes", consisting of a 2D or 3D array of physical qubits representing one logical qubit, are an active area of research.

25.11 Quantum Threshold Theorem

What about errors that might occur in the QEC circuit itself? A theorem in quantum computing, called the quantum threshold theorem or quantum fault-tolerance theorem, states that quantum error correction can reduce the logical error rate to arbitrarily low levels, providing an error rate below a certain threshold can be achieved. The probability of an error occurring must be below 1 in 100–1000 gate operations; i.e., the gate accuracy (called fidelity) must be at least ~99.0–99.9% accurate. Reaching this fidelity is the goal of every quantum computing platform but may require 1000s of physical qubits per logical qubit. When this occurs, it will be possible to detect and correct the errors faster than they occur. The goal is to demonstrate a logical qubit with a lifetime that is longer than any physical component. This goal was reached with a superconducting transmon qubit with a quantum-error-corrected lifetime (T_2) of 320 μs, 20 times longer than the lifetime of the physical transmon [8].

References

1. F. Bloch, Phys. Rev. **105**, 1206 (1957)
2. A.G. Redfield, IBM J. Res. Dev. **1**, 19 (1957)
3. T.D. Ladd et al., Nature **464**, 45 (2010)
4. P.W. Shor, Phys. Rev. A **52**, R2493 (1995)
5. A.M. Steane, Phys. Rev. Lett. **77**, 793 (1996)
6. R. Laflamme et al., Phys. Rev. Lett. **77**, 198 (1996)
7. S.J. Devitt, W.J. Munro, K. Nemoto, Rep. Prog. Phys. **76**, (2013)
8. N. Ofek et al., Nature **536**, 441 (2016)

Chapter 26
Topological Quantum Computing

Quipu is an ancient Incan method of storing information in knots (Fig. 26.1). Topological quantum computing is a method of quantum information storage and processing using "topological" knots. In this chapter, we will examine topological quantum computing.

26.1 Exchange Statistics: Bosons, Fermions and Anyons

Consider the exchange in position of two particles, as shown in Fig. 26.2. If the two particles are identical, there should be no change in the wavefunction after the exchange. There are two possibilities: the wavefunction could be symmetric under particle exchange:

$$\psi(\mathbf{r}_2, \mathbf{r}_1) = +\psi(\mathbf{r}_1, \mathbf{r}_2) \tag{26.1}$$

or anti-symmetric under particle exchange:

$$\psi(\mathbf{r}_2, \mathbf{r}_1) = -\psi(\mathbf{r}_1, \mathbf{r}_2) \tag{26.2}$$

In either case, the probability $|\psi|^2$ is the same before and after the exchange, as expected for identical particles. Particles that obey Eq. (26.1) are called bosons (e.g., photons), while particles that obey Eq. (26.2) are called fermions (e.g., electrons). This wavefunction symmetry is responsible for many phenomena in condensed matter such as the Pauli exclusion principle, superfluidity, superconductivity, Bose-Einstein condensation, etc.

We can generalize Eqs. (26.1) and (26.2) as follows:

$$\psi(\mathbf{r}_2, \mathbf{r}_1) = e^{i\theta}\psi(\mathbf{r}_1, \mathbf{r}_2) \tag{26.3}$$

© The Author(s), under exclusive license to Springer Nature Switzerland AG 2021
R. LaPierre, *Introduction to Quantum Computing*, The Materials Research Society Series,
https://doi.org/10.1007/978-3-030-69318-3_26

Fig. 26.2 The exchange in
position of two particles

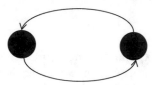

where $0 < \theta < \pi$. Particles that obey Eq. (26.3) are called anyons. When one anyon
is moved around another anyon, its state acquires a phase of $e^{i\theta}$. The probability
is the same before and after the exchange, since $|\psi(\mathbf{r}_2, \mathbf{r}_1)|^2 = |e^{i\theta}\psi(\mathbf{r}_1, \mathbf{r}_2)|^2 = |\psi(\mathbf{r}_1, \mathbf{r}_2)|^2$. Bosons and fermions may be considered as a special case of anyons
with $\theta = 0$ and $\theta = \pi$, respectively. Anyons therefore generalize the statistics of the
commonly known bosons and fermions. Particle exchanges are depicted in Fig. 26.3.
Anyons obeying Eq. (26.3) are called Abelian anyons, meaning that the order of
operations does not matter.

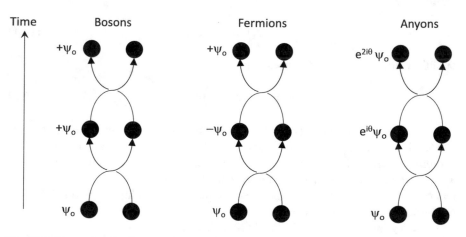

Fig. 26.3 The exchange of two particles, defining bosons, fermions and Abelian anyons

26.2 Topological Quantum Computing

Anyons that result in a trivial phase upon particle exchange, as in Eq. (22.3), are called Abelian anyons. These are not of interest for quantum computation. However, another class of anyons are called non-Abelian anyons, in which particle exchange results in a non-trivial unitary operation and the order of operations matters (the operations are non-commutative). A common example of a non-Abelian operation is rotation in three-dimensional space. For example, rotating an object 90° around one axis and then 90° around a different axis is not the same if the order of operations is swapped.

In 2003, Alexei Kitaev proposed non-Abelian anyons as a means of quantum computing [2]. Quantum computing using non-Abelian anyons is performed by controlled particle exchanges called "braiding", like those shown in Fig. 26.3. A qubit is formed by pairs of anyons. The quantum gates are implemented by the braiding of pairs of anyons, which look like entangled strings [3]. The unitary transformation implemented by the motion of the anyons depends only on the braid and is independent of the details of the trajectories. This type of quantum computing is known as "topological quantum computing".

Roughly speaking, topology is the science and mathematics of "shapes". For example, what is the difference between a sphere, a donut and coffee cup? The donut and coffee cup have the same topology because their shapes can be continuously deformed into each other without tearing or punching a hole in the object. However, the sphere is topologically distinct from the donut or coffee cup because it cannot be deformed into those shapes without punching a hole in it. Michael Kosterlitz, David Thouless, and Duncan Haldane (Fig. 26.4) won the Nobel Prize in Physics in 2016 for their theoretical explanations of strange states of matter in two-dimensional materials, known as topological phases.

In the case of non-Abelian anyons, the quantum information is encoded in the braid's topology; i.e., the way the anyons are braided. Like a knot, local noise in

Fig. 26.4 From left: David Thouless [4], Duncan Haldane [5] and Michael Kosterlitz [6] (from left); Nobel Prize in Physics in 2016. *Credit* Wikimedia Commons [4–6]

the environment cannot undo the braids, so the quantum information encoded in the braids is not perturbed. The quantum information is thus said to be "topologically protected" and "delocalized". Hence, the braiding of non-Abelian anyons should allow quantum computation without the need for quantum error correction [2]. But do non-Abelian anyons exist?

26.3 Fractional Quantum Hall Effect

The fractional quantum Hall effect (fQHE) is a variation of the familiar Hall effect. The Hall effect, discovered by Edwin Hall in 1879, involves the measurement of a voltage (V_y), transverse to an applied electric current (I_x) and a transverse applied magnetic field (B_z), as shown in Fig. 26.5. An important parameter is the Hall resistance, defined as the ratio of the transverse voltage and applied current, $R_H = V_y/I_x$. The Hall resistance can be derived classically (Exercise 26.1) as:

$$R_H = B_z/ent \qquad (26.4)$$

where n is the charge carrier density, t is the sample thickness, and e is the fundamental charge.

Exercise 26.1 Referring to Fig. 26.5, derive Eq. (26.4).

If the conductor is a high mobility two-dimensional electron gas (2DEG) at low temperatures and strong magnetic fields, then something unusual happens. At low magnetic fields (less than a few Tesla), R_H follows Eq. (26.4) with n (charge per unit volume) replaced by nt (charge per unit area) for the 2DEG. Therefore, at low

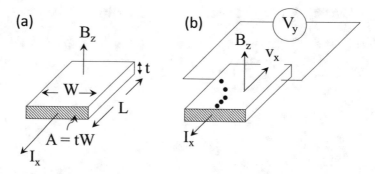

Fig. 26.5 **a** In the Hall effect, a current I_x is applied along a sample with a perpendicular magnetic field B_z. Sample dimensions are indicated. **b** Due to B_z, the charge carriers (electrons depicted as filled circles) with velocity v_x are deflected by the Lorentz force, producing a Hall voltage V_y

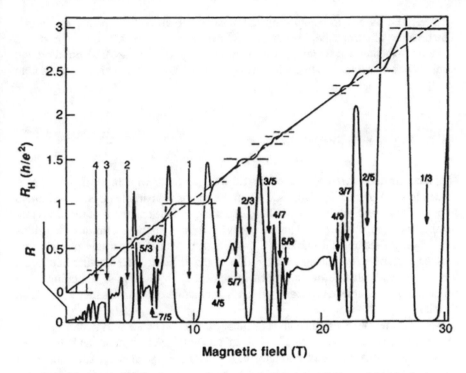

Magnetic field (T)

Fig. 26.6 Fractional quantum Hall effect showing Hall resistance, R_H, and longitudinal resistance, R. At low magnetic fields, R_H follows Eq. (26.4). However, at large magnetic fields, R_H follows Eq. (26.5) with quantized plateaus. At the plateaus, the longitudinal resistance drops to zero. Reprinted from H.L. Stormer, Physica B 177 (1992) 401, https://doi.org/10.1016/0921-452 6(92)90138-I, Copyright © 1992, with permission from Elsevier [7]

magnetic fields, R_H is proportional to the field as seen in Fig. 26.6. However, at high magnetic fields, R_H exhibits plateaus or quantized values of resistance. At these plateaus,

$$R_H = \left(h/e^2\right)/\nu \tag{26.5}$$

where h/e^2 is the resistance quantum or von Klitzing constant (~25.8 kΩ) and ν is either an integer, $n = 1, 2, \ldots$ (called the "integer quantum Hall effect" (IQHE)) or a rational fraction p/q such as 1/3, 2/5, 3/7, etc. (called the "fractional quantum Hall effect" (fQHE)) [7]. At the plateaus of R_H, the longitudinal resistance, R, along the applied current direction drops to zero. An example of R_H and R, demonstrating the fQHE, is shown in Fig. 26.6 [7]. The 1985 Nobel Prize in Physics was awarded to Klaus von Klitzing for his discovery of the IQHE. The 1998 Nobel Prize in Physics was awarded to Robert Laughlin, Horst Störmer and Daniel Tsui for their discovery of the fQHE.

The fQHE is due to charge-flux composites known as "composite fermions". The braiding of composite fermions associated with the $\nu = 5/2$ state is predicted to be non-Abelian and therefore useful for quantum computing. The braiding and non-Abelian character of composite fermions in the fQHE is an active research field [8, 9].

26.4 Majorana Fermions

Majorana fermions were proposed by Ettore Majorana, who mysteriously disappeared in 1946 (Fig. 26.7) [10]. In 1937, in the context of particle physics, Majorana hypothesized a real solution to the Dirac equation, and the existence of fermions that are their own antiparticle [11]. These are now known as Majorana fermions. Neutrinos are the only fundamental particle of the Standard Model of particle physics that may be Majorana fermions. However, due to the difficulty of detecting neutrinos, we are presently unsure of their particle statistics. Physicists are searching for a process called "double beta decay", which is a signature of neutrinos being Majorana fermions.

Majorana fermions can also arise as quasiparticles in condensed matter systems. Quasiparticles are elementary excitations of a many-particle system that behave like particles (e.g., holes in a semiconductor, or phonons in a crystal lattice). In condensed matter, Majorana fermions come in pairs, with a regular fermion (Dirac fermion; e.g., an electron) comprised of two Majorana fermions (Fig. 26.8). Conversely, each Majorana quasiparticle can be considered as an equal superposition of an electron and a hole. Hence, in Fig. 26.7, each Majorana quasiparticle is represented by a half-filled circle.

Majorana fermions are predicted to form as localized bound states (edge states) at the ends of a one-dimensional superconductor (a nanowire) [13], or at certain defects

Fig. 26.7 Ettore Majorana (1906–1938). *Credit* Wikimedia Commons [12]

Fig. 26.8 In condensed matter, Majorana fermions come in pairs, with a regular (Dirac) fermion comprised of two Majorana fermions

in the superconductor. Hence, the Majorana quasiparticles are also called "Majorana bound states". Furthermore, the superconductor must be a p-wave superconductor, meaning the superconducting gap is dependent on the wavevector, **k**. This characteristic is believed to form in certain nanowires with large spin-orbit interaction and large magnetic fields. Under these conditions, Majorana fermions form at the ends of the nanowire. It is as though we split each electron into two halves and placed each half at the ends of the nanowire (Fig. 26.9). The electron becomes delocalized.

The Majoranas occupy a state at zero energy in the superconductor and are protected by an energy gap Δ (Fig. 26.10). This zero-energy state is referred to as a "Majorana zero mode" (MZM). Unlike regular fermions, the MZMs are degenerate, permitting the occupation of many fermions (Majorana pairs). In this system, $|0\rangle$ and $|1\rangle$ are not represented by a ground and excited state. Instead, $|0\rangle$ and $|1\rangle$ are represented by each mode being empty or filled (absence or presence of a fermion). Excitations are not possible, so the energy relaxation time, T_1, is irrelevant. Also, local noise cannot destroy the MZM state and should exhibit long coherence times (T_2). Since Majoranas are their own anti-particle, they have no charge or spin, making them hard to detect but also unsusceptible to local noise. Most importantly, Majoranas are believed to obey non-Abelian statistics, making them useful for quantum computing. Hence, the name "Majorana fermion" in this context is inaccurate, since Majorana quasiparticles obey non-Abelian statistics, not fermion statistics; however, the name Majorana fermion is still used sometimes in the literature.

The two Majoranas comprising a single fermion (electron) are spatially separated and can form a non-local qubit. Every MZM can be empty or it can be occupied by a fermion (Majorana pair), giving us two possible degenerate quantum states, $|0\rangle$ and $|1\rangle$, for each pair of Majoranas. Hence the quantum state, $|0\rangle$ or $|1\rangle$, is the

Fig. 26.9 Majorana fermions at the ends of a one-dimensional superconductor (nanowire)

Fig. 26.10 Energy diagram for Majoranas in a 1D superconductor

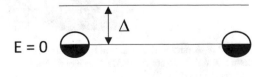

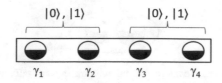

$|0\rangle, |1\rangle$ $|0\rangle, |1\rangle$

γ_1 γ_2 γ_3 γ_4

Fig. 26.11 Four Majorana fermions along a nanowire, denoted γ_1 to γ_4. Each pair of Majoranas (γ_1 and γ_2, γ_3 and γ_4) form a regular (Dirac) fermion (e.g., an electron)

fermion occupation number. We will have 2^n possible quantum states for n pairs of Majoranas. For example, consider the state of four Majoranas (two pairs), denoted γ_1 to γ_4 in Fig. 26.11. The positions of the Majorana pairs are controlled by barriers created with electrostatic gates. The most general quantum state is a superposition: $|\psi\rangle = \alpha_{00}|00\rangle + \alpha_{01}|01\rangle + \alpha_{10}|10\rangle + \alpha_{11}|11\rangle$. Using the vector representation, we have $|\psi\rangle = \begin{pmatrix} \alpha_{00} \\ \alpha_{01} \\ \alpha_{10} \\ \alpha_{11} \end{pmatrix}$. The unitary transformation associated with the braiding will be 4×4 matrices.

A method is required to braid two Majoranas without resulting in their collision. As described below, collision of two Majoranas is called fusion and results in a regular fermion, thereby losing the quantum coherence between the two Majoranas. In principle, the braiding is achieved by forming a nanowire T-junction or cross (Fig. 26.12), allowing the particle exchange (braiding) to take place without collision of the two Majoranas [14]. For example, Fig. 26.13a shows a braiding operation of Majoranas labelled γ_1 and γ_2. γ_1 is first moved down one "leg" of the T-junction, γ_2 is then moved to the previous position of γ_1, and finally γ_2 is moved to the previous position of γ_1, thereby completing a braiding of two Majoranas without creating a collision. The corresponding diagram for the braiding of γ_1 and γ_2 is shown in Fig. 26.13b. During the braiding process, the Majoranas remain in the ground state. The braiding process is implemented by moving the barriers formed by electrostatic gates.

The states that result from a braiding process can be described by a unitary transformation. Suppose the unitary 4×4 matrix for the braiding of γ_1 and γ_2 is U_{12}.

Fig. 26.12 A nanowire T-junction for Majorana braiding. Reprinted by permission from Springer Nature: J. Alicea, Nature Nanotechnology 8 (2013) 623, https://doi.org/10.1038/nnano.2013.178, Copyright © 2013 [15]

Fig. 26.13 a Consecutive steps for braiding two Majoranas, γ_1 and γ_2, using a T-junction. **b** Corresponding braiding diagram

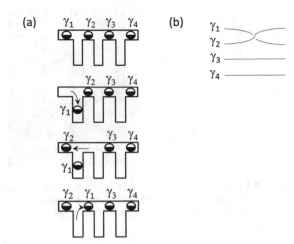

Similarly, the braiding of γ_2 and γ_3 leads to a different quantum operation described by the unitary transformation U_{23}, and the braiding of γ_3 and γ_4 is described by U_{34}. With the appropriate braiding operations (described by U_{12}, U_{23} and U_{34}) on the correct pairs of Majoranas, we can perform quantum gates. The exact path followed by the Majoranas during each braiding process does not affect the transformation. The exchange paths are topologically equivalent. However, the operations are non-commutative (for example, $U_{23}U_{12} \neq U_{12}U_{23}$), so the order of the braiding matters. These unusual properties classify the Majoranas as non-Abelian anyons. The braiding of Majoranas is a discrete process; i.e., the braiding either happens or does not happen. Therefore, the quantum operations are not affected by local noise—they are topologically protected.

The quantum information is read-out in a process called fusion. Two Majoranas comprise a regular fermion (an electron). At the end of the braiding process, we combine (fuse) pairs of Majoranas, resulting in an empty or filled state (absence or presence of fermions; i.e., $|0\rangle$ or $|1\rangle$), for each fusion pair. The state, $|0\rangle$ or $|1\rangle$, can then be read out by charge sensors.

26.5 Outlook

A Majorana fermion is believed to have been observed in 2012 in an InSb nanowire partially covered by a superconducting NbTiN film (Fig. 26.14) [16], and also in an InAs nanowire covered by a superconducting Al film [17]. The nanowire material is normally not a superconductor. However, due to the tunneling of Cooper pairs, the superconductor lying on top of the nanowire can induce superconductivity in the nanowire. This is known as the proximity effect. Majorana fermions are then believed to form at the two ends of the nanowire in the presence of a strong magnetic field. Majoranas are zero energy states (Fig. 26.10), so we expect to see a peak in the

Fig. 26.14 A nanowire (grey) with two leads (green) and partially covered in a superconductor (orange). Reprinted by permission from Springer Nature: J. Alicea, Nature Nanotechnology 8 (2013) 623, https://doi.org/10.1038/nnano.2013.178, Copyright © 2013 [15]

conductance of the nanowire at zero voltage due to the MZMs. Indeed, this is what is observed experimentally in conductance measurements (Fig. 26.15). Since then, Majoranas are believed to have been formed in other experimental systems [18], and there is a huge experimental effort to perform braiding of Majoranas to demonstrate their non-Abelian statistics. This effort has large support from Microsoft Quantum Research.

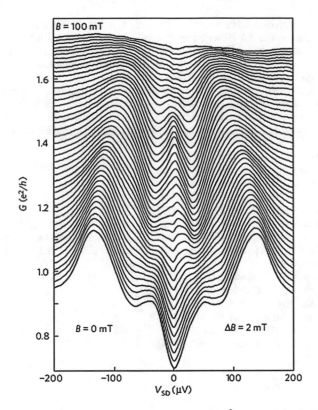

Fig. 26.15 Nanowire conductance (G), measured in units of e^2/h, versus applied bias (V_{SD}). A small conductance peak at zero volts (the so-called "zero bias peak") in the presence of a strong magnetic field is a signature of MZMs. Each curve represents a different magnetic field from 0 to 100 mT in steps of 2 mT. Reprinted by permission from Springer Nature: A. Das et al., Nature Physics 8 (2012) 887, https://doi.org/10.1038/nphys2479, Copyright © 2012 [17]

References

1. This file is licensed under the Creative Commons Attribution-Share Alike 4.0 International license (https://creativecommons.org/licenses/by-sa/4.0/deed.en). File: Quipo in the Museo Machu Picchu, Casa Concha, Cusco.jpg. (2020, October 28). *Wikimedia Commons, the free media repository*. Retrieved 19:03, December 8, 2020 from https://commons.wikimedia.org/w/index.php?title=File:Quipo_in_the_Museo_Machu_Picchu,_Casa_Concha,_Cusco.jpg&oldid=505376745
2. A. Yu Kitaev, Annals Phys. **303**, 2 (2003)
3. G.P. Collins, Sci. Amer. 57 (2006)
4. Attribution: Mary Levin/University of Washington. This file is licensed under the Creative Commons Attribution-Share Alike 4.0 International license (https://creativecomm ons.org/licenses/by-sa/4.0/deed.en). File: DavidThouless 1995 UW.jpg. (2020, September 20). *Wikimedia Commons, the free media repository*. Retrieved 19:10, December 8, 2020 from https://commons.wikimedia.org/w/index.php?title=File:DavidThouless_1995_UW.jpg&oldid=465406312

5. Author: Bengt Nyman from Vaxholm, Sweden. This file is licensed under the Creative Commons Attribution 2.0 Generic license (https://creativecommons.org/licenses/by/2.0/deed.en). File: Duncan Haldane.jpg. (2020, September 15). *Wikimedia Commons, the free media repository*. Retrieved 19:13, December 8, 2020 from https://commons.wikimedia.org/w/index.php?title=File:Duncan_Haldane.jpg&oldid=459619406
6. Attribution : Brown University (http://www.brown.edu/). This file is licensed under the Creative Commons Attribution-Share Alike 3.0 Unported license (https://creativecommons.org/licenses/by-sa/3.0/deed.en). File: Jkosterl.jpg. (2020, October 24). *Wikimedia Commons, the free media repository*. Retrieved 19:16, December 8, 2020 from https://commons.wikimedia.org/w/index.php?title=File:Jkosterl.jpg&oldid=498961821
7. H.L. Stormer, Phys. B **177**, 401 (1992)
8. M. Freedman, C. Nayak, K. Walker, Phys. Rev. B **73**, (2006)
9. A. Stern, Annals Phys. **323**, 204 (2008)
10. B.R. Holstein, J. Phys: Conf. Ser. **173**, (2009)
11. E. Majorana, Nuovo Cimento **14**, 171 (1937)
12. Unknown author (Mondadori Publishers). File: Ettore Majorana.jpg. (2018, September 4). *Wikimedia Commons, the free media repository*. Retrieved 19:43, December 8, 2020 from https://commons.wikimedia.org/w/index.php?title=File:Ettore_Majorana.jpg&oldid=318536900
13. A. Yu Kitaev, Phys. Usp. **44**, 131 (2001)
14. S.R. Plissard et al., Nat. Nanotech. **8**, 859 (2013)
15. J. Alicea, Nat. Nanotech. **8**, 623 (2013)
16. V. Mourik et al., Science **336**, 1003 (2012)
17. A. Das et al., Nat. Phys. **8**, 887 (2012)
18. S. Nadj-Perge et al., Science **346**, 602 (2014)

Further Reading

1. E.G. Rieffel, W.H. Polak, *Quantum Computing: A Gentle Introduction* (MIT Press, 2011)
2. C. Bernhardt, *Quantum Computing for Everyone* (MIT Press, 2019)
3. M.A. Nielsen, I.L. Chuang, *Quantum Computation and Quantum Information* (Cambridge University Press, 2017)
4. P. Kaye, R. Laflamme, M. Mosca, *An Introduction to Quantum Computing* (Oxford University Press, 2010)
5. N. David Mermin, *Quantum Computer Science: An Introduction* (Cambridge University Press, 2016)
6. B. Schumacher, M. Westmoreland, *Quantum Processes, Systems and Information* (Cambridge University Press, 2010)
7. J.D. Hidary, *Quantum Computing: An Applied Approach* (Springer, 2019)

© Materials Research Society, under exclusive license to Springer Nature Switzerland AG 2021

R. LaPierre, *Introduction to Quantum Computing*, The Materials Research Society Series, https://doi.org/10.1007/978-3-030-69318-3

Index

A

Abelian anyons, 348, 349
Adiabatic Quantum Computing (AQC), 323–325
Adiabatic theorem, 323, 324
Alferov, Zhores, 260
Amplitude amplification, 167
Angular frequency, 3, 8, 19, 30, 197, 198, 200, 201, 247, 259, 298, 299
Anharmonic potential, 296, 301, 302, 306, 307, 337
Anti-crossing, 227, 270, 313, 318, 320
Anyons, 347–349
Aspect, Alain, 88
Avoided crossing, 224, 227, 310
Azimuthal quantum number, 31

B

Bardeen, Cooper and Schrieffer (BCS), 286
Basis states, 27–30, 44, 60, 82, 96, 106, 109, 111, 122, 128, 134, 139, 140, 142, 143, 151, 165, 173, 183, 221
BB84, 95
Beam splitter, 8, 9, 42, 44, 94, 331–335
Bell, John, 79
Bell state circuit, 122, 123, 128, 136
Bell states, 75
Bell test, 79, 89
Bennett, Charles, 95, 98, 106, 129
Big O, 145, 163
Birthday paradox, 180
Bit flips, 108, 109, 343, 344
Bits, 15, 59, 94, 96, 106, 121, 142, 159, 172, 342, 343
Bloch, Felix, 54, 249

Bloch sphere, 54, 55, 66, 67, 83–85, 105, 113, 124, 193, 194, 197, 202, 205, 206, 211–213, 229, 237, 240, 252, 329, 332, 338–340
Born, Max, 19, 20
Born rule, 19, 25
Bosons, 42, 287, 347, 348
Bound states, 25, 35, 352
Bra, 61, 62, 64, 80, 107
Brassard, Gilles, 95

C

Charge qubit, 223, 224, 226, 262, 269, 270, 285, 304, 311–317
CHSH inequality, 86–88, 138
Church, Alonzo, 147
Church-Turing thesis, 147
Chu, Steven, 278
Cirac, Ignacio, 275, 276
Circuit model, 103, 240, 300
Circuit QED (cQED), 318–321, 335
Cohen-Tannoudji, Claude, 278
Complex conjugate, 27, 29, 62, 104, 106, 107
Composite fermions, 352
Composite state, 73, 74
Computational basis, 60, 63, 64, 98, 111, 149, 202
Computational complexity, 139, 144–146, 178, 324
Cooper pair box, 311, 314
Copenhagen interpretation, 9, 12, 122
Coplanar waveguide, 318
Coulomb blockade, 263, 264, 266, 268, 269
CPHASE gate, 117
Cryptography, 91, 178
Current-phase relation, 297

© Materials Research Society, under exclusive license to Springer Nature Switzerland AG 2021
R. LaPierre, *Introduction to Quantum Computing*, The Materials Research Society Series,
https://doi.org/10.1007/978-3-030-69318-3

D

Davisson, Clinton, 10–12
DC-SQUID, 303, 304, 311, 312, 314, 316, 319–321
De Broglie waves, 10
De Broglie, Louis, 10, 11
De Broglie wavelength, 10, 259, 265
Decoherence, 14, 241, 243, 255, 257, 270, 271, 315, 320, 338
Decoherence time, 338
Dephasing, 254, 302, 338, 339, 341
Deutsch algorithm, 157, 159
Deutsch, David, 142, 143, 149
Deutsch-Jozsa algorithm, 149, 159–161
Diffusion transform, 167–169, 171, 173–175
Dirac notation, 46, 49, 59, 63, 65, 72, 108–111, 113, 115, 118–120, 141, 172, 203
Dirac, Paul, 46, 47
Discrete Fourier Transform (DFT), 180–182, 189
DiVincenzo criteria, 241–244, 337
DiVincenzo, David, 241
Doppler cooling, 278
Doppler effect, 278
Double slit interference, 1, 6–9, 11, 13
Dual rail representation, 328, 333, 334
D-Wave, 325

E

E91, 97
Eigenfunction, 23, 28, 230
Eigenvalue, 23, 28, 33, 34, 48–50, 52, 107, 202, 218, 224, 225, 293
Einstein, Albert, 5, 6, 76, 78
Ekert, Arthur, 97
Electron confinement, 259
Electron Spin Resonance (ESR), 209, 211–213, 217–219, 222, 229, 250, 259, 266, 268, 301
Electrostatic quantum dots, 267
Entanglement, 74, 76–79, 86, 97, 123, 124, 127–130, 144, 156, 159, 269, 325
EPR paradox, 78
Exchange statistics, 347
Excited state, 19, 26, 28–30, 35, 59, 203, 217, 241, 279–281, 291, 295, 318, 337, 338, 353
Expectation value, 33, 107, 138, 224

F

Fermions, 42, 245, 347, 348, 352, 353, 355

Feynman, Richard, 14, 15
First Josephson relation, 297
Fluorescence, 242, 270, 276, 281
Flux quantization, 289, 290, 296, 302, 303
Flux quantum, 288, 298, 304, 310
Flux qubit, 285, 304, 308–312, 314, 315, 319, 323, 325
Flying qubits, 244, 327
Fourier, Jean-Baptiste Joseph, 180, 181
Fractional Quantum Hall Effect (fQHE), 268, 350, 351
Fredkin gate, 119, 120
Free Induction Decay (FID), 253, 255
Free particle, 21, 23, 259
Free precession, 229, 237, 238

G

Germer, Lester, 11
G-factor, 43, 198, 203
Global phase factor, 114, 156, 186, 198, 205
Goudsmit, Samuel, 41, 42
Greenberger-Horne-Zeilinger, 123
Ground state, 19, 26, 28, 29, 35, 203, 217, 242, 277–281, 291, 295, 302, 306, 308
Grover algorithm, 163, 164, 167, 172, 174–176
Grover, Lov, 164
Gyromagnetic ratio, 43, 198, 200, 234, 247, 248

H

Hadamard gate, 111, 112, 123, 128, 130, 139–141, 154–157, 165, 171, 187, 198, 279, 332, 334
Hahn Echo, 340, 341
Haldane, Duncan, 349
Hall, Edwin, 350
Hall effect, 268, 350, 351
Hall resistance, 350, 351
Hamiltonian, 21–23, 107, 111, 197, 201, 202, 209, 211, 213, 215–217, 223–226, 229, 233–235, 291, 293–295, 300, 301, 306, 311–313, 320, 323, 324
Hermitian operators, 34, 106, 107
Heterostructure, 260, 261, 265, 267
Hidden variables, 79, 81
Hydrogen atom, 31, 35, 36

I

Identity operator, 112, 115, 149, 150, 194, 195
Infinite quantum well, 23–30, 46, 223
Initialization, 165, 241, 242, 279
Inner product, 62, 64, 65, 80, 105, 141
Integer Quantum Hall Effect (IQHE), 268, 351
Interaction Hamiltonian, 218, 235
Interference, 1, 3, 4, 6–12, 31, 46, 47, 77, 78, 144, 303, 304, 308, 327
Inversion about the mean, 167–169
Ising model, 148, 235, 324

J

Jaynes-Cummings Hamiltonian, 320
J-coupling, 234, 236–239, 251
Jönsson, Claus, 11, 12
Josephson, Brian, 296, 297
Josephson energy, 300, 306, 309, 316
Josephson inductance, 301
Josephson junction, 296, 298–306, 308, 309, 311, 313, 314, 316, 317, 319, 321
Josephson relations, 294, 297–300, 305

K

Kelvin, Lord William Thomson, 1
Kerr effect, 318, 334
Kerr, John, 334
Ket, 46, 61, 62, 64, 67, 73, 80, 107
Kilby, Jack, 260
Kitaev, Alexei, 349
Kosterlitz, Michael, 349
Kroemer, Herbert, 260
Kronecker delta, 27

L

Landauer, Rolf, 106
Larmor, Joseph, 200, 201
Laughlin, Robert, 268, 351
LC circuit, 227, 289–293, 295, 301, 318, 320
Leakage, 296, 337, 338
Level repulsion, 227, 311
Lewis, Gilbert, 6
Linear transformations, 103
Longitudinal coherence time, 338
Longitudinal relaxation, 253, 255, 337–339

M

Mach-Zehnder interferometer, 8, 9

Macroscopic quantum tunneling, 307, 308
Magnetic dipole moment, 37, 38
Magnetic Resonance Imaging (MRI), 255
Majorana, Ettore, 352
Majorana fermions, 352–355
Majorana Zero Modes (MZMs), 353, 356, 357
Mathieu equation, 301, 313
Maxwell, James Clerk, 4
Mirror, 8, 9, 104, 318, 319, 321, 329, 330, 333, 335

N

Neutral atom, 38, 277, 282
Newton, Isaac, 1, 2
Niels Bohr, 36
No cloning theorem, 96, 124, 129
Noisy Intermediate-Scale Quantum computing (NISQ), 148
Non-Abelian anyons, 349, 350, 355
Nondeterministic Polynomial (NP), 146, 163, 172, 324
Non-linear inductance, 303
Non-locality, 76, 78
Normalization, 27, 60, 64, 67, 70, 71, 173, 185, 191
NOT gate, 106, 108, 109, 120, 121, 152, 170, 212, 333, 344
Nuclear Magnetic Resonance (NMR), 165, 191, 245, 248, 250, 251, 255–257, 338, 340, 341
Nuclear spin, 245, 247, 248, 250–252, 254, 270–272
Number-phase uncertainty relation, 315

O

Observables, 14, 23, 33, 34, 48, 72, 87, 94, 106–108, 263, 287
One-time pad, 93, 94, 97
Operator, 21, 23, 33, 34, 48–50, 52, 59, 65, 71, 101–104, 106–108, 115, 117, 122, 124, 135, 137, 138, 152, 156, 166–168, 172, 183, 184, 194, 195
Optical lattice, 282
Optical pumping, 242, 279
Optical qubit, 277
Oracle, 152–154, 161, 166–168, 172–174, 190
Orbital angular momentum, 31, 35, 37, 38, 40, 42, 43, 277, 279, 327
Orbital angular momentum quantum number, 31

Orthonormal states, 27, 55
Outer product, 64, 65, 109–112, 119

P

Partial measurement rule, 69
Path encoding, 327, 328
Pauli gates, 111
Pauli spin matrices, 50, 51
Pauli, Wolfgang, 46, 47, 111, 275
Paul trap, 275–277
Period finding, 185
Persistent current, 285, 289, 308
Phase coherence time, 338
Phase flips, 344
Phase gate, 108, 113, 122, 187, 329
Phase kick-back, 156, 157, 159
Phase particle, 306, 307
Phase qubit, 285, 304–309, 315
Phase shifter, 327–329, 332, 333, 335
Phillips, William, 278
Photoelectric effect, 5, 6
Photon, 5–12, 19, 31, 42, 46, 59, 76–78, 88,
        89, 93–98, 103, 124, 129, 217, 241,
        244, 266, 278, 281, 287, 318, 320,
        327, 328, 331, 332, 334, 335, 347
Planck constant, 6
Podolsky, Boris, 76, 78
Polarization encoding, 327
Precession, 193, 198–200, 202, 205, 206,
        209–211, 213, 229, 237, 247, 249,
        251, 252, 271, 338, 340
Preskill, John, 147, 148
Prime factors, 91, 145, 177–179, 189, 191
Principal quantum number, 31, 282
Probability amplitude, 8, 9, 11, 19, 20, 26,
        28, 29, 34, 47, 48, 53, 60, 64, 67, 68,
        72, 94, 96, 101, 109, 122, 127, 133,
        139, 140, 144, 164–167, 174, 182,
        221, 228, 229, 327, 328, 332
Projection operators, 65
Proximity effect, 355
Pure dephasing time, 338

Q

Quantization, 19, 26, 31, 35, 37, 38, 40, 260,
        289, 293
Quantum advantage, 148
Quantum annealing, 324
Quantum Dot (QD), 88, 260–271, 340
Quantum eraser, 77, 78
Quantum Error Correction (QEC), 148, 243,
        337, 341–345, 350

Quantum fault-tolerance theorem, 345
Quantum Fourier Transform (QFT), 112,
        148, 182–191
Quantum gates, 61, 101, 102, 108, 124, 144,
        157, 193, 197, 209, 217, 241, 243,
        269, 279, 333, 349, 355
Quantum harmonic oscillator, 289
Quantum Key Distribution (QKD), 91, 93–
        98, 124, 129, 178
Quantum metrology, 78
Quantum money, 98
Quantum parallelism, 15, 72, 139, 142–144,
        152–154, 157, 167, 180
Quantum supremacy, 148
Quantum threshold theorem, 345
Quantum well, 23, 26, 28, 30, 223, 259, 261,
        268, 271
Quantum wire, 260, 261
Qubits, 15, 59, 66, 68, 70–75, 89, 93, 98,
        101–103, 117, 118, 120, 122–124,
        128–130, 133, 138–142, 144, 148,
        154, 159, 161, 164, 165, 171, 182,
        184, 187, 191, 192, 219, 221, 241–
        245, 251, 255–257, 266, 270, 271,
        275, 277, 279, 282, 285, 296, 297,
        299, 302, 304, 312, 315, 317, 318,
        320, 324, 337, 340–345

R

Rabi formula, 227, 230
Rabi frequency, 211, 217, 220, 224, 228–
        230, 251
Rabi, Isidor, 209
Rabi oscillations, 212, 221, 223, 230, 233,
        266, 270, 276, 279, 295, 296, 307,
        308, 311, 314, 316, 320, 338
Ramsey experiment, 340
Ramsey, Norman, 340
RCSJ model, 305
Read-out, 241, 242, 250, 252, 267, 281, 304,
        308, 314, 320, 355
Realism, 78, 79, 89
Redundant encoding, 342
Refocusing, 257, 341
Relaxation, 253, 255, 271, 338
Relaxation time, 242, 253, 295, 311, 338,
        339, 353
Resistance quantum, 351
Resistively and capacitively shunted junc-
        tion model, 305
Reversibility, 101, 106
RF qubit, 277

Rivest, Shamir and Adleman (RSA), 91, 177, 178, 180, 192
Rosen, Nathan, 76, 78
Rotating Wave Approximation (RWA), 211, 213, 220
Rotation matrices, 193, 194, 202, 212, 222
Rydberg atom, 282

S
Schrodinger equation, 20, 21, 37, 225, 291, 292, 298, 301, 306, 313
Schrodinger, Erwin, 20
Second Josephson relation, 294, 298
Separable state, 74, 77, 144
Separation of variables, 22, 28
Shannon, Claude, 93
Shor algorithm, 178, 282
Shor, Peter, 177–179, 344
Sideband cooling, 278
Sifted key, 96, 97
Sign flips, 344
Simple harmonic oscillator, 292, 293, 295, 296, 301
Single electron transistor, 264, 266
Single-qubit gates, 144, 193, 217, 238, 239, 250, 251
Smolin, John, 98
Spin, 21, 35, 42–55, 59, 63, 76, 77, 79–83, 86–88, 103, 108, 111, 112, 138, 193, 197–200, 202, 203, 205, 206, 211, 223, 224, 226, 233–237, 241, 242, 245, 247–251, 253–255, 257, 259, 262, 266–268, 270–272, 277, 286, 287, 310, 311, 314, 320, 323, 324, 327, 340, 353
Spin operator, 48, 49
Spin projection quantum number, 42, 53
Spin quantum number, 246, 247
Spin qubits, 219, 255, 259, 266, 270, 272
Stationary qubits, 244
Stationary states, 25, 29–31, 203, 205, 218, 225, 226
Stern-Gerlach experiment, 38, 39
Stern, Otto, 38
Störmer, Horst, 268, 351
Superconductivity, 285–288, 320, 347, 355
Supercurrent, 285, 288, 289, 296, 297, 299, 300, 303, 305, 307, 309–312
Superdense coding, 129, 130
Superposition, 1, 2, 8, 9, 14, 15, 28–30, 33–35, 46, 53, 55, 57, 59, 60, 68, 70–72, 75, 78, 81, 94, 97, 101, 108, 109, 111, 112, 114, 121, 125, 127, 139, 140, 142, 153–155, 157, 159, 164–166, 173, 174, 189, 190, 205, 206, 212, 218, 221, 241, 242, 255, 266, 270, 311–314, 324, 325, 332, 338, 339, 341, 342, 344, 352, 354
Surface codes, 345
SWAP gate, 118–120, 187
Switching curves, 307, 308

T
Teleportation, 78, 127–130
Tensor, 115, 133, 134, 139, 159, 160
Tensor product, 65, 109, 115, 133–137, 144, 172, 173
Thomson, George Paget, 11, 12
Thomson, J.J., 10, 11
Thouless, David, 349
Time-dependent Schrodinger equation, 20–22, 28, 29, 202, 209, 215, 216, 218, 219
Time-independent Schrodinger equation, 23, 31, 197, 202, 215, 217, 235
Toffoli gate, 120, 121
Toffoli, Tommaso, 121
Topological quantum computing, 268, 347, 349
Topology, 349
Torque, 198–200
Transmon qubit, 285, 315–317, 319–321, 341, 345
Transverse relaxation, 254, 339
Transverse relaxation time, 254
Trapped Ion Quantum Computing (TIQC), 129, 275, 282
Tsui, Daniel, 268, 351
Turing, Alan, 147
Two-qubit gates, 114, 123, 144, 237, 251, 282, 334, 335

U
Uhlenbeck, George, 41, 42
Unary operator, 149–151
Uncertainty principle, 14
Unitary, 104, 105, 108–112, 114, 116, 121, 122, 124, 125, 142, 149–152, 154, 169, 184, 193, 194, 197, 213, 222, 237, 349, 354
Universal gates, 124, 233, 242, 243

V

Vacuum Rabi oscillations, 318
Vector axioms, 61, 65, 66, 103
Vernam, Gilbert, 93
Voltage-phase relation, 294, 298
Von Klitzing constant, 351
Von Klitzing, Klaus, 268, 351

W

Walther Gerlach, 38
Washboard potential, 306, 307
Wavefunction, 8–10, 12, 14, 19, 21, 33, 34,
         36, 46, 63, 77, 122, 185, 186, 190,
         206, 223, 226, 229, 259, 270–272,
         288, 289, 292, 296–298, 300, 306,
         307, 323, 347
Wave-particle duality, 12, 14
Wavevector, 3, 8, 19, 24, 259, 327, 353

Wineland, David, 280

X

X gate, 108, 109, 344

Y

Y gate, 109
Young, Thomas, 1, 2

Z

Zeeman effect, 202, 203, 247, 250, 267
Zeeman, Pieter, 203, 204
Zeilinger, Anton, 98
Z gate, 110, 111, 129, 130, 279, 334
Zoller, Peter, 275, 276

Printed in the United States
by Baker & Taylor Publisher Services